30107 005 150 997

Fundamentals and Applications of

ULTRASONIC WAVES

CRC SERIES *in* PURE *and* APPLIED PHYSICS

Dipak Basu
Editor-in-Chief

PUBLISHED TITLES

Handbook of Particle Physics
M. K. Sundaresan

High-Field Electrodynamics
Frederic V. Hartemann

Fundamentals and Applications of Ultrasonic Waves
J. David N. Cheeke

FORTHCOMING TITLES

Introduction to Molecular Biophysics
Jack Tuszynski

Physics of Semiconductor Electron Devices
Pratul K. Ajmera

Fundamentals and Applications of ULTRASONIC WAVES

J. David N. Cheeke
Concordia University
Montreal, Canada

CRC PRESS

Boca Raton London New York Washington, D.C.

Cover Design: Polar diagram (log scale) for a circular radiator with radius/wavelength of 10. (Diagram courtesy of Zhaogeng Xu.)

Library of Congress Cataloging-in-Publication Data

Cheeke, J. David N.
Fundamentals and applications of ultrasonic waves / David Cheeke
p.; cm. (CRC series in pure and applied physics)
Includes bibliographical references and index.
ISBN 0-8493-0130-0 (alk. paper)
1. Ultrasonic waves. 2. Ultrasonic waves–Industrial applications. I. Title. II. Series.
QC244 .C47 2002
534.5′5—dc21 2002018807

This book contains information obtained from authentic and highly regarded sources. Reprinted material is quoted with permission, and sources are indicated. A wide variety of references are listed. Reasonable efforts have been made to publish reliable data and information, but the author and the publisher cannot assume responsibility for the validity of all materials or for the consequences of their use.

Neither this book nor any part may be reproduced or transmitted in any form or by any means, electronic or mechanical, including photocopying, microfilming, and recording, or by any information storage or retrieval system, without prior permission in writing from the publisher.

The consent of CRC Press LLC does not extend to copying for general distribution, for promotion, for creating new works, or for resale. Specific permission must be obtained in writing from CRC Press LLC for such copying.

Direct all inquiries to CRC Press LLC, 2000 N.W. Corporate Blvd., Boca Raton, Florida 33431.

Trademark Notice: Product or corporate names may be trademarks or registered trademarks, and are used only for identification and explanation, without intent to infringe.

Visit the CRC Press Web site at www.crcpress.com

© 2002 by CRC Press LLC

No claim to original U.S. Government works
International Standard Book Number 0-8493-0130-0
Library of Congress Card Number 2002018807
Printed in the United States of America 1 2 3 4 5 6 7 8 9 0
Printed on acid-free paper

Preface

This book grew out of a semester-long course on the principles and applications of ultrasonics for advanced undergraduate, graduate, and external students at Concordia University over the last 10 years. Some of the material has also come from a 4-hour short course, "Fundamentals of Ultrasonic Waves," that the author has given at the annual IEEE International Ultrasonics Symposium for the last 3 years for newcomers to the field. In both cases, it was the author's experience that despite the many excellent existing books on ultrasonics, none was entirely suitable for the context of either of these two courses.

One reason for this is that, except for a few specialized institutions, acoustics is no longer taught as a core subject at the university level. This is in contrast to electricity and magnetism, where, in nearly every university-level institution, there are introductory (college), intermediate (mid- to senior-level undergraduate), and advanced (graduate) courses. In acoustics the elementary level is covered by general courses on waves, and there are many excellent books aimed at the senior graduate (doctoral) level, most of which are cited in the references. Paradoxically, there are precious few books that are suitable for the nonspecialized beginning graduate student or newcomers to the field. For the few acoustics books of this nature, ultrasonics is only of secondary interest. This situation provided the specific motivation for writing this book.

The end result is a book that addresses the advanced intermediate level, going well beyond the simple, general ideas on waves but stopping short of the full, detailed treatment of ultrasonic waves in anisotropic media. The decision to limit the present discussion to isotropic media allows us to reduce the mathematical complexity considerably and put the emphasis on the simple physics involved in the relatively wide range of topics treated. Another distinctive feature of the approach lies in putting considerable emphasis on applications, to give a concrete setting to newcomers to the field, and to show in simple terms what one can do with ultrasonic waves. Both of these features give the reader a solid foundation for working in the field or going on to higher-level treatises, whichever is appropriate.

The content of the book is suitable for use as a text for a one-semester course in ultrasonics at the advanced B.Sc. or M.Sc. level. In this context it has been found that material for 8 to 9 weeks can be selected from the fundamental part (Chapters 1 through 10), and material for applications can be selected from the remaining chapters.

The following sections are recommended for the semester-long fundamental part: 3.1, 3.2, 4.1, 4.2, 4.3, 4.5, 5.1, 5.2, 6.1, 6.3, 7.1, 7.3, 7.4, 8.1, 8.2, 9.1, 10.1, and 10.2. Many of the sections omitted from this list are more specialized and can be left for a second or subsequent reading, such as Sections 4.4, 8.3.1, and 10.5. For each of these chapters, a summary has been given at the end where the principal concepts have been reviewed. Students should be urged to read these summaries to ensure that the concepts are well understood; if not, the appropriate section should be reread until comprehension has been achieved. A number of questions/problems have also been included to assist in testing comprehension or in developing the ideas further.

There is more than adequate material in the remaining chapters to use the rest of the semester to study selected applications. It has been the author's practice to assign term papers or open-ended experimental/computational projects during this stage of the course. In this connection, Chapters 11 and 12 have been provided as useful swing chapters to enable a transition from the more formal early text to the practical considerations of the applications chapters.

J. David N. Cheeke
Physics Department
Concordia University
Montreal, Canada

BARCODE No. 5150997
CLASS No. 534.55 CHE
BIB CHECK
21 JAN 2005
PROC CHECK
FINAL CHECK 1105 JB.
OS SYSTEM No.
LOAN CATEGORY NL

Acknowledgments

It has been said that a writer never completes a book but instead abandons it. This must have some truth in that, if nothing else, the publisher's deadline puts an end to activities. In any case, the completion of what has turned into a major project is in large part due to the presence of an enthusiastic support group, and it is a pleasure to thank them at this stage.

My graduate students over the last 10 years have been at the origin of much of the work, and I would particularly like to thank Martin Viens, Xing Li, Manas Dan, Steve Beaudin, Julien Banchet, Kevin Shannon, and Yuxing Zhang for many enjoyable working hours together. Over the years, my close colleagues Cheng-Kuei Jen and Zuoqing Wang have joined me in many pleasant hours of discussion of acoustic paradoxes and interpretation of experimental results. I would like to thank Camille Pacher for her help with the text, equations, and figures. Zhaogeng Xu made a significant and much-appreciated contribution with the numerical calculations for many of the figures, including Figures 6.3, 6.4, 6.6 through 6.8, 7.5, 7.6, 8.3, 9.1, 9.3, 9.4, 10.3, 10.5, and 10.6. Joe Shin has made a constant and indispensable contribution, with his deep understanding of the psyche of computers, and I also thank him for bailing me out of trouble so many times. Lastly, my wife Guerda has been a constant source of motivation and encouragement.

I wish to thank John Wiley & Sons for permission to use material from my chapter, "Acoustic Microscopy," in the *Wiley Encyclopedia of Electrical and Electronics Engineering*, which makes up a large part of Chapter 14. I also thank the *Canadian Journal of Physics* for permission to use several paragraphs from my article, "Single-bubble sonoluminescence: bubble, bubble, toil and trouble" (*Can. J. Phys.*, 75, 77, 1997), and the IEEE for permission to use several paragraphs from Viens, M. et al., "Mass sensitivity of thin rod acoustic wave sensors" (*IEEE Trans. UFFC*, 43, 852, 1996). I thank Larry Crum and EDP Sciences, Paris (Crum, L.A., *J. Phys. Colloq.*, 40, 285, 1979), for their permission to use Larry's magnificent photo of an imploding bubble in the preface.

This work was done during a sabbatical leave from the Faculty of Arts and Science of Concordia University, Montreal, and that support is gratefully acknowledged.

Finally, I would like to thank Nora Konopka, Helena Redshaw, Madeline Leigh, and Christine Andreasen of CRC Press for providing such a pleasant and efficient working environment during the processing of the manuscript.

The Author

J. David N. Cheeke, Ph.D., received his bachelor's and master's degrees in engineering physics from the University of British Columbia, Vancouver, Canada, in 1959 and 1961, respectively, and his Ph.D. in low temperature physics from Nottingham University, U.K., in 1965. He then joined the Low Temperature Laboratory, CNRS, Grenoble, France, and also served as professor of physics at the University of Grenoble.

In 1975, Dr. Cheeke moved to the Université de Sherbrooke, Canada, where he set up an ultrasonics laboratory, specialized in physical acoustics, acoustic microscopy, and acoustic sensors. In 1990, he joined the physics department at Concordia University, Montreal, where he is currently head of an ultrasonics laboratory. He was chair of the department from 1992 to 2000. He has published more than 120 papers on various aspects of ultrasonics. He is senior member of IEEE, a member of ASA, and an associate editor of *IEEE Transactions on Ultrasonics, Ferroelectrics, and Frequency Control.*

Contents

1

Ultrasonics: An Overview

1.1 Introduction

Viewed from one perspective, one can say that, like life itself, ultrasonics came from the sea. On land the five senses of living beings (sight, hearing, touch, smell, and taste) play complementary roles. Two of these, sight and hearing, are essential for long-range interaction, while the other three have essentially short-range functionality. But things are different under water; sight loses all meaning as a long-range capability, as does indeed its technological counterpart, radar. So, by default, sound waves carry out this long-range sensing under water. The most highly developed and intelligent forms of underwater life (e.g., whales and dolphins) over a time scale of millions of years have perfected very sophisticated range-finding, target identification, and communication systems using ultrasound. On the technology front, ultrasound also really started with the development of underwater transducers during World War I. Water is a natural medium for the effective transmission of acoustic waves over large distances; and it is indeed, for the case of transmission in opaque media, that ultrasound comes into its own.

We are more interested in ultrasound in this book as a branch of technology as opposed to its role in nature, but a broad survey of its effects in both areas will be given in this chapter. Human efforts in underwater detection were spurred in 1912 by the sinking of *RMS Titanic* by collision with an iceberg. It was quickly demonstrated that the resolution for iceberg detection was improved at higher frequencies, leading to a push toward the development of ultrasonics as opposed to audible waves. This led to the pioneering work of Langevin, who is generally credited as the father of the field of ultrasonics. The immediate stimulus for his work was the submarine menace during World War I. The U.K. and France set up a joint program for submarine detection, and it is in this context that Langevin set up an experimental immersion tank in the Ecole de Physique et Chimie in Paris. He also conducted large-scale experiments, up to 2 km long, in the Seine River. The condenser transducer was soon replaced by a quartz element, resulting in a spectacular improvement in performance, and detection up to a distance of 6 km was obtained.

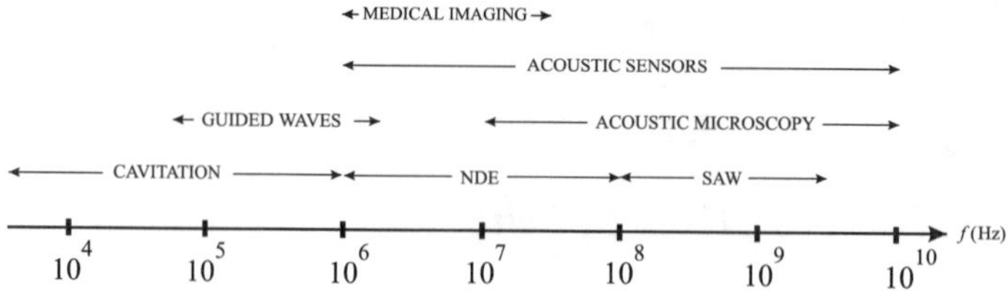

FIGURE 1.1
Common frequency ranges for various ultrasonic processes.

With Langevin's invention of the more efficient sandwich transducer shortly thereafter the subject was born. Although these developments came too late to be of much use against submarines in that war, numerous technical improvements and commercial applications followed rapidly.

But what, after all, is ultrasonics? Like the visible spectrum, the audio spectrum corresponds to the standard human receptor response function and covers frequencies from 20 Hz to 20 kHz, although, with age, the upper limit is reduced significantly. For both light and sound, the "human band" is only a tiny slice of the total available bandwidth. In each case the full bandwidth can be described by a complete and unique theory, that of electromagnetic waves for optics and the theory of stress waves in material media for acoustics.

Ultrasonics is defined as that band above 20 kHz. It continues up into the MHz range and finally, at around 1 GHz, goes over into what is conventionally called the hypersonic regime. The full spectrum is shown in Figure 1.1, where typical ranges for the phenomena of interest are indicated. Most of the applications described in this book take place in the range of 1 to 100 MHz, corresponding to wavelengths in a typical solid of approximately 1 mm to 10 μm, where an average sound velocity is about 5000 m/s. In water—the most widely used liquid—the sound velocity is about 1500 m/s, with wavelengths of the order of 3 mm to 30 μm for the above frequency range.

Optics and acoustics have followed parallel paths of development from the beginning. Indeed, most phenomena that are observed in optics also occur in acoustics. But acoustics has something more—the longitudinal mode in bulk media, which leads to density changes during propagation. All of the phenomena occurring in the ultrasonic range occur throughout the full acoustic spectrum, and there is no theory that works only for ultrasonics. So the theory of propagation is the same over the whole frequency range, except in the extreme limits where funny things are bound to happen. For example, diffraction and dispersion are universal phenomena; they can occur in the audio, ultrasonic, or hypersonic frequency ranges. It is the same theory at work, and it is only their manifestation and relative importance that change. As in the world of electromagnetic waves, it is the length scale that counts. The change in length scale also means that quite different technologies must be used to generate and detect acoustic waves in the various frequency ranges.

Why is it worth our while to study ultrasonics? Alternatively, why is it worth the trouble to read (or write) a book like this? As reflected in the structure of the book itself, there are really two answers. First, there is still a lot of fundamentally new knowledge to be learned about acoustic waves at ultrasonic frequencies. This may involve getting a better understanding of how ultrasonic waves occur in nature, such as a better understanding of how bats navigate or dolphins communicate. Also, as mentioned later in this chapter, there are other fundamental issues where ultrasonics gives unique information; it has become a recognized and valuable tool for better understanding the properties of solids and liquids. Superconductors and liquid helium, for example, are two systems that have unique responses to the passage of acoustic waves. In the latter case they even exhibit many special and characteristic modes of acoustic propagation of their own. A better understanding of these effects leads to a better understanding of quantum mechanics and hence to the advancement of human knowledge.

The second reason for studying ultrasonics is because it has many applications. These occur in a very broad range of disciplines, covering chemistry, physics, engineering, biology, food industry, medicine, oceanography, seismology, etc. Nearly all of these applications are based on two unique features of ultrasonic waves:

1. Ultrasonic waves travel slowly, about 100,000 times slower than electromagnetic waves. This provides a way to display information in time, create variable delay, etc.
2. Ultrasonic waves can easily penetrate opaque materials, whereas many other types of radiation such as visible light cannot. Since ultrasonic wave sources are inexpensive, sensitive, and reliable, this provides a highly desirable way to probe and image the interior of opaque objects.

Either or both of these characteristics occur in most ultrasonic applications. We will give one example of each to show how important they are. Surface acoustic waves (SAW) are high-frequency versions of the surface waves discovered by Lord Rayleigh in seismology. Due to their slow velocity, they can be excited and detected on a convenient length scale (cm). They have become an important part of analog signal processing, for example, in the production of inexpensive, high-quality filters, which now find huge application niches in the television and wireless communication markets. A second example is in medical applications. Fetal images have now become a standard part of medical diagnostics and control. The quality of the images is improving every year with advances in technology. There are many other areas in medicine where noninvasive acoustic imaging of the body is invaluable, such as cardiac, urological, and opthalmological imaging. This is one of the fastest growing application areas of ultrasonics. It is not generally appreciated that ultrasonics occurs in nature in quite a few different ways—both as sounds emitted and received by animals, birds, and fish, but also in the form of acoustic emission from inanimate objects. We will discuss the two cases in turn.

One of the best-known examples is ultrasonic navigation by bats, the study of which has a rather curious history [1]. The Italian natural philosopher Lazzaro Spallanzani published results of his work on the subject in 1794. He showed that bats were able to avoid obstacles when flying in the dark, a feat that he attributed to a "sixth sense" possessed by bats. This concept was rejected in favor of a theory related to flying by touch. In the light of further experimental evidence, Spallanzani modified his explanation to one based on hearing. Although this view was ultimately proven to be correct, it was rejected and the touch theory was retained. The subject was abandoned; it was only in the mid-20th century that serious research was done in the subject, principally by Griffin and Pye. The acoustic theory was retained, and considerable experimental work was carried out to characterize the pulse width, the repetition rate, and the frequency spectrum. It was found that at long range the repetition rate was quite low (10 pps) and it increased significantly at close range (100 pps), which is quite understandable from a signal processing point of view. In fact, many of the principles developed for radar and ultrasonic pulse echo work in the laboratory have already been used by bats. For example, Pye showed that the frequency changes monotonically throughout the pulse width, similar to the chirp signal described in Chapter 12, which is used in pulse compression radar. There is also evidence that bats make use of beat frequencies and Doppler shifting. There is evidence that the bat's echolocation system is almost perfectly optimized; small bats are able to fly at full speed through wire grid structures that are only slightly larger than their wingspans.

It is also fascinating that one of the bat's main prey, the moth, is also fully equipped ultrasonically. The moth can detect the presence of a bat at great distances—up to 100 ft—by detecting the ultrasonic signal emitted by the bat. Laboratory tests have shown that the moth then carries out a series of evasive maneuvers, as well as sending out a jamming signal to be picked up by the bat! Several types of birds use ultrasonics for echolocation, and, of course, acoustic communication between birds is highly developed. Of the major animals, the dog is the only one to use ultrasonics. Dogs are able to detect ultrasonic signals that are inaudible to humans, which is the basis of the silent dog whistle. However, dogs do not need ultrasonics for echolocation, as these functions are fully covered by their excellent sight and sense of smell for long- and short-range detection.

In passing to the use of ultrasonics under water, the seal is an interesting transition story. The seal provides nature's lesson in acoustic impedance, as it has two sets of ears—one set for use in air, centered at 12 kHz, and the other for use under water, centered at 160 kHz. These frequencies correspond to those of its principal predators. As will be seen for dolphins and whales, the ultrasonic frequencies involved are considerably higher than those in air; this is necessary to get roughly similar spatial resolution in the two cases, as the speed of sound in water is considerably higher than in air.

Next to bats, dolphins (porpoises) and whales are the best-known practitioners of ultrasound under water. Their ultrasonic emissions have been studied extensively, and the work is ongoing. It is believed that dolphins have a well-defined vocabulary. Some of the sounds emitted are described by graphic terms such as mewing, moaning, rasping, whistling, and clicking, all with characteristic ultrasonic properties. The latter two are the most frequent. The whistle is a low-frequency sound in pulses about a second long and frequencies in the range 7 to 15 kHz. The clicks are at considerably higher frequencies, up to 150 kHz, at repetition rates up to several hundred per second. The widths of the clicks are sufficiently short so that there is no cavitation set up in the water by the high amplitudes that are generated. High-amplitude clicks are also produced by another well-studied denizen, the snapping shrimp.

It is not often realized that natural events can give rise to ultrasonic waves. Earthquakes emit sound, but it is in the very-low-frequency range, below 20 Hz, which is called *infrasound*. The much higher ultrasonic frequencies are emitted in various processes that almost always involve the collapse of bubbles, which is described in detail in Chapter 17. The resonance of bubbles was studied by Minnaert, who calculated the resonance frequency and found that it varied inversely with the bubble size. Hence, very small bubbles have very high resonance frequencies, well into the ultrasonic range. Bubbles and many other examples of physics in nature are described in a charming book, *Light and Color in the Open Air*, by Minnaert [2].

The babbling brook is a good example of ultrasonic emission in nature as the bubbles unceasingly form and collapse. Leighton [3] measured a typical spectrum to be in the range of 3 to 25 kHz. Waterfalls give rise to the high-frequency contact, while low frequencies are produced by the water as it flows over large, round boulders. Another classic example is rain falling on a puddle or lake. The emitted sound can easily be measured by placing a hydrophone in the water. Under usual conditions a very wide spectrum, 1 to 100 kHz, is obtained, with a peak around 14 kHz. The source of the spectrum is the acoustic emission associated with impact of the water drop on the liquid surface and the entrainment of bubbles. It turns out that the broad spectrum is due to impact and the peak at 14 kHz to the sum of acoustic resonances associated with the bubble formation. An analogous effect occurs with snowflakes that fall on a water surface, apparently giving rise to a deafening cacophony beneath the surface.

Easily the largest source of ultrasound is the surface of an ocean, where breaking waves give rise to a swirly mass of bubbles and agitated water. The situation is, of course, very complicated and uncontrolled, with single bubbles, multibubbles, and fragments thereof continually evolving. This situation has been studied in detail by oceanographers. The effect is always there, but like the tree falling in the forest, there is seldom anyone present to hear it.

While ultrasonics in nature is a fascinating study in its own right, of far greater interest is the development of the technology of ultrasonic waves that is studied in the laboratory and used in industry. Ultrasonics developed as part of acoustics—an outgrowth of inventions by Langevin. There were, of course, a number of precursors in the 19th and early 20th centuries. In what follows we summarize the main developments from the beginning until about 1950; this discussion relies heavily on the excellent review article by Graff [1]. After 1950, the subject took off due to a happy coincidence of developments in materials, electronics, industrial growth, basic science, and exploding opportunities. There were also tremendous synergies between technology and fundamental advances. It would be pointless to describe these developments chronologically, so a sectorial approach is used.

A number of high-frequency sources developed in the 19th century were precursors of the things to come. They included:

1. The Savant wheel (1830) can be considered to be the first ultrasonic generator. It worked up to about 24 kHz.
2. The Galton whistle (1876) was developed to test the upper limit of hearing of animals. The basic frequency range was 3 to 30 kHz. Sounds at much higher frequencies were produced, probably due to harmonic generation, as the operation was poorly understood and not well controlled.
3. Koenig (1899) developed tuning forks that functioned up to 90 kHz. Again, these experiments were poorly understood and the conclusions erroneous, almost certainly due to nonlinear effects.
4. Various high-power sirens were developed, initially by Cagniard de la Tour in 1819. These operated below ultrasonic frequencies but had an important influence on later ultrasonic developments.

In parallel with the technological developments mentioned above, there was an increased understanding of acoustic wave propagation, including velocity of sound in air (Paris 1738), iron (Biot 1808), and water (Calladon and Sturm 1826)—the latter a classic experiment carried out in Lake Geneva. The results were reasonably consistent with today's known values—perhaps understandably so, as the measurement is not challenging because of the low value of the velocity of sound compared with the historical difficulties of measuring the velocity of light. Other notable advances were the standing wave approach for gases (Kundt 1866) and the stroboscopic effect (Toepler 1867), which led to Schlieren imaging.

One of the key events leading directly to the emergence of ultrasonics was the discovery of piezoelectricity by the Curie brothers in 1880; in short order they established both the direct and inverse effect, i.e., the conversion of an electrical to a mechanical signal and vice versa. The 20th century opened with the greatest of all acousticians, Lord Rayleigh (John W. Strutt). Rayleigh published what was essentially the principia of acoustics, *The Theory of Sound*, in 1889 [4]. He made definitive studies and discoveries in acoustics, including atomization, acoustic surface (Rayleigh) waves, molecular relaxation, acoustic pressure, nonlinear effects, and bubble collapse.

The sinking of the *Titanic* and the threat of German submarine attacks led to Langevin's experiments in Paris in 1915—the real birth of ultrasonics. On the one hand, his work demonstrated the practicality of pulse echo work at high frequencies (150 kHz) for object detection. The signals were so huge that fish placed in the ultrasonic immersion tank were killed immediately when they entered the ultrasonic beam. On the other hand, the introduction of quartz transducers and then the sandwich transducer (steel-quartz-steel) led to the first practical and efficient use of piezoelectric transducers. Quite surprisingly, almost none of Langevin's work on ultrasonics was published. His work was followed up by Cady, which led to the development of crystal-controlled oscillators based on quartz.

Between the wars, the main thrust was in the development of high-power sources, principally by Wood and Loomis. For example, a very-high-power oscillator tube in the range 200 to 500 kHz was developed and applied to a large number of high-power applications, including radiation pressure, etching, drilling, heating, emulsions, atomization, chemical and biological effects, sonoluminescence, sonochemistry, etc. Supersonic was the key buzzword, and high-power ultrasonics was applied to a plethora of industrial processes. However, this was mainly a period of research and development; and it was only in the period following this that definitive industrial machines were produced. This period, 1940–1955, was characterized by diverse applications, some of which include:

1. New materials, including poled ceramics for transduction
2. The Mason horn transducer (1950) for efficient concentration of ultrasonic energy by the tapered element
3. Developments in bubble dynamics by Blake, Esche, Noltink, Neppiras, Flynn, and others
4. Ultrasonic machining and drilling
5. Ultrasonic cleaning; GE produced a commercial unit in 1950
6. Ultrasonic soldering and welding, advances made mainly in Germany
7. Emulsification: dispersal of pigments in paint, cosmetic products, dyes, shoe polish, etc.
8. Metallurgical processes, including degassing melts

From the 1950s onward there were so many developments in so many sectors that it is feasible to summarize only the main developments by sector. Of course, the list is far from complete, but the aim is to give examples of the explosive growth of the subject rather than provide an encyclopedic coverage of the developments. The proceedings of annual or biannual conferences on the subject, such as the IEEE Ultrasonics Symposium and Ultrasonics International, are good sources of progress in many of the principal directions.

1.2 Physical Acoustics

A key element in the explosive growth of ultrasonics for electronic device applications and material characterization in the 1960s and beyond was the acceptance of ultrasonics as a serious research and development (R&D) tool by the condensed matter research community. Before 1950, ultrasonics would not have been found in the toolkit of mainline condensed matter researchers, who relied mainly on conductivity, Hall effect, susceptibility, specific heat, and other traditional measurements used to characterize solids. However, with developments in transducer technology, electronic instrumentation, and the availability of high-quality crystals it then became possible to carry out quantitative experiments on velocity and attenuation as a function of magnetic field, temperature, frequency, etc., and to compare the results with the predictions of microscopic theory. The trend continued and strengthened, and ultrasonics soon became a choice technique for condensed matter theorists and experimentalists. A huge number of sophisticated studies of semiconductors, metals, superconductors, insulators, magnetic crystals, glasses, polymers, quantum liquids, phase transitions, and many others were carried out, and unique information was provided by ultrasonics. Some of this work has become classic. Two examples will be given to illustrate the power of ultrasonics as a research tool.

Solid state and low-temperature physics underwent a vigorous growth phase in the 1950s. One of the most spectacular results was the resolution of the 50-year-old mystery of superconductivity by the Bardeen, Cooper, and Schrieffer (BCS) theory in 1957. The BCS theory proposed that the conduction electrons participating in superconductivity were coupled together in pairs with equal and opposite momentum by the electron–phonon interaction. The interaction with external fields involves so-called coherence factors that have opposite signs for electromagnetic and acoustic fields. The theory predicted that at the transition temperature there would be a peak of the nuclear spin relaxation time and a straight exponential decrease of the ultrasonic attenuation with temperature. This was confirmed by experiment and was an important step in the widespread acceptance of the BCS theory. The theory of the ultrasonic attenuation was buttressed on the work of Pippard, who provided a complete description of the interaction of ultrasonic waves with conduction electrons around the Fermi surface of metals.

A second example is provided by liquid helium, which undergoes a transition to the superfluid state at 2.17 K. Ultrasonic experiments demonstrated a change in velocity and attenuation below the transition. Perhaps more importantly, further investigation showed the existence of other ways of propagating sound in the superfluid state in different geometries—so that one talks of a first (ordinary), second, third, and fourth sound in such systems. These acoustics measurements went a long way to providing a fuller understanding of the superfluid state. The case of He^3 was even more fruitful for acoustic studies. The phase diagram was much more complicated, involving the magnetic field, and many new hydrodynamic quantum modes were discovered. Recently, even purely propagating transverse waves were found in this superfluid medium.

This and other fundamental work led to attempts to increase the ultrasonic frequency. Coherent generation by application of microwave fields at the surface of piezoelectrics raised the effective frequency well into the hypersonic region above 100 GHz. Subsequently, the superconducting energy gap of thin films was used to generate and detect high-frequency phonons at the gap frequency, extending the range to the THz region. Heat pulses were used to generate very-high-frequency broadband pulses of acoustic energy. In another approach, the development of high-flux nuclear reactors led to measurement of phonon dispersion curves over the full high-frequency range, and ultrasonics became a very useful tool for confirming the low-frequency slope of these curves. In summary, all of this work in physical acoustics gave new legitimacy to ultrasonics as a research tool and stimulated development of ultrasonic technologies.

1.3 Low-Frequency Bulk Acoustic Wave (BAW) Applications

This main focus of our discussion on the applications of ultrasonics provides some of the best examples of ultrasonic propagation. The piezoelectric transducer itself led to some of the earliest and most important applications. The quartz resonator was used in electronic devices starting in the 1930s. The quartz microbalance became a widely used sensor for detection of the mass loading of molecular species in gaseous and aqueous media and will be fully described in Chapter 13. Many other related sensors based on this principle were developed and applied to many problems such as flow sensing (including Doppler), level sensing, and propagation (rangefinders, distance, garage door openers, camera rangefinders, etc.). A new interest in propagation led to the development of ultrasonic nondestructive evaluation (NDE). Pulse echo techniques developed during World War II for sonar and radar led to NDE of materials and delay lines using the same principles and electronic instrumentation. Materials NDE with shorter pulse and higher frequencies was made possible with the new electronics developed during the war, particularly radar. A first ultrasonic flaw detection patent was issued in 1940. From 1960 to the present there have been significant advances in NDE technology for detecting defects in multilayered, anisotropic samples, raising ultrasonics to the status of a major research tool, complementary to resistivity, magnetization, x-rays, eddy currents, etc.

One of the most important areas in low-frequency BAW work was the development of ultrasonic imaging, which started with the work of Sokolov. By varying the position and angle of the transducer, A (line scan), B (vertical cross-section), and C (horizontal cross-section) scans were developed. C scan has turned out to be the most commonly used, where the transducer is translated in the x-y plane over the surface of a sample to be inspected so that surface and subsurface imaging of defects can be carried out. Realization by Quate in the early 1970s that microwave ultrasonics waves in water have optical wavelengths led to the development of the scanning acoustic microscope (SAM) by Lemons and Quate in 1974. This is covered in detail in Chapter 14 because it is a textbook example of the design of an ultrasonic instrument. The SAM provides optical resolution for frequencies in the GHz range, high intrinsic contrast, quantitative measure of surface sound velocities, and subsurface imaging capability. In more recent developments the atomic force microscope (AFM), also developed by Quate, has been used to carry out surface, near-surface, and near-field imaging with nanometer resolution. In parallel, much progress has been made in acoustic imaging with phased arrays. Recent developments include time-reversal arrays and the use of high-performance micromachined capacitive transducer arrays.

1.4 Surface Acoustic Waves (SAWs)

The SAW was one of the modes discovered very early on by Lord Rayleigh in connection with seismology studies. In the device field it remained a scientific curiosity with few applications until the development of the interdigital transducer (IDT) by White and Voltmer in the 1960s. This breakthrough allowed the use of planar microelectronic technology, photolithography, clean rooms, etc. for the fabrication of SAW devices in large quantities. A second breakthrough was a slow but ultimately successful development of sputtering of high-quality ZnO films on silicon, which liberated device design from bulk piezoelectric substrates and permitted integration of ultrasonics with silicon electronics. Since the 1960s, there has been a huge amount of work on the fundamentals and the technology of SAW and its application to signal processing, NDE, and sensors. The SAW filter has been particularly important commercially in mass consumer items such as TV filters and wireless communications. There is presently a push to very-high-frequency devices (5 to 10 GHz) for communications applications.

The above topics are the main ones covered in the applications sections. Of course, there are many other extremely important areas of ultrasonics, but a selection was made of those topics that seemed best suited as examples of the basic theory and which the author was qualified to address. Some of the important areas omitted (and the reasons for omission) include piezoelectric materials, transducers, medical applications (specialized and technical), high-power ultrasonics (lacks a well-developed theoretical base), underwater acoustics, and seismology (more acoustics than ultrasonics and lacking in unity with the other topics). In these cases, a brief summary of some of the highlights is given to complete the introductory survey of this chapter.

1.5 Piezoelectric Materials

Much of the remarkable progress made in ultrasonics is due to the synergy provided by new high-performance materials and improved electronics. This is perhaps best exemplified in the work of Langevin in applying quartz to transduction and then developing the composite transducer. A second major step forward occurred in the 1940s with the development of poled ceramic transducers of the lead zirconate (PZT) family, which were relatively inexpensive, rugged, high performance, and ideally suited to field work. For the laboratory, more expensive but very high-performance new crystals such as lithium niobate entered into widespread use. A third wave occurred with piezoelectric films. After a false start with CdS, ZnO (and also AlN to some extent) became the standard piezoelectric film for device applications such as SAW. The development of polyvinylidine (PVDF) and then copolymers based on it was important for many niche applications—particularly in medical ultrasonics, as the acoustic impedance is very well matched to water. Other favorable properties include flexibility and wide bandwidth. They are, however, very highly attenuating, so they are not suitable for SAW or high-frequency applications.

More recently, the original PZT family has been improved by the use of finely engineered piezocomposites for general BAW applications. New SAW substrates are still under development, particularly with the push to higher frequencies. Microelectromechanical (MEMS) transducers are under a stage of intense development as they have potential for high-quality, real-time, mass-produced acoustic imaging systems.

1.6 High-Power Ultrasonics

This was one of the first areas of ultrasonics to be developed, but it has remained poorly developed theoretically. It involves many heavy-duty industrial applications, and often the approach is semi-empirical. Much of the early work was carried out by Wood and Loomis, who developed a high-frequency, high-power system and then used it for many applications. One of the problems in the early work was the efficient coupling of acoustic energy into the medium, which limited the available power levels. A solution was found with the exponential horn; a crude model was developed by Wood and Loomis, and this was perfected by Mason using an exponential taper in 1950. The prestressed ceramic sandwich transducers also were important in raising the acoustic power level. Another problem, which led in part to the same limitation, was cavitation. Once cavitation occurs at the transducer or horn surface, the transfer of acoustic energy is drastically reduced due to the acoustic impedance mismatch introduced by the air.

However, work on cavitation gradually led to it becoming an important subject in its own right. Ramification of the process led to operations such as drilling, cutting, and ultrasonic cleaners. Other applications of cavitation included sonochemistry and sonoluminescence. High-power ultrasonics also turned out to be a useful way to supply large amounts of heat, leading to ultrasonic soldering and welding of metals and plastics.

1.7 Medical Ultrasonics

From a purely technical ultrasonic standpoint, there are many similarities between NDE and medical ultrasonics. Basically, one is attempting to locate defects in an opaque object; the same technological approaches are relevant, such as discriminating between closely spaced echoes and digging signals out of the noise. So it is not surprising that many developments on one side have been applied to problems on the other. Of course, there are differences: one is that inspection of *in vivo* samples is an important part of medical ultrasonics. Respiratory effects, blood flow, and possible tissue damage are issues that are totally absent in NDE. This has led to much R&D on induced cavitation and cavitation damage as well as development of very sophisticated Doppler schemes for monitoring blood flow.

Historically, during the 1940s and 1950s, there was strong emphasis on therapy. This declined in the 1950s when the current dominant theme of medical imaging started. There was much work on the brain, followed by applications in urology, ophthalmology, and vital organs (heart and liver). Certainly the most celebrated application of ultrasonic imaging in medicine is fetal imaging; images of tremendous detail and clarity can be obtained in real time. High-resolution *in vitro* imaging has been carried out in the same way. Current trends for *in vivo* imaging include phased arrays for real-time imaging and nonlinear imaging using contrast agents as well as harmonic imaging of basic tissue.

1.8 Acousto-Optics

The interaction of light and sound was discovered early in the history of ultrasonics. Brillouin suggested the existence of Brillouin scattering in 1922, which was followed by low-frequency diffraction (Debye-Sears 1932 and Raman-Nath 1935). Schlieren visualization of ultrasonic fields has long been a useful tool for exploring scattering and propagation phenomena. Bragg cells for acousto-optic modulators are important components in optical communication systems. An important developing area is that of laser ultrasonics.

It has been known since the 1960s that absorption of a laser beam can lead to generation of ultrasonic waves by the thermoelastic effect. The mode generated can be partly controlled by the surface condition. An all-optical system can be made by using a Michelson interferometer to monitor surface displacement. A special application of laser ultrasonics is described in Chapter 16.

1.9 Underwater Acoustics and Seismology

Fascinating as they are, underwater acoustics and seismology cannot be properly put under the umbrella of ultrasonics as almost all of the work in these areas is done in the audio or infrasonic frequency range. It is only the tail end, as it were, of a few graphs that penetrate into the ultrasonic regime. Nevertheless, the basic theory is the same, and only the length scale is much larger. Also, the acoustic phenomena of interest are in many cases identical. One needs only cite the names of Rayleigh, Love, and Sezawa waves in the earth's crust, longitudinal and transverse wave propagation in the bulk of the earth, and multilayer and reflection and transmission phenomena in the case of seismology. For underwater acoustics we have again reflection and transmission phenomena, guided waves in channels due to stratified layers caused by temperature gradients, scattering of acoustic waves by targets of all sorts, bubble phenomena, acoustic imaging, sonar, and the list goes on. In both cases we have the inverse problem that is at the base of a large chunk of NDE. One of the advantages of the situation, at least in principle, is that it should be relatively easy for experts in ultrasonics to work on problems in these other fields and vice versa.

2

Introduction to Vibrations and Waves

2.1 Vibrations

The general objective of this chapter is to give an introduction to vibrations and waves (see, e.g., [5]). More specifically, the chapter also has the goal of recalling the basic mathematical apparatus necessary to read the book and to introduce the simple physical ideas and analogies that will be useful throughout the book. The model system used will be a simple oscillator, a mass connected to a spring, although a simple pendulum or any other similar system could have been used. For small displacements it will be seen that the oscillations are sinusoidal at a single frequency, so-called simple harmonic motion.

Looking at Figure 2.1, we easily see that the motion will be periodic. If the mass is displaced initially there will be a restoring force due to the spring. For small displacements, Hooke's law applies, so that the restoring force is given by $F = -kx$. This is in fact the leading term in a Taylor's expansion of the force in terms of the displacement. Hooke's law is ubiquitous in mechanical problems of vibrations and waves. For example, it is this approximation that is used to define the elastic constants of crystals and that is also at the basis of the theory of elasticity of solids. If Hooke's law is not obeyed then things become much more complicated, mathematically and physically, and we enter the realm of nonlinear acoustics. Except where stated otherwise, we will always remain in the linear regime described by Hooke's law.

Hooke and Newton were great English scientists of the 17th century and there was ill-concealed tension between them. It is thus somewhat ironic that the basic equation for the simple oscillator and the wave equation are both obtained by a happy combination of Hooke's law and Newton's equation of motion. For the mass-spring system this can be written

$$F = m\frac{d^2x}{dt^2} \tag{2.1}$$

or

$$\frac{d^2x}{dt^2} + \frac{k}{m}x = 0 \tag{2.2}$$

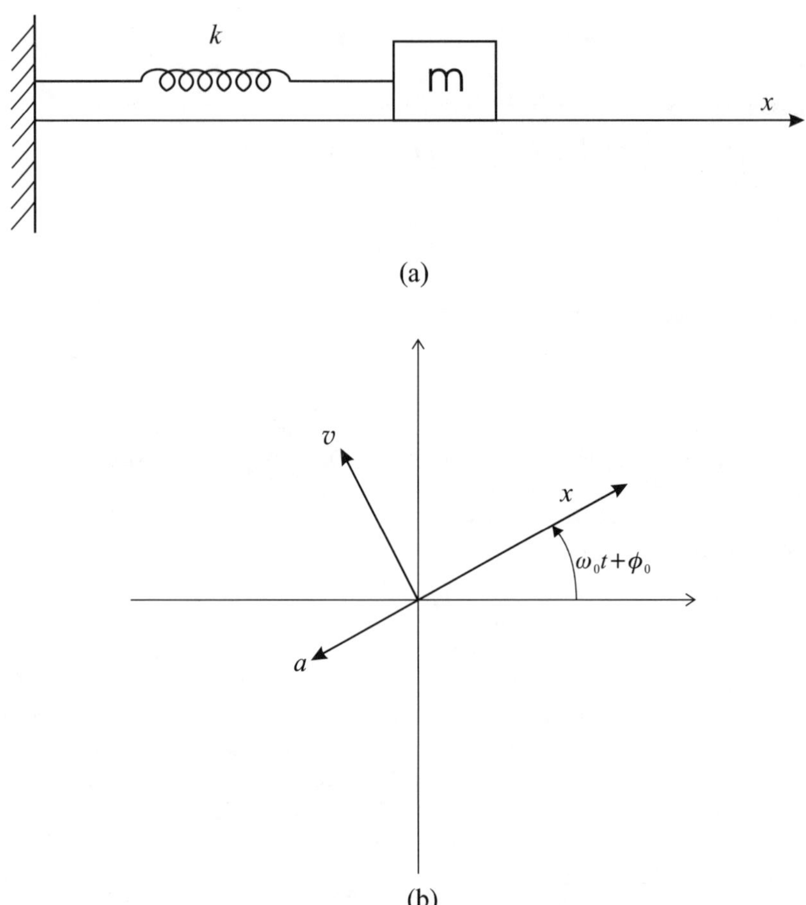

FIGURE 2.1
(a) Mass-spring oscillator. (b) Phasor diagram for simple harmonic motion.

Physically, this equation provides the solution $x(t)$ for the displacement of the mass. Once the mass is released at $t = 0$, it is pulled in the $-x$ direction by the spring, which is in turn compressed by the movement of the mass. At the moment of maximum compression, all of the energy of the system is stored as potential energy in the spring. The mass is then repelled to the right by the spring and at the instant where the spring extension is zero, the potential energy is also zero and all of the energy of the system is now in the form of kinetic energy of the mass. If there is no dissipative force, the process will be periodic with exchange from kinetic to potential energy and vice versa and will continue ad infinitum. If there is dissipation, for example, friction with the supporting surface, the motion will be progressively damped and will finally come to a halt. Finally, it should be noted that this is a fixed, isolated vibrator that undergoes periodic motion. There is no wave propagated here: that aspect will be discussed in Section 2.2.

Returning to Equation 2.1, this can be clearly identified as the harmonic equation, with harmonic solutions. Defining the angular frequency $\omega_0^2 = k/m$, these solutions are of the form

$$x = A_1 \cos \omega_0 t + A_2 \sin \omega_0 t \tag{2.3}$$

For this second-order homogeneous differential equation the solution has two arbitrary constants to be determined by the initial conditions. Alternatively, the solution can be written

$$x = A \sin(\omega_0 t + \phi_0) \tag{2.4}$$

where ϕ_0 is an initial phase angle. The frequency f and the period T are determined by

$$f_0 = \frac{\omega_0}{2\pi} \tag{2.5}$$

$$T = \frac{1}{f_0} \tag{2.6}$$

The subscript zero is used as this is a simple undamped oscillator.

The complete solution can be found using the initial conditions. At $t = 0$, we define the initial displacement x_0 and the initial velocity v_0, from which we immediately find

$$A = \left[x_0^2 + \left(\frac{v_0}{\omega_0}\right)^2\right]^{\frac{1}{2}} \tag{2.7}$$

$$\phi_0 = \tan^{-1}\left(\frac{-v_0}{\omega_0 x_0}\right) \tag{2.8}$$

which completely determines the displacement from Equation 2.4. The velocity v and acceleration a are immediately found as

$$v = v_m \cos(\omega_0 t + \phi_0) \tag{2.9}$$

and

$$a = -\omega_0 v_m \sin(\omega_0 t + \phi_0) \tag{2.10}$$

From these solutions we can deduce that the displacement and velocity are in phase quadrature (displacement lags by $\pi/2$), and the displacement and acceleration are π out of phase. This type of analysis will be found to be important for waves.

2.1.1 Vibrational Energy

For a mechanical system in general the total energy U is the sum of the potential energy U_P and the kinetic energy U_K. These are readily calculated for our model system. U_P is determined by the work done to compress the spring:

$$U_P = \int_0^x kx\,dx = \frac{1}{2}kx^2 = \frac{1}{2}kA^2\sin^2(\omega_0 t + \phi_0) \tag{2.11}$$

The kinetic energy is determined by the usual mechanical formula for a mass m:

$$U_K = \frac{1}{2}mv_m^2 = \frac{1}{2}mv_m^2\cos^2(\omega_0 t + \phi_0)$$

Hence, the total energy is given by

$$U = U_P + U_K = \frac{1}{2}m\omega_0^2 A^2 \tag{2.12}$$

Alternatively, as could have been deduced from the discussion of energy exchange during a cycle, the total energy is simply equal to the maximum potential or kinetic energy:

$$U = \frac{1}{2}kA^2 = \frac{1}{2}mv_m^2 \tag{2.13}$$

2.1.2 Exponential Solutions: Phasors

The previous results for x, v, and a were obtained using the real trigonometric functions sine and cosine to represent the periodic variation with time. There is an alternative representation that is conceptually simple and mathematically more economic than the use of real trigonometric functions. This is the use of complex exponentials, which is almost universally employed in research papers. In the complex plane, it is well known that we can represent sine and cosine functions in the complex plane by using Euler's rule

$$e^{j\theta} = \cos\theta + j\sin\theta$$

where $j = \sqrt{-1}$. Generally, j is used in engineering practice and i in mathematics and physics, but this is not universal. When they are not used as an index, the scalars i or j always represent $\sqrt{-1}$. We may use them interchangeably. In the complex plane the x axis represents the "real" part and the y axis represents the "imaginary" part of a variable $z = x + iy = re^{i\theta}$. When a physical quantity is represented by a complex variable z, by convention its physically significant part is given by Re(z). This is pure convention; since the real and imaginary parts contain redundant information, the imaginary part could equally well have been chosen. The semantics have been chosen to reinforce the conventional choice.

Complex exponential notation is ideally suited for the representation of harmonic vibrations. Thus, instead of describing a physical displacement as $x = A\cos\omega t$, we can represent it by the quantity $x = Ae^{j\theta} = Ae^{j\omega t}$. The radius vector A is real and it rotates at constant angular velocity $\dot{\theta} = \omega$. Thus, the projection on the x axis, the real part, traces out the variation $x = A\cos\omega t$ with time. The polar representation is called the phasor representation (A is a "phasor"). Phasors are a simple graphical way to represent vibrations and they are particularly useful when several different vibrations are added and one wishes to calculate the resultant. As before, two quantities must be given to specify a phasor, namely the amplitude (radius vector) and the phase (angle θ). Another analytical advantage of the use of complex numbers and phasors is that multiplication by j corresponds to an advance in phase by 90° (rotation from the real to the imaginary axis). Similarly, multiplication by $-j$ retards the phase by $\pi/2$. Thus, phase relationships can be deduced instantly from analytical formulae by identifying the imaginary terms and their sign.

2.1.3 Damped Oscillations

A simple undamped oscillator is, of course, an academic simplification. In the real world, there are always frictional and resistive effects that eventually damp out an oscillator's movement unless it is maintained by an external force. In this section we examine the damping effects and then study the forced, damped oscillator in the subsequent section.

Most if not all damping mechanisms provide an opposing force that is proportional to the velocity or current. Frictional forces and the potential drop across a resistor are two common examples. The force can be written

$$F = -R_m\frac{dx}{dt} \tag{2.14}$$

where the subscript m stands for mechanical, to distinguish R_m from an electrical resistance R. In a mass-spring system, R_m is often represented as a dashpot that slows the movement of the mass. The equation of motion can now be written

$$\frac{d^2x}{dt^2} + \frac{R_m}{m}\frac{dx}{dt} + \omega_0^2 x = 0 \tag{2.15}$$

using a trial solution $x = Ae^{\gamma t}$

$$\left(\gamma^2 + \frac{R_m}{m}\gamma + \omega_0^2\right)x = 0$$

leading to a condition on γ

$$\gamma = -\alpha \pm \sqrt{\alpha^2 - \omega_0^2} \tag{2.16}$$

where $\alpha = R_m/2m$.

For typical mechanical systems of interest, the oscillation persists for at least several cycles so that $\alpha < \omega$ for this case. We then define a frequency $\omega_1^2 = \omega_0^2 - \alpha^2$ for the damped oscillator, so that finally

$$x = e^{-\alpha t}(A_1 e^{j\omega t} + A_2 e^{-j\omega t}) = Ae^{-\alpha t}e^{j(\omega_1 t + \phi)} \tag{2.17}$$

2.1.4 Forced Oscillations

In practice, virtually all oscillators are forced, either by external amplifiers or by feedback. Hence, the frequency response is of prime importance; depending on the application, the objective may be to excite the oscillator at a particular frequency or over a wide bandwidth. We start by establishing the system response at a single driving frequency and then extend these results to the response for an arbitrary frequency.

For an applied force $Fe^{j\omega t}$, the differential equation can be written

$$\frac{d^2x}{dt^2} + \frac{R_m}{m}\frac{dx}{dt} + \omega_0^2 x = Fe^{j\omega t} \tag{2.18}$$

Physically, in the steady state, the system must respond at the applied frequency, so we look for solutions of the form $x = Ae^{j\omega t}$. Substitution in Equation 2.18 gives

$$x = \frac{1}{j\omega}\frac{Fe^{j\omega t}}{R_m + j\left(\omega m - \frac{k}{\omega}\right)} \tag{2.19}$$

and

$$v = \frac{dx}{dt} = \frac{Fe^{j\omega t}}{R_m + j\left(\omega m - \frac{k}{\omega}\right)} \tag{2.20}$$

Equation 2.20 has the form of Ohm's law for an electrical AC circuit. A formal analogy can be established by defining the mechanical impedance

$$Z_m = R_m + jX_m \tag{2.21}$$

where the mechanical reactance $X_m = \omega m - k/\omega$ follows from Equation 2.20. Analogous to Ohm's law, we then have impedance = force/velocity. This analogy is also valid for acoustic waves and the concept of acoustic impedance will be used throughout this book.

Analogous to electrical circuits, the real and imaginary parts of the impedance can be represented by a vector diagram, corresponding to the complex plane, with phase angle $\tan\theta = [\omega m - k/\omega]/R_m$. The real values of displacement and velocity are given by

$$x = \left(\frac{F}{\omega Z_m}\right)\sin(\omega t - \theta) \tag{2.22}$$

$$v = \left(\frac{F}{Z_m}\right)\cos(\omega t - \theta) \tag{2.23}$$

Thus the velocity lags the applied force by a phase angle θ. As in an AC circuit this will affect the power transferred to the oscillator as the force and velocity are, in general, not in phase. The power transferred at time t is

$$P(t) = F(t)v(t) = \left(\frac{F^2}{Z_m}\right)\cos\omega t\cos(\omega t - \theta) \tag{2.24}$$

Of more importance is the average power transferred over a cycle

$$\begin{aligned} P_0 &= \langle P(t)\rangle = \frac{1}{T}\int_0^T P(t)dt \\ &= \frac{F^2}{2Z_m}\cos\theta = \frac{F^2 R_m}{2Z_m^2} \end{aligned} \tag{2.25}$$

The maximum power transferred occurs when the mechanical reactance vanishes ($\theta = 0$) and the impedance Z_m takes its minimum value R_m, which occurs at $\omega = \omega_0$. This is called the resonance frequency of the system. The power as a function of frequency is shown in Figure 2.2. An important parameter of the

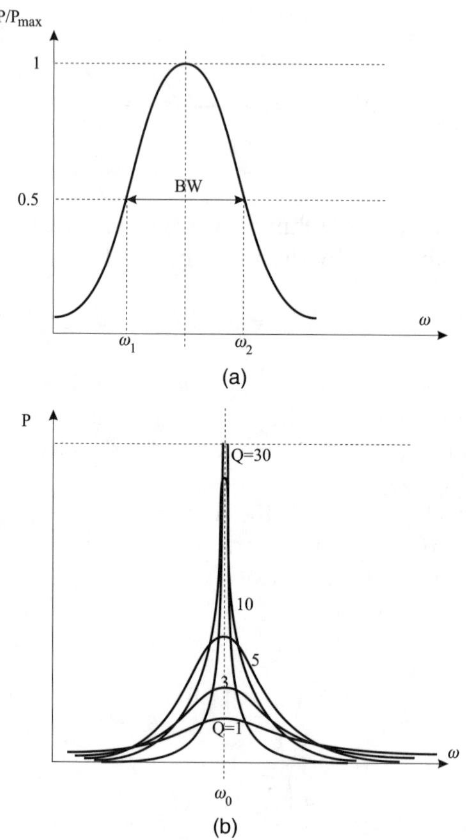

FIGURE 2.2
(a) Mean power input as a function of frequency to show the sharpness of the resonance curve. (b) Mean power absorbed by a forced oscillator as a function of frequency in units of $F^2/2m\omega_0$.

power curve $P_0(\omega)$ is the relative width of the curve around the resonance. Like the equivalent electrical system, the width is described by the Q or quality factor. There are various ways to define and describe the Q of the system and these are summarized as follows:

1. The Q can be defined as the resonance frequency divided by the bandwidth $BW \equiv$ frequency difference between the upper and lower frequencies for which the power has dropped to half of its maximum value:

$$Q = \frac{\omega_0}{BW} \tag{2.26}$$

   Hence high Q corresponds to a sharp resonance with a narrow bandwidth.

2. The above form for Q can be rewritten in terms of mechanical constants. For the two half power points $Z_m^2 = 2R_m^2$. Using $X_m = \omega m - k/\omega$, this gives

$$Q = \frac{\omega_0 m}{R_m} \tag{2.27}$$

   Thus high Q corresponds to small R_m or low loss.

3. In terms of the decay time τ of the free oscillator, which is the time for the amplitude to fall to $1/e$ of its initial value, $\tau = 1/\alpha$ from Equation 2.17, $\alpha = R_m/2m$,

$$Q = \frac{1}{2}\omega_0 \tau \tag{2.28}$$

   This means that a high Q oscillator when used as a free oscillator will "ring" for a long time, of the order of τ, before the amplitude falls to zero.

4. Finally, a formal definition of Q, equivalent to the above, is

$$Q = \frac{\text{stored energy}}{\text{total energy dissipated}} \tag{2.29}$$

   Again, a high Q oscillator is a low loss system.

5. Q can also be seen as an amplification factor. As R decreases the displacement-frequency curve gets sharper and the amplitude at resonance A_0 increases significantly. Direct calculation of Q from the definition leads to

$$Q = A_0\left(\frac{k}{F_0}\right) \tag{2.30}$$

F_0/k is the amplitude at asymptotically low frequencies, so Equation 2.30 means that the amplitude at resonance is a factor of Q greater than at low frequencies. This is the physical basis for the demonstrably high displacements attainable in mechanical systems at resonance. The same principle is routinely exploited in high Q electrical circuits, for example, in RF receivers.

The full analogy between electrical and mechanical quantities is displayed in Table 2.1, together with a list of key formulae. Physically, by Lenz's law, inductance corresponds to the inertia (mass) of the system to change in current. The condenser stores the potential energy as does the compressed spring in the mechanical system. The resistance corresponds to the dissipated energy in both cases. Care must be taken in what quantities are held constant when comparing electrical circuits to mechanical configurations. For example, in Figure 2.3(a) the source voltage is held constant and the same current flows

TABLE 2.1

Comparison of Equivalent Electrical and Mechanical Resonant Circuits

Electrical	Mechanical
Charge Q	Displacement x
Current I	Velocity v
Applied voltage V	Applied force F
Resistance R	Mechanical resistance R_m
Inductance L	Mass m
Capacitance C	Spring compliance $C = 1/k$
Impedance	Mechanical impedance
$Z = R + j(\omega L - 1/\omega C)$	$Z_m = R_m + j(\omega m - k/\omega)$
Differential Equation	
$L\frac{d^2Q}{dt^2} + R\frac{dQ}{dt} + \frac{Q}{C} = V_0 e^{j\omega t}$	$m\frac{d^2x}{dt^2} + R_m\frac{dx}{dt} + kx = F_0 e^{j\omega t}$
Solution	
$Q = \frac{1}{j\omega}\frac{V}{Z}$	$x = \frac{1}{j\omega}\frac{F}{Z_m}$
Resonant Frequency	
$\omega_0 = \sqrt{1/LC}$	$\omega_0 = \sqrt{k/m}$
Energy	
$U_K = \frac{1}{2}LI^2$	$U_K = \frac{1}{2}mv^2$
$U_P = \frac{1}{2}CV^2 = \frac{Q^2}{2C}$	$U_P = \frac{1}{2}kx^2$
Phase Angle	
$\phi = \tan^{-1}\left(\frac{(\omega L - 1/\omega C)}{R}\right)$	$\phi = \tan^{-1}\left(\frac{(\omega m - k/\omega)}{R_m}\right)$

through all elements in the electrical circuit. This clearly corresponds to the mechanical configuration shown in Figure 2.3(b), where all elements have the same velocity and amplitude if the force is constant.

2.1.5 Phasors and Linear Superposition of Simple Harmonic Motion

A phasor has amplitude and orientation (phase angle) and as such is a vector. If two phasors have the same frequency then they can be added vectorially. Graphically they can be drawn head to tail to give a resultant phasor with

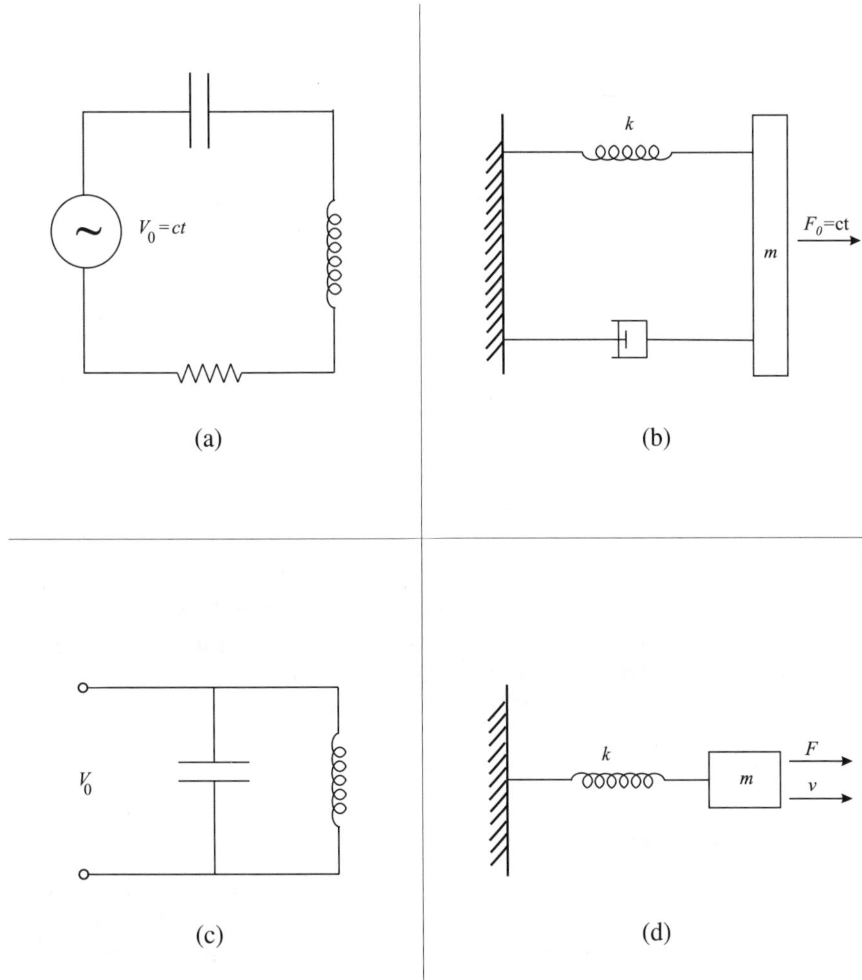

FIGURE 2.3
(a) Series electrical circuit and (b) its mechanical equivalent. (c) Parallel electrical circuit and (d) its mechanical equivalent.

components as shown in Figure 2.4. For n such phasors we have

$$A = \left[\left(\sum A_n \cos\phi_n\right)^2 + \left(\sum A_n \sin\phi_n\right)^2\right]^{\frac{1}{2}}$$
$$\tan\phi = \frac{\Sigma A_n \sin\phi}{\Sigma A_n \cos\phi} \tag{2.31}$$

For $n \to \infty$ and equal contribution for each constituent, the polygonal locus becomes an arc of a circle. In this way, interference and diffraction patterns in acoustics and optics can be constructed.

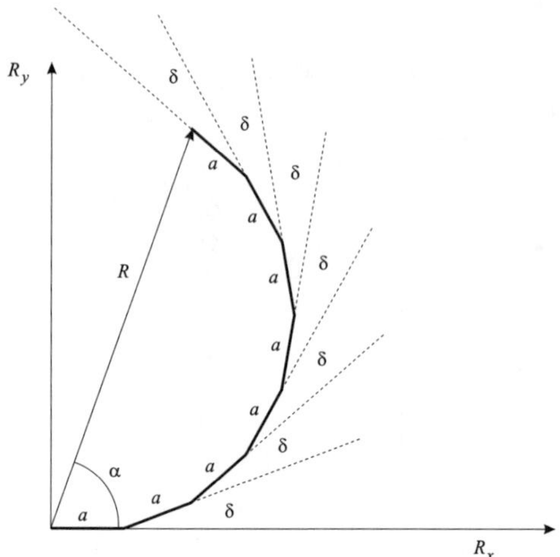

FIGURE 2.4
Addition of phasors of equal amplitude and phase difference.

The above results are for superposition of vibrations at the same frequency. If the frequencies are different the motion becomes complicated and aperiodic, even if there are only two components. In the case of two vibrations with frequencies very close together, "beats" can be observed at the difference frequency. The question will be taken up for the case of waves and the formation of wave packets later in the chapter.

2.1.6 Fourier Analysis

We now turn to what is in some respects the inverse problem to the addition of phasors presented in the last section. If we start with an arbitrary periodic function, Fourier showed that it can be represented as an infinite sum of sine and cosine (i.e., harmonic) terms. The subject, together with that of Fourier transforms for nonperiodic functions, has been treated in numerous texts and we only summarize some of the main results here.

We consider an anharmonic (nonsinusoidal) periodic function of time, such as a square wave. Then Fourier's theorem states that it can be represented as a Fourier series

$$f(t) = \frac{A_0}{2} + \sum_{n=1}^{\infty} A_n \cos n\omega t + \sum_{n=1}^{\infty} B_n \sin n\omega t \tag{2.32}$$

where

$$A_n = \frac{2}{T}\int_0^T f(t)\cos n\omega t \, dt$$

$$B_n = \frac{2}{T}\int_0^T f(t)\sin n\omega t \, dt$$

The symmetry or lack thereof of the function to be analyzed can lead to important simplifications. For example, suppose that the origin has been chosen so that the square wave in question has odd symmetry. Since sine waves have odd symmetry ($\sin t = -\sin(-t)$) and cosine waves are even ($\cos t = \cos(-t)$), the Fourier series of this square wave can have only sine terms. After only three terms, the general shape of the square wave is reproduced, but clearly it will take many terms (in principle an infinite number) to reproduce the vertical front.

2.1.7 Nonperiodic Waves: Fourier Integral

The previous results on Fourier analysis (synthesis) can be extended from periodic functions to nonperiodic functions (for example, single pulses) by a simple artifice. If we extend the period T in Equation 2.32 to $T \to \infty$ then we effectively have a single pulse or more generally a transient disturbance $f(t)$ that we can describe by a simple generalization of the series

$$f(t) = \frac{1}{\pi}\left[\int_0^\infty A(\omega)\cos\omega t\,d\omega + \int_0^\infty B(\omega)\sin\omega t\,d\omega\right] \qquad (2.33)$$

where

$$A(\omega) = \int_{-\infty}^{\infty} f(t)\cos\omega t\,dt$$

$$B(\omega) = \int_{-\infty}^{\infty} f(t)\sin\omega t\,dt$$

As an example for a square pulse (see Figure 2.5)

$$\begin{aligned} f(t) &= E_0 \quad |t| < \frac{T}{2} \\ &= 0 \quad |t| > \frac{T}{2} \end{aligned} \qquad (2.34)$$

which is an even function, the sine term is zero, and

$$\begin{aligned} A(\omega) &= E_0 T \frac{\sin\left(\frac{\omega T}{2}\right)}{\frac{\omega T}{2}} \\ &= \operatorname{sinc}\left(\frac{\omega T}{2}\right) \end{aligned} \qquad (2.35)$$

which is also shown in Figure 2.5. This is a very familiar result in optics when variables t and ω are replaced by x and k. It corresponds to diffraction by a single slit.

It is more economical and standard practice to rewrite Equation 2.33 in complex notation to obtain a Fourier transform pair

$$\begin{aligned} f(t) &= \int_{-\infty}^{\infty} g(\omega)e^{j\omega t}d\omega \\ g(\omega) &= \frac{1}{2\pi}\int_{-\infty}^{\infty} f(t)e^{-j\omega t}dt \end{aligned} \tag{2.36}$$

where the negative frequency, by Euler's theorem, is nothing more than a way to write the complex conjugate

$$e^{j(\pm\omega t)} = \cos\omega t \pm j\sin\omega t \tag{2.37}$$

It is readily seen that dimensionally the members of the Fourier transform pair are the inverse of each other. Moreover, if the pulse is very narrow in t space, it is very wide in ω space and vice versa. Two important examples are the slit function, already shown as having a sine Fourier transform, and the Gaussian, both shown in Figure 2.5. The Gaussian transform can easily

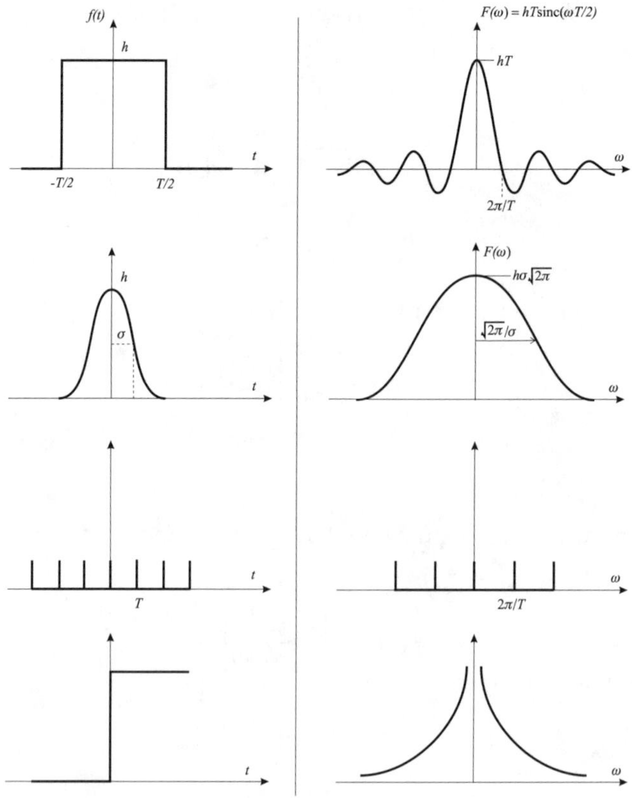

FIGURE 2.5
Some common Fourier transform pairs.

be changed into another Gaussian. As a limiting case consider the Dirac delta function

$$\begin{aligned} \delta(t) &= 0 \quad \text{for } t \neq 0 \\ \int_{-\infty}^{\infty} \delta(t)dt &= 1 \end{aligned} \tag{2.38}$$

which is an infinite spike of unit area at $t = 0$. Then the Fourier transform

$$g(\omega) = \frac{1}{2\pi}\int_{-\infty}^{\infty} \delta(t)e^{-j\omega t}dt = \frac{1}{2\pi} \tag{2.39}$$

is a constant, independent of frequency.

The δ function results are a direct demonstration of the bandwidth theorem, which states that

$$\Delta\omega\,\Delta t \sim 1 \tag{2.40}$$

Applied to a single pulse, the theorem states that the narrower the pulse the wider the associated frequency spectrum and vice versa consistent with the results for the Fourier transform of the Gaussian. We return to the bandwidth theorem in the next section to generalize it to the case of waves and wave packets.

2.2 Wave Motion

Waves are universal, presenting themselves in different guises in nature, and they are ubiquitous in the physics and engineering laboratory. They are in fact so common in different areas of science (acoustics, optics, electromagnetics, etc.) that wave motion is usually taught as a subject in its own right in elementary physics courses. What follows is not a substitute for these elementary treatments but rather a summary that enables us to collect the main results in one place, establish notation, and emphasize certain concepts that are important for this book, such as phase and group velocity.

The first question is: What is a wave? In fact, simple intuitive answers to this question can be reformulated in precise mathematical language to provide a test for a given function to decide if it corresponds to wave propagation or not. For the moment we avoid pathological problems such as strongly scattering, highly dispersive media, etc., and concentrate on the linear regime in simple, nondispersive media. In this spirit we then define a wave "as the self-sustaining propagation at constant velocity of a disturbance without change of shape." We can represent the shape of the disturbance by the function $f(x,t)$, a Gaussian $f(x,0)$ at $t = 0$. The pulse is then propagated at constant velocity V, and at time t we can describe the same profile in a moving reference frame x' as $f(x')$. Since $x' = x - Vt$ by inspection and there has been no change in shape, we have $f(x,t) = f(x - Vt)$ for any time t. This form $f(x - Vt)$ is characteristic of a wave traveling to the right, or in the forward direction. It is easy to see, for the same coordinate system, that a wave propagating to the left would be described by $f(x + Vt)$. This simple rule has a functionality that will become clear throughout the book. For example, according to it, $\sin(\omega t - kx)$ is indeed a wave and $\sin \omega t$ is not; in fact, the latter is clearly an example of harmonic motion of a fixed oscillator, as discussed in Section 2.1.

As for the case of simple harmonic motion for a mechanical oscillator, we determine the equation of motion of the mechanical system under study by combining Hooke's law with Newton's equation of motion. One of the simplest possible examples is that of the transverse vibrations of a string or a cord (see Figure 2.6). For simplicity we consider the string to be under a certain

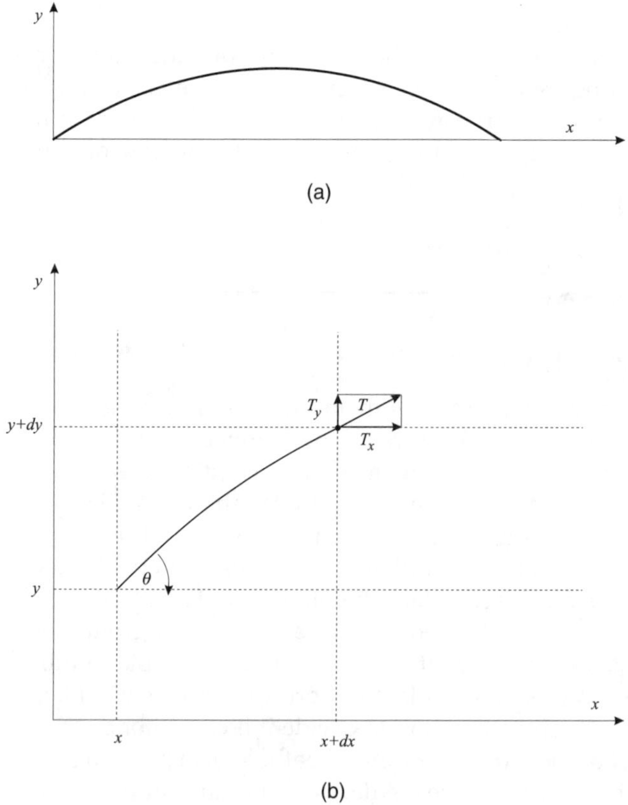

FIGURE 2.6
(a) Vibrating string with fixed end points. (b) Forces on a string element.

LANCASHIRE LIBRARY

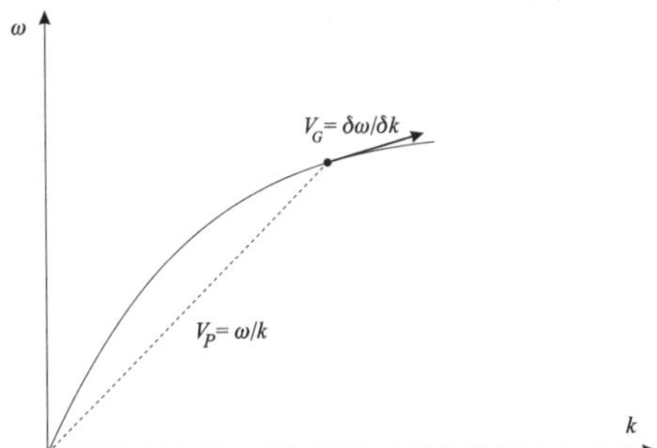

FIGURE 2.7
Typical dispersion curve showing phase velocity and group velocity for one point on the curve.

tension T and to be infinite in length. While the tension T is constant along the string, this is not true for the y component, due to the curvature of the string.

From Figure 2.6 for an element dx

$$dF_y = (T\sin\theta)_{x+dx} - (T\sin\theta)_x \tag{2.41}$$

Doing a Taylor's expansion for F_y

$$F(x+dx) = F(x) + \left(\frac{\partial F}{\partial x}\right)dx \tag{2.42}$$

we have

$$dF_y = \left[(T\sin\theta)_x + \frac{\partial(T\sin\theta)}{\partial x}dx + \cdots\right] - (T\sin\theta)_x$$

so that

$$dF_y = \frac{\partial(T\sin\theta)}{\partial x}dx \tag{2.43}$$

For small displacement (θ) of the string, $\sin\theta \sim \frac{\partial y}{\partial x}$ so that

$$dF_y = T\frac{\partial^2 y}{\partial x^2}dx \tag{2.44}$$

From Newton's law for a string of mass per unit length ρ_l

$$dF_y = \rho_l dx\frac{\partial^2 y}{\partial t^2} \tag{2.45}$$

Combining Equations 2.44 and 2.45 we have the one-dimensional wave equation

$$\frac{\partial^2 y}{\partial x^2} = \frac{1}{V_0^2}\frac{\partial^2 y}{\partial t^2} \tag{2.46}$$

where

$$V_0^2 = \frac{T}{\rho_l} \tag{2.47}$$

The form of the wave equation, Equation 2.46, is in fact completely general for all types of waves and the form of V_0^2 is typical for that sort of mechanical system. The tension T that can be applied is proportional to the mechanical stiffness of the system, and this fact can be used to obtain *a priori* estimates of the sound velocity in a given system. For example, for a given value of ρ_l, the mechanical wave (sound) velocity in a steel cord is going to be much higher than that in a cord made of cooked spaghetti.

2.2.1 Harmonic Waves

For a general wave motion, we write $\psi(x,t) = f(x,t)$ so that the wave equation for ψ is

$$\frac{\partial^2 \psi}{\partial x^2} = \frac{1}{V_0^2}\frac{\partial^2 \psi}{\partial t^2} \tag{2.48}$$

with a general solution of the form

$$\psi = C_1 f(x - V_0 t) + C_2 g(x + V_0 t) \tag{2.49}$$

In order to summarize the basic wave parameters, we consider a wave profile

$$\psi = A \sin kx \tag{2.50}$$

and if this is propagating to the right then from before

$$\psi = A \sin k(x - V_0 t + \phi) \tag{2.51}$$

and the well-known wave parameters are:

- initial phase angle ϕ
- wavelength $\lambda = 2\pi/k$
- wave number k

- period $T = 1/f$
- frequency $f = \omega/2\pi$

all of which lead to

$$\omega = V_0 k \quad \text{or} \quad V_0 = \lambda f \tag{2.52}$$

and

$$\psi = A\sin(kx - \omega t + \phi) \tag{2.53}$$

Let us look in more detail at the velocity. We define the phase of the wave as the argument of the harmonic function

$$\varphi \equiv kx - \omega t + \phi \tag{2.54}$$

Then the phase velocity is defined as the velocity of propagation of constant phase, e.g., that of a wave crest.

Then

$$\varphi = kx - \omega t + \phi = \text{constant} \tag{2.55}$$

Hence

$$kdx - \omega\, dt = 0 \quad \text{for } \varphi = \text{constant}$$

so

$$V_P \equiv \left(\frac{dx}{dt}\right)_\varphi = \frac{\omega}{k} \tag{2.56}$$

Alternatively, this result can be obtained using the chain law for partial derivatives from thermodynamics

$$\left(\frac{\partial x}{\partial t}\right)_\varphi = \frac{-\left(\frac{\partial \varphi}{\partial t}\right)_x}{\left(\frac{\partial \varphi}{\partial x}\right)_t} = \frac{\omega}{k} \tag{2.57}$$

Finally, it is common practice to describe wave motion using the complex exponential. As outlined previously,

$$\psi(x,t) = \mathrm{Re}[A\exp i(\omega t - kx + \phi)] = A\cos(\omega t - kx + \phi) \tag{2.58}$$

In physics, it is common to use the above notation, e.g.,

$$\psi(x,t) = A\exp i(kx - \omega t)$$

and in engineering it is more common to use the complex conjugate

$$\psi(x,t) = A\exp j(\omega t - kx)$$

where both i and j represent $\sqrt{-1}$.

Both notations are encountered frequently in the literature. For uniformity, we arbitrarily adopt the form exp $j(\omega t - kx)$ in the rest of the book.

2.2.2 Plane Waves in Three Dimensions

We adopt a three-dimensional coordinate system (x, y, z) with propagation in the direction of the propagation vector $\vec{k} = (k_x, k_y, k_z)$. The wavefront is the locus of points of constant phase at a given time t, so for plane waves it can be represented as a series of parallel planes. If $\vec{r}$ is a position vector from the origin to a point on the wavefront at time t then the equation of the wavefront is

$$\vec{k} \cdot \vec{r} = \text{constant} \tag{2.59}$$

and so in complex notation we describe the plane wave by

$$\psi(x,t) = A\exp j(\omega t - \vec{k} \cdot \vec{r} + \phi) \tag{2.60}$$

By a simple generalization of the one-dimensional case we can show directly from the above that, as before,

$$V_P = \frac{\omega}{k} \tag{2.61}$$

The solution for propagation in an arbitrary direction $\vec{k}$ can be written

$$\psi(x,y,z,t) = A\exp j(\omega t - [k_x x + k_y y + k_z z]) \tag{2.62}$$

or in terms of direction cosines n_x, n_y, n_z for $\vec{k}$

$$\psi(x,y,z,t) = A\exp j(\omega t - k[n_x x + n_y y + n_z z]) \tag{2.63}$$

where

$$k^2 = k_x^2 + k_y^2 + k_z^2 \tag{2.64}$$

and

$$n_x^2 + n_y^2 + n_z^2 = 1 \tag{2.65}$$

The plane wave equation in three dimensions is usually written in terms of the Laplacian

$$\nabla^2 = \frac{\partial}{\partial x^2} + \frac{\partial}{\partial y^2} + \frac{\partial}{\partial z^2} \tag{2.66}$$

so that

$$\nabla^2 \psi = \frac{1}{V_0^2}\frac{\partial^2 \psi}{\partial t^2} \tag{2.67}$$

2.2.3 Dispersion, Group Velocity, and Wave Packets

Up to now we have considered the simplest possible model for wave propagation: isotropic, homogeneous, linear, and dispersionless. Some of these simplifications will be removed later but dispersion is appropriate to consider now. Dispersion basically means that the phase velocity varies with the frequency. In optics, dispersion manifests itself in the splitting of white light into its spectral components by a prism or a raindrop. In that case, the dispersion is due to the frequency-dependent movement of the atomic mass. In acoustics, the same effects happen at very high frequencies or with thermal phonons near the Brillouin zone boundaries, and the resulting dispersion curves can be measured directly by neutron scattering. However, in acoustics the relevant length scale is 10^5 times larger than in optics, so that for the relatively low ultrasonic frequency range the wavelengths are quite large, of the order of 100 μm to 1 mm. This is of the same order of magnitude as the critical dimension of the films, plates, wires, etc. used to guide ultrasonic waves, so we can expect to encounter dispersion in such structures on purely geometrical grounds. Hence, it is essential that we appreciate the consequences of dispersion right from the beginning.

For waves of all types, no information whatsoever is transmitted by the "pure" sinusoidal carrier wave, apart from its characteristic frequency. To transmit information we need to modulate the carrier with other frequencies, and it is appropriate to consider the velocity of propagation of this modulation, and thus, more generally, the velocity of propagation of information and of energy. The simplest case to consider is that of the wave packet, treated in detail in all standard texts on waves. If several neighboring frequencies are linearly superimposed, they form a wave packet with finite extension in space and a corresponding finite Fourier frequency spectrum. The modulation is somewhat analogous to the beats for simple harmonic motion considered earlier. The modulation travels at the velocity of this wave packet. For a simple model of two waves with a small difference in frequency

$$\begin{aligned} \psi_1 &= \cos(\omega_1 t - k_1 x) \\ \psi_2 &= \cos(\omega_2 t - k_2 x) \end{aligned} \tag{2.68}$$

the superposition of ψ_1 and ψ_2 gives

$$\psi = 2\cos\left[\frac{(\omega_1+\omega_2)}{2}t - \frac{(k_1+k_2)}{2}x\right]\cos\left[\frac{(\omega_1-\omega_2)}{2}t - \frac{(k_1-k_2)}{2}x\right] \tag{2.69}$$

A whole wave packet can be built up by the superposition of such pairs with a center frequency $\omega_0 = (\omega_1 + \omega_2)/2$ and a modulation frequency $\omega_m = (\omega_1 - \omega_2)/2$. For this simple example, the modulation has velocity $(\omega_1 - \omega_2)/(k_1 - k_2)$. For $\omega_1 \to \omega_2 \to \omega_0$ this goes to $V_G = \partial\omega/\partial k$, the group velocity. Two different forms of V_G for calculation purposes are

$$V_G = \frac{\partial\omega}{\partial k} = V_P + k\frac{dV_P}{dk} \tag{2.70}$$

and

$$\frac{1}{V_G} = \frac{\partial k}{\partial\omega} = \frac{1}{V_P} - \frac{\omega}{V_P^2}\frac{dV_P}{d\omega} \tag{2.71}$$

The bandwidth theorem of simple harmonic motion can be generalized to waves by considering conjugate variables x, k in addition to ω, t. Thus,

$$\Delta\omega\Delta t \sim 1 \qquad \Delta x\Delta k \sim 1 \tag{2.72}$$

where the latter relation becomes evident from Figure 2.5. The bandwidth relation has its most famous application to wave packets in quantum mechanics where $p = h/\lambda = \hbar k$ is the particle momentum and $E = \hbar\omega$ the particle energy. Thus, we have

$$\Delta x\Delta p \sim \hbar \quad \text{and} \quad \Delta E\Delta t \sim \hbar \tag{2.73}$$

the celebrated Heisenberg uncertainty principle.

Summary

Simple harmonic motion refers to harmonic vibrations at a frequency f of a point mass about an equilibrium point. The movement is in general maintained by an external force and is damped by frictional forces.The motion is governed by Newton's second law and Hooke's law, giving a natural frequency $\omega^2 = k/m$.

Phasor is a representation of the vibration in the complex plane. The phasor rotates at the phase angle $\theta = \omega t$ and the radius vector is the amplitude of vibration. Two or more phasors can be added algebraically in the complex plane.

Resonance occurs when the imaginary part of the impedance is zero. The resonance can be described by the Q or quality factor, which is a measure of the sharpness of the resonance. There are five different ways of expressing the Q, each with a different but complementary physical interpretation.

Fourier series is a way of representing a periodic function as a sum of sines or cosines with argument an integral multiple of the fundamental frequency. The sine series is used for functions of odd symmetry, the cosine series for even functions.

Fourier integral is a generalizaton of Fourier series to the representation of pulses by a frequency spectrum. A Fourier transform pair links Fourier representations of a pulse in the time and frequency domain or quantities in spatial and wave number space.

Traveling waves correspond to the self-sustaining propagation of a disturbance in space at constant velocity without change of shape. A progressive or traveling wave is characterized by a functional form $f(kx - Vt)$.

Harmonic wave is a wave at a single frequency ω, described by a sine or cosine function, or in complex notation by exp $j(\omega t - kx)$.

Phase velocity $V_P = \omega/k$ is the velocity of a wavefront of constant phase.

Group velocity $V_G = \delta\omega/\delta k$ is the velocity of propagation of a wave packet. For lossless or low loss media it is also the velocity of propagation of energy.

Dispersion describes a situation in which the phase velocity varies with frequency; it occurs in dispersive media.

Questions

1. Draw a diagram to show how to add two phasors graphically, to determine their total amplitude and phase angle. Determine analytical expressions for the latter.
2. Make a graph of the displacement and velocity for a forced simple harmonic oscillator as a function of frequency. Draw the corresponding phasor diagram. Compare results for oscillators where $R \to 0$ and $R \to \infty$.
3. Consider a triangular waveform as a function of time. Define the amplitude and period. Choose an origin and sketch the first three Fourier components. Comment on the use of sine or cosine functions.

4. Draw two limiting cases (width going to zero or infinity) for the Fourier transform of a Gaussian pulse.
5. Draw the vector diagram corresponding to $\tan\theta$ for simple harmonic motion.
6. Decide which of the following are traveling waves and calculate the appropriate phase velocity:
   i. $f(x, t) = (ax - bt)^2$
   ii. $f(x, t) = (ax + bt + c)^2$
   iii. $f(x, t) = 1/(ax^2 + b)$
   a, b, and c are positive constants.
7. Consider a harmonic wave with given ω and k. Give V_P, T, and λ in terms of these quantities.
8. Consider the dispersion curve $w(k) = A\,|\sin ka|$. Plot $w(k)$ over the range $-\pi/a \le k \le \pi/a$. Make plots of $V_P(k)$ and $V_G(k)$. Do likewise for $V_P(\omega)$ and $V_G(\omega)$.
9. Plot Equation 2.69 for the case where $\omega_1 \gg \omega_2$. Comment on the pertinence of this case for communications.
10. Calculate the group velocity for the following cases where the phase velocity is known:
    i. Transverse elastic wave in a rod

$$V_P = A/\lambda$$

    ii. Deep water waves

$$V_P = A\sqrt{\lambda}$$

    iii. Surface waves in a liquid

$$V_P = A/\sqrt{\lambda}$$

    iv. Electromagnetic waves in the ionosphere

$$V_P = \sqrt{c^2 + A^2\lambda^2}$$

    where c is the velocity of light.

3

Bulk Waves in Fluids

This chapter makes an extension of the introductory material of Chapter 2 to the simplest acoustic case of interest to us here, namely the propagation of bulk waves in liquids and gases. Formally, this case is much simpler than that of solids; fluids in equilibrium are always isotropic and only longitudinal (compressional) waves can propagate. Hence, there is no polarization to specify, and scalar wave theory can be applied. From another point of view, ultrasonic waves in liquids are sufficiently different from those in solids that a separate discussion is required. Finally, these results on liquids form a good basis for extending the theory to solids. A good discussion of waves in liquids is given in [6] and [7].

In terms of notation, V_i with a subscript i will be used for sound velocity, V_0 for bulk waves in liquids, V_L and V_S for longitudinal and shear waves in solids, V_P and V_G for phase and group velocity, etc. When the symbol V stands alone, it normally represents the thermodynamic variable for volume V.

3.1 One-Dimensional Theory of Fluids

We consider bulk fluids that are homogeneous, isotropic, and compressible with equilibrium pressure p_0 and density ρ_0. As for the case with waves in strings in Chapter 2 we apply Newton's law to an element of volume, and we need an additional equation relating a pressure increase to change in volume of the fluid, which will be provided by the definition of the compressibility.

Considering a simple volume element, a wave will be provided in the following way. If a pressure increase is applied at $t = 0$ to the plate at the origin, this will cause an increase in pressure and density in the layer of fluid next to it relative to the layer at the right. Hence, particles will flow to the right, leading to an increase in pressure and density, and the disturbance will then flow as a series of alternative compressions and rarefactions.

Considering the volume element between x and $x + dx$ we have

$$dF_x = \left[P(x) - \left\{P(x) + \frac{\partial P}{\partial x}dx\right\}\right]A = -\frac{\partial P}{\partial x}dxA \tag{3.1}$$

Applying Newton's law to the element of mass $\rho_0 dxA$

$$\frac{\partial P}{\partial x} = -\rho_0 \frac{\partial^2 u}{\partial t^2} \tag{3.2}$$

Here P and u are the instantaneous pressure and displacement, respectively.

For simplicity, we distinguish between the equilibrium pressure P_0 and the instantaneous pressure P to the excess, acoustic pressure p by

$$p = P - P_0$$

so

$$\frac{\partial p}{\partial x} = -\rho_0 \frac{\partial^2 u}{\partial t^2} \tag{3.3}$$

To link the applied pressure to the compression of the liquid, we define the compressibility

$$\chi = -\frac{1}{V}\left(\frac{\partial V}{\partial p}\right) \tag{3.4}$$

and the compression of the liquid will be described by the dilatation S

$$S \equiv \frac{\Delta V}{V} \tag{3.5}$$

During a compression of the volume $dV = Adx$ at pressure p on the left to $dV = A(1 + \partial u/\partial x)dx$ at pressure $p + dp$ on the right

$$S = \frac{\Delta V}{V} = \frac{\partial u}{\partial x} \tag{3.6}$$

From the definition of the compressibility

$$p = -\frac{S}{\chi} = -\frac{1}{\chi}\frac{\partial u}{\partial x} \tag{3.7}$$

Hence, the equation of motion can be rewritten

$$\frac{\partial^2 u}{\partial t^2} = V_0^2 \frac{\partial^2 u}{\partial x^2} \tag{3.8}$$

where

$$V_0^2 \equiv \frac{1}{\rho_0 \chi} \tag{3.9}$$

The compressibility can be rewritten

$$\chi = -\frac{1}{V}\frac{\partial V}{\partial p} = \frac{1}{\rho_0}\frac{\partial \rho}{\partial p} \tag{3.10}$$

which gives a more general form

$$V_0^2 = \frac{\partial P}{\partial \rho} \tag{3.11}$$

Since pressure is only proportional to density in first order, this highlights the fact that V_0 = constant only to first order. In other words, since the pressure-density relation is nonlinear in an exact theory, linear acoustics, corresponding to V_0 = constant, does not exist as such but is only an approximation.

Summarizing from the previous, the wave equation can be written in the form

$$\frac{\partial^2 u}{\partial t^2} = V_0^2 \frac{\partial^2 u}{\partial x^2}$$

or

$$\frac{\partial^2 p}{\partial t^2} = V_0^2 \frac{\partial^2 p}{\partial x^2} \tag{3.12}$$

or

$$\frac{\partial^2 v}{\partial t^2} = V_0^2 \frac{\partial^2 v}{\partial x^2} \tag{3.13}$$

where

$v = \partial u/\partial t$ = particle velocity,
S = dilatation = $\partial u/\partial x$, and
$p = -\rho_0 V_0^2 S$

All of these three forms of the wave equation are equivalent by the above relations in the linear approximation. We will focus on the solutions for the displacement $u(x, t)$. These can be written

$$u = A\exp j(\omega t - kx) + B\exp j(\omega t + kx) = u_+ + u_- \tag{3.14}$$

where $A(u_+)$ is the amplitude (displacement) of the wave in the forward $(+x)$ direction and $B(u_-)$ is the amplitude (displacement) of the wave in the backward $(-x)$ direction.

Then p, S, and v can also be written in the form

$$p = -\rho_0 V_0^2 \frac{\partial u}{\partial x} = j\rho_0 \omega V_0 (u_+ - u_-) \tag{3.15}$$

$$S = \frac{\partial u}{\partial x} = jk(-u_+ + u_-) \tag{3.16}$$

$$v = \frac{\partial u}{\partial t} = j\omega (u_+ + u_-) \tag{3.17}$$

One immediate consequence of these equations is that they provide the phase relations between pressure, displacement, dilatation, and velocity. These can best be displayed on a complex phasor diagram as shown in Figure 3.1. From a practical viewpoint the relation for the pressure and the velocity are most important. For the forward wave, the pressure and velocity lead the displacement by $\pi/2$; for the backward wave, the velocity leads by $\pi/2$ and the pressure lags by $\pi/2$. The change in phase relationship with propagation direction comes about because pressure and dilatation are scalar quantities while displacement and velocity are vectorial.

3.1.1 Sound Velocity

As seen by the form of the solutions the sound velocity $V_0 = \sqrt{\partial P/\partial \rho} = \omega/k$ is the phase velocity of the wave. For bulk waves in infinite media, it is a constant for a given medium but is dependent on all of the thermodynamic parameters such as compressibility, density, external pressure, temperature, etc. Within the present context it is independent of frequency (infinite media) and amplitude (linear regime) but in general this is, of course, not the case. In fact, the analysis of the velocity is quite different for gases and liquids so these two cases will be treated separately.

3.1.1.1 Gases

The approximation of an ideal gas will be made: $PV = n_0RT$ or $P = (RT/M)\rho$, where n_0 = number of moles. Since sound propagation in a gas is known to be essentially an adiabatic process, the relation PV^γ = constant is also applicable.

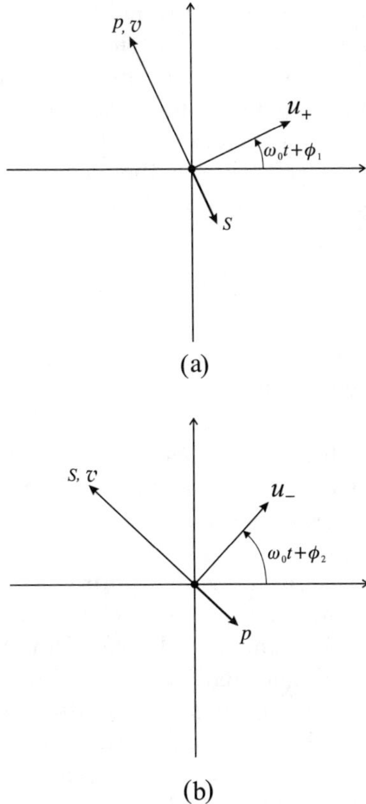

FIGURE 3.1
Phasor representation for an acoustic wave in a fluid. (a) Forward wave. (b) Backward wave.

This can be written in the form

$$\frac{P}{\rho^{\gamma}} = \text{constant} \tag{3.18}$$

so that

$$\frac{\partial P}{\partial \rho} = \frac{\gamma P}{\rho} \tag{3.19}$$

and for equilibrium conditions P_0, ρ_0

$$V_0 = \sqrt{\frac{\gamma P_0}{\rho_0}} = \sqrt{\frac{\gamma RT}{M}} \tag{3.20}$$

For air (diatomic) at room temperature (20°C) $\gamma = 1.4$, $P_0 = 1.01 \times 10^5$ Pa giving $V_0 \sim 343$ m/s in good agreement with experiment.

In the present treatment of fluids, the first implicit assumption of local thermodynamic equilibrium has been made, in that only under this condition can local values of P, T, ρ, etc. be assigned. In the case of a gas, the length scale for thermodynamic equilibrium is the mean free path l of the gas particles, i.e., the mean distance between collisions of the molecules. It is standard that

$$l = v_0 \tau \tag{3.21}$$

where

τ = mean time between collisions

v_0 = thermodynamic particle velocity of the molecules

l can be inferred from transport measurements on the gas and v_0 is well-known from the kinetic theory of gases. In order of magnitude $v_0 = \sqrt{3RT/M} \propto 300$ m/s at 20°C.

The second implicit assumption is that in order to obtain wave propagation conditions, the thermodynamic parameters must be well defined over distances much shorter than the wavelength. Otherwise, the propagating quantities such as pressure and density would simply not be defined with respect to the wave. This then gives the condition $\lambda \gg l$, which must be respected for a wave description to apply. This implies an upper frequency limit for wave propagation in a gas, for example, for air at STP $l \propto 10^{-5}$ cm, leading to a critical frequency $f \propto 1$ GHz.

It should be noted that the same conditions apply for liquids and solids but the critical frequencies are much higher and do not have any practical consequence for ultrasonic waves.

3.1.1.2 Liquids

It is relatively easy to find simple models for the limiting cases of sound propagation in gases and solids. Liquids, however, constitute an intermediate case and it is more difficult to find a simple model connecting the sound velocity V_0 to the molecular constants. The few available models will be outlined briefly.

A semi-empirical approach, similar to that for gases, gives

$$V_0 = \sqrt{\frac{\gamma K_T}{\rho_0}} \tag{3.22}$$

where K_T is the isothermal bulk modulus. Another semi-empirical approach is Rao's rule, of the form

$$V_0^{1/3} V = R_a \tag{3.23}$$

where V is the molar volume and R_a is a constant for a given liquid. It was pointed out by Rao that R_a undergoes regular increments among the members of a homologous series of liquids so that

$$R_a = AM + B \tag{3.24}$$

where M is the molecular weight.

One of the few relations between V_0 and the liquid structure was provided by the early study of Schaaffs [8]. He assumed that although a realistic equation of state for the liquid was too complicated, some properties of organic liquids such as the sound velocity could be deduced from the van der Waals equation

$$\left(P + \frac{a}{V^2}\right)(V - b) = RT \tag{3.25}$$

where R is the universal gas constant, a = constant, and b = excluded volume.

Schaaffs obtained for organic liquids

$$V_0 = \sqrt{\gamma RT\left(\frac{M}{3(M-\rho b)^2} - \frac{2}{M-\rho b}\right)} \tag{3.26}$$

Actual comparisons were made by solving for b

$$b = \frac{M}{\rho}\left[1 - \frac{RT}{MV_0^2}\left(\left(1 + \frac{MV_0^2}{3RT}\right)^{\frac{1}{2}} - 1\right)\right] \tag{3.27}$$

Excellent agreement was obtained by comparing $b = 4V_{\text{molecule}}$ with molecular volumes determined by other means. Further discussion of other semi-empirical approaches is given by Beyer and Letcher [7], including that for the sound velocity in liquid mixtures. Values for representative liquids are given in Table 3.1.

3.1.2 Acoustic Impedance

Using the electromechanical analogy developed in Chapter 2, we define the specific acoustic impedance Z of an acoustic wave

$$Z = \frac{p}{v} \tag{3.28}$$

Z carries a sign as v can be either in the positive or negative direction. The absolute value of Z for plane waves, useful to characterize the bulk (infinite) medium, is called the characteristic impedance of the liquid, $Z_0 = \rho_0 V_0$. A third variant, the normal acoustic impedance, will be introduced in Chapter 7 for reflection and transmission analysis.

TABLE 3.1

Acoustic Properties of Representative Liquids

Liquid	V_L (km-s^{-1})	ρ (10^3 kg-m^{-3})	Z_0 (MRayls)
Acetone	1.17	0.79	1.07
Liquid argon (87 K)	0.84	1.43	1.20
Methanol	1.1	0.79	0.87
Gallium (30 K)	2.87	6.10	17.5
Glycerin	1.92	1.26	2.5
Liquid He^4 (2 K)	0.228	0.145	0.033
Mercury	1.45	13.53	19.6
Liquid nitrogen (77 K)	0.86	0.85	0.68
Silicone oil	1.35	1.1	1.5
Seawater	1.53	1.02	1.57
Water (20°C)	1.48	1.00	1.483

Using the previous notation we can determine the acoustic impedance for forward and backward propagation

$$Z_+ = \frac{p_+}{v_+} = \frac{j\rho_0 \omega V_0 u_+}{j\omega u_+} = \rho_0 V_0 \tag{3.29}$$

$$Z_- = \frac{p_-}{v_-} = \frac{-j\rho_0 \omega V_0 u_-}{j\omega u_-} = -\rho_0 V_0 \tag{3.30}$$

Acoustic impedance is a highly useful concept in ultrasonics. From Chapter 2, it is the direct analogy of impedance in electrical circuits. In the latter case, it is well known that there is maximum power transfer between two circuits when the impedances are matched. In the ultrasonic case, this corresponds to maximum transmission of an ultrasonic wave from one medium to another when the characteristic impedances are equal. Characteristic acoustic impedances for some liquids are shown in Figure 3.2 in a representation that is useful for choosing liquids with prescribed density and sound velocity.

3.1.3 Energy Density

The energy density is the total energy per unit volume, comprised of the sum of the kinetic and potential energy. By definition, the kinetic energy density is

$$u_K = \frac{1}{2}\rho_0 \dot{u}^2 \tag{3.31}$$

For the potential energy, we consider a volume element V changed to V' by the passage of the acoustic wave.

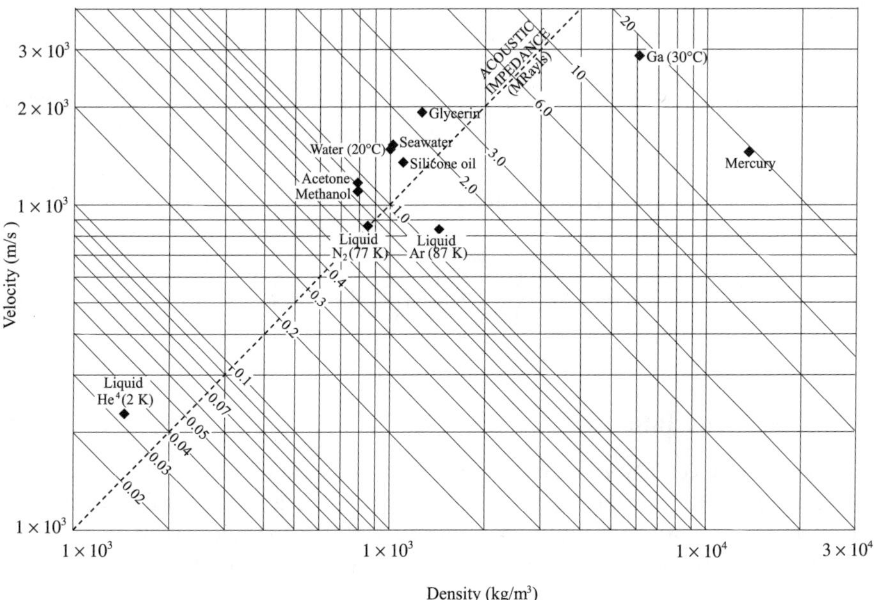

FIGURE 3.2
Density-sound velocity/characteristic acoustic impedance relation on a log-log scale for various liquids. (Based on a graph by R. C. Eggleton, described in Jipson, V. B., Acoustical Microscopy at Optical Wavelengths, Ph.D. thesis, E. L. Ginzton Laboratory, Stanford University, Stanford, CA, 1979.)

Since $S = \frac{\partial u}{\partial x}$ from Equation 3.6

$$V' = V\left(1 + \frac{\partial u}{\partial x}\right)$$
$$= V\left(1 - \frac{p}{\rho_0 V_0^2}\right) \tag{3.32}$$

and the change in potential energy is

$$\Delta U_P = -\int p dV' \tag{3.33}$$

From Equation 3.32

$$dV' = -\frac{V dp}{\rho_0 V_0^2} \tag{3.34}$$

Hence,

$$\Delta U_P = \frac{V}{\rho_0 V_0^2}\int_0^p p dp = \frac{1}{2}\frac{p^2 V}{\rho_0 V_0^2} \tag{3.35}$$

Finally,

$$\Delta U_{tot} = \Delta U_K + \Delta U_P = \frac{1}{2}\rho_0\left(\dot{u}^2 + \frac{p^2}{\rho_0^2 V_0^2}\right)V \tag{3.36}$$

so that the acoustic energy density

$$u_a = \frac{\Delta U_{tot}}{V} = \frac{1}{2}\rho_0\left(\dot{u}^2 + \frac{p^2}{\rho_0^2 V_0^2}\right) \tag{3.37}$$

3.1.4 Acoustic Intensity

The acoustic intensity I is the average flux of acoustic energy per unit area per unit time. For a plane wave, it is clear that for a tube element of area A and length $V_0 dt$, all of the acoustic energy dU_a inside the cylindrical element will traverse the end face and leave the cylinder in time dt.

Hence,

$$dU_a = u_a A V_0 dt$$

so that

$$I \equiv \frac{\overline{dU_a}}{A dt} = \overline{u}_a V_0 \tag{3.38}$$

3.2 Three-Dimensional Model

The previous results can be generalized immediately to three dimensions. Displacement u and velocity v now become explicitly vectors $\vec{u}$ and $\vec{v}$ while the acoustic pressure p remains a scalar. Hence the 3D description of the acoustic properties of fluids is usually carried out in terms of the pressure; not only is this the simplest choice, but pressure is also the variable that is usually measured in the laboratory.

For a surface element $\overrightarrow{dA}$ with displacement $\vec{u}$ the associated volume is $dV = \vec{u} \bullet \overrightarrow{dA}$. By Gauss' theorem

$$\Delta V = \oint_S \vec{u} \cdot \overrightarrow{dA} = \int_V (\vec{\nabla} \cdot \vec{u}) dV \equiv \int_V S(\vec{r}) dV \tag{3.39}$$

where $S(\vec{r})$ is the dilatation.

Hence,

$$S(\vec{r}) = \vec{\nabla} \bullet \vec{u} \equiv \frac{\partial u_x}{\partial x} + \frac{\partial u_y}{\partial y} + \frac{\partial u_z}{\partial z} \tag{3.40}$$

Newton's law in three dimensions is

$$\rho_0 \frac{\partial^2 \vec{u}}{\partial t^2} = -\vec{\nabla} p \tag{3.41}$$

where $-\vec{\nabla} p$ is the net force on the element.

We want to change to a simple set of variables so that $\vec{u}$ on the left-hand side should be expressed in terms of the pressure. This can be done by using $S(\vec{r}) = \vec{\nabla} \bullet \vec{u}$ and then using Equation 3.7, the relation between the dilatation and the pressure. Applying those steps to Equation 3.41, we obtain

$$\rho_0 \frac{\partial^2 S}{\partial t^2} = -\Delta(p) \tag{3.42}$$

where

$$\Delta = \vec{\nabla} \bullet \vec{\nabla} = \frac{\partial^2}{\partial x^2} + \frac{\partial^2}{\partial y^2} + \frac{\partial^2}{\partial z^2} \equiv \text{Laplacian}$$

and finally the wave equation

$$\Delta(p) = \frac{1}{V_0^2} \frac{\partial^2 p}{\partial t^2} \tag{3.43}$$

where

$$V_0^2 = \frac{1}{\rho_0 \chi} \tag{3.44}$$

In analogy with Equation 3.43 the 3D wave equations for $\vec{u}$ and $\vec{v}$ are

$$\nabla^2 \vec{u} = \frac{1}{V_0^2} \frac{\partial^2 \vec{u}}{\partial t^2} \tag{3.45}$$

$$\nabla^2 \vec{v} = \frac{1}{V_0^2} \frac{\partial^2 \vec{v}}{\partial t^2} \tag{3.46}$$

and the solutions for $\vec{u}$ are

$$\vec{u} = \vec{u}_0 \exp j(\omega t - \vec{k} \bullet \vec{r}) \tag{3.47}$$

where $\vec{k}$ is the propagation vector whose direction gives the direction of propagation and whose magnitude is

$$|\vec{k}| = \frac{2\pi}{\lambda} \tag{3.48}$$

3.2.1 Acoustic Poynting Vector

In the presence of applied volume forces $\vec{f}$ per unit volume Equation 3.42 becomes

$$\rho_0 \frac{\partial \vec{v}}{\partial t} = -\vec{\nabla}(p) + \vec{f} \tag{3.49}$$

If this force represents the force by the adjoining fluid on an element dV, then the work done per unit volume in time dt is

$$\begin{aligned} dw &= \vec{f} \bullet \overrightarrow{du} = \vec{f} \bullet \vec{v} dt \\ &= \rho_0 \vec{v} \bullet \overrightarrow{dv} + \overrightarrow{\nabla p} \bullet \overrightarrow{du} \quad \text{by Equation (3.49)} \\ &= d\left(\frac{1}{2}\rho_0 \vec{v}^2\right) - p dS + \vec{\nabla} \bullet (p\overrightarrow{du}) \end{aligned} \tag{3.50}$$

Referring to the one-dimensional model, we immediately identify the first two terms as the variation of the kinetic and potential energy per unit volume, respectively.

Hence,

$$u_K = \frac{1}{2}\rho_0 v^2 \tag{3.51}$$

and

$$u_P = -\int_0^S p\, dS = \int_0^S \frac{S dS}{\chi} = \frac{1}{2}\frac{S^2}{\chi} \quad \text{by Equation 3.7}$$

We define the acoustic Poynting vector

$$\vec{\mathcal{P}} \equiv p\vec{v} \tag{3.52}$$

and taking the time derivative of Equation 3.50

$$\frac{dw}{dt} = \frac{d}{dt}(u_K + u_P) + \vec{\nabla} \bullet \vec{\mathcal{P}}$$

For a finite system, integrating over the volume

$$\frac{dw}{dt} = \frac{d}{dt}(U_K + U_P) + \oint_S \vec{\mathcal{P}} \bullet \overrightarrow{dA} \tag{3.53}$$

where $\vec{\mathcal{P}}$ is the instantaneous acoustic power per unit area radiated from the system through the surface S. This equation represents the law of conservation of energy at a given time.

The average value of $\vec{\mathcal{P}} \equiv \vec{I}$ then corresponds to the average flux density carried by the acoustic wave. For a system with no absorption $\vec{I}$ = constant and by Equation 3.53 the net acoustic power radiated from a closed element in the steady state is zero.

3.2.2 Attenuation

Up to now we have assumed perfectly lossless reversible behavior of the fluid. In practice, there are losses or absorption of acoustic energy by the medium. These losses are normally attributable to viscosity and thermal conductivity leading to the so-called classical attenuation. In addition, there are molecular processes where acoustic energy is transformed into internal molecular energy. The finite time for these processes leads to relaxation and loss effects.

In fact, all of the loss effects in fluids can be described by a phase lag between acoustic pressure and the medium response (density or volume change). A classical example from thermodynamics is that of the *P-V* diagram, which can be used to display the work done on the medium due to a pressure change.

The situation is shown in Figure 3.3 on the usual *P-V* diagram for compression and expansion of a gas. Let us suppose that changes in P and V are due to an acoustic wave. The work done or supplied by the system is given by

$$W = -\int P dV$$

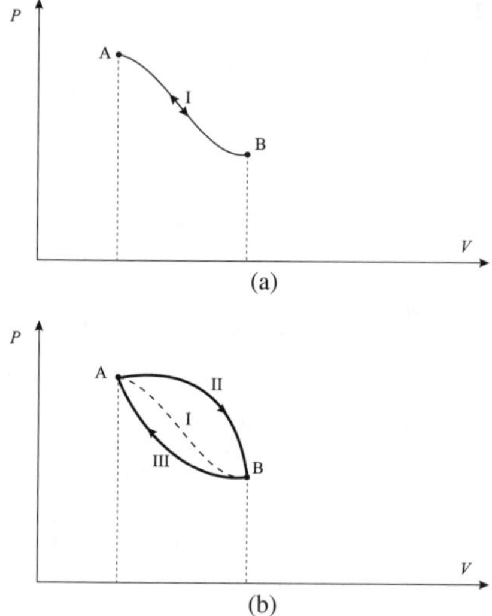

FIGURE 3.3
(a) Reversible transformation from A to B and from B to A in a lossless medium. (b) Transformation from A to B and from B to A in a lossy medium.

for the appropriate process. It is well known that the area enclosed by the curve for a cycle is the net work done on the system. In the lossless case, the system evolves along the same path *I* during expansion from *A* to *B* ($\int_A^B$) and compression from *B* back to *A* ($\int_B^A$). These two amounts of work are of opposite sign so the net amount of work absorbed by the system from the acoustic wave is zero. On the other hand, if the system does not respond immediately then intuitively volume change will tend to lag that for the reversible case for both expansion (II) and compression (I), leading to a net amount of work per cycle by the acoustic wave on the medium, leading to absorption of energy.

3.2.2.1 *Decibel Scale of Attenuation*

If we consider the displacement u of the wave as

$$u = u_0 \exp j(\omega t - kx)$$

for the wave without dissipation, then $I \propto u^2$ for plane waves. If now we add dissipation, the only effect is that the wave vector $\vec{k}$ becomes complex, i.e., $k \rightarrow \beta - j\alpha$, where α is seen to be the attenuation coefficient for the amplitude of the wave, as now

$$u = u_0 \exp j(\omega t - \beta x)\exp(-\alpha x) \tag{3.54}$$

In plane wave conditions, which are standard for attenuation measurement, $I \propto u^2$, so that the acoustic intensity decays as $\exp(-2\alpha x)$. The factor of two comes from the difference in attenuation between the amplitude and the intensity due to the quadratic term. In practice, care must be taken as to what is being measured (and calculated) to avoid confusion on this point.

In practice the attenuation factor for the amplitude is measured by determining the amplitude ratio r_{12} of the wave at two different positions x_1 and x_2. Hence,

$$r_{12} = \exp \alpha(x_2 - x_1)$$

The attenuation in nepers $\equiv \ln(r_{12}) = \alpha(x_2 - x_1)$, so that α is measured in Np/m.

It is more common to use the decibel (dB) scale to compare acoustic intensity level; the attenuation in dB is defined as

$$\begin{aligned} \text{attenuation (dB)} &= 10\log_{10}(r_{12})^2 \\ &= 20(\log_{10}e)\alpha(x_2 - x_1)\,\text{dB} \end{aligned} \tag{3.55}$$

where α is in dB/m.

Hence, the relation between the two units is

$$\alpha\ (\text{dB/m}) = 20(\log_{10}e)\alpha\ (\text{Np/m}) = 8.686\alpha\ (\text{Np/m}) \tag{3.56}$$

3.2.2.2 Relaxation Time Formulation for Viscosity

Stokes' classic treatment includes a time-dependent term in the pressure-condensation relation [6]

$$p = \rho_0 V_0^2 s + \eta \frac{\partial s}{\partial t} \tag{3.57}$$

where η is a viscosity coefficient and $s = -S$ is the relative density change or condensation. For an applied pressure $p_a = p_{a0} \exp(j\omega t)$, if we assume a response for the condensation $s = s_0 \exp(j\omega t)$, direct substitution yields

$$s_0 = \frac{p_0}{\rho_0 V_0^2 + j\omega\eta} \tag{3.58}$$

Clearly, the density change lags the applied pressure by a phase angle ϕ where

$$\tan\phi = \frac{\omega\eta}{\rho_0 V_0^2} \tag{3.59}$$

If a step function pressure change Δp_{a0} is applied at $t = 0$, the solution is

$$s = \frac{\Delta p_0}{\rho_0 V_0^2}\left(1 - \exp\left(-\frac{\rho_0 V_0^2 t}{\eta}\right)\right) \tag{3.60}$$

and if a step function pressure is suddenly removed

$$s = \frac{\Delta p_0}{\rho_0 V_0^2}\exp\left(-\frac{\rho_0 V_0^2 t}{\eta}\right) \tag{3.61}$$

Recalling the electromechanical analogy, it is readily seen that these solutions are identical to those for the current in an *L-R* circuit when a potential difference is suddenly applied or removed. That process is described by a relaxation time $\tau = L/R$. By analogy, we define a viscous relaxation time

$$\tau = \frac{\eta}{\rho_0 V_0^2}. \tag{3.62}$$

3.2.2.3 Attenuation Due to Viscosity

The effects of attenuation are normally incorporated by using a complex wave number

$$k \equiv \beta - j\alpha \tag{3.63}$$

Then

$$u = u_0 \exp j(\omega t - (\beta - j\alpha)x) = u_0 e^{-\alpha x} \exp j(\omega t - \beta x) \tag{3.64}$$

using the Stokes term for the pressure, the wave equation is

$$\frac{\partial^2 u}{\partial t^2} = V_0^2 \frac{\partial^2 u}{\partial x^2} + \frac{\eta}{\rho_0}\frac{\partial^2 u}{\partial x \partial \tau} \tag{3.65}$$

substituting for u and separating real and imaginary parts

$$\alpha^2 = \frac{\omega^2}{2V_0^2}\left(\frac{1}{\sqrt{1+\omega^2\tau^2}} - \frac{1}{1+\omega^2\tau^2}\right) \tag{3.66}$$

$$\beta^2 = \frac{\omega^2}{2V_0^2}\left(\frac{1}{\sqrt{1+\omega^2\tau^2}} + \frac{1}{1+\omega^2\tau^2}\right) \tag{3.67}$$

and

$$V_P^2 \equiv \frac{\omega^2}{\beta^2} = \frac{2V_0^2(1+\omega^2\tau^2)}{1+\sqrt{1+\omega^2\tau^2}} \tag{3.68}$$

For most fluids at ultrasonic frequencies at room temperature, $\omega\tau \ll 1$; hence,

$$\alpha \sim \frac{\omega^2 \tau}{2V_0} = \frac{\omega^2 \eta}{2\rho_0 V_0^3} \tag{3.69}$$

and the modified phase velocity

$$V_P = V_0\left(1 + \frac{3}{8}\omega^2\tau^2\right) \tag{3.70}$$

The important result here is that in this limit $\alpha \sim \omega^2$. This means that α rises rapidly with frequency; this will have important implications for acoustic devices and NDE. The change in the velocity is small and is neglected in most cases in practice.

3.2.2.4 Attenuation Due to Thermal Conduction

In simple descriptions of sound propagation, perfect adiabaticity is usually assumed. This is only strictly true if the thermal conductivity $\kappa \equiv 0$. In fact, there is always a finite κ, so heat will be transported from the hot regions (compressions) to the cooler regions (rarefactions) created by the sound wave. As for viscous effects the temperature change will lag the applied pressure, leading to additional attenuation, described by a relaxation time

$$\tau = \frac{\kappa}{\rho_0 V_0 C_p} \tag{3.71}$$

The corresponding attenuation in the limit $\omega\tau \ll 1$ when added to the viscous term that Equation 3.69 gives for the so-called classical attenuation coefficient of liquids

$$\alpha = \frac{\omega^2}{2\rho_0 V_0^3}\left(\frac{4\eta}{3} + \frac{\kappa(\gamma - 1)}{C_p}\right) \tag{3.72}$$

It is interesting to compare this classical attenuation to that actually observed experimentally in liquids, which is done in Table 3.2 for liquids and gases. Excellent quantitative agreement is obtained for inert gases (He and Ar) and in cases where the viscous term dominates (glycerin). Otherwise, the experimental value exceeds the classical one sometimes significantly. This is due to molecular relaxation phenomena.

3.2.2.5 Molecular Relaxation

This is a subject of physical chemistry in itself, which could easily fill a large book. However, since the subject is now well understood and is not of current research interest, only a brief overview will be given. A detailed discussion has been given by Herzfeld and Litovitz [9] and Beyer and Letcher [7].

TABLE 3.2

Acoustic Absorption in Fluids

All Data for $T = 20°C$ and $\mathcal{P}_0 = 1$ atm	α/f^2 ($Np \cdot s^2 \cdot m^{-1}$)			
	Shear Viscosity	Thermal Conductivity	Classical	Observed
Gases		*Multiply All Values by 10^{-11}*		
Argon	1.08	0.77	1.85	1.87
Helium	0.31	0.22	0.53	0.54
Oxygen	1.14	0.47	1.61	1.92
Nitrogen	0.96	0.39	1.35	1.64
Air (dry)	0.99	0.38	1.37	α/f peaks at 40 Hz
Carbon dioxide	1.09	0.31	1.40	α/f peaks at 30 kHz
Liquids		*Multiply All Values by 10^{-15}*		
Glycerin	3000.0	—	3000.0	3000.0
Mercury	—	6.0	6.0	5.0
Acetone	6.5	0.5	7.0	30.0
Water	8.1	—	8.1	25.0
Seawater	8.1	—	8.1	α/f peaks at 1.2 kHz and 136 kHz

Source: Data from Kinsler, L.E. et al., *Fundamentals of Acoustics*, John Wiley & Sons, New York, 2000.

To see how molecular structure can contribute to relaxation effects, let us look briefly at the simple physics of relaxation, which will also give insight into the viscosity and thermal conduction contributions. Consider a physical system at constant temperature that is excited to a higher energy state by energy absorbed from an incident ultrasonic wave. The system will attempt to return to equilibrium by giving up this energy to surrounding regions at a rate determined by a temperature dependent relaxation time. Let us now slowly increase the ultrasonic frequency from zero.

In the regime $\omega\tau \ll 1$, the variations of the applied field are so slow that the process is approximately reversible, the system follows in phase the applied field, there is little excess absorbed energy, and the attenuation is tiny. This is less true as the frequency is raised, leading to an increase in attenuation.

In the opposite limit with $\omega\tau \gg 1$ at sufficiently high frequencies, the ultrasonic field varies so fast that the system cannot follow it. Hence, there is almost no absorbed energy and the attenuation is again very small. As the frequency is reduced, the system progressively starts to follow the field and absorb energy; hence, the attenuation starts to increase.

Clearly, there is an optimal situation at which the system absorbs a large amount of energy from the field and dumps it efficiently and irreversibly as heat to the surroundings, thus giving a high value of attenuation. This optimal situation occurs at $\omega\tau \sim 1$, at which there is a well-defined peak in the attenuation. Experimentally, one can also observe the peak at constant frequency by varying the temperature (or other parameter) which makes τ sweep the critical region around the peak. Another point is that experimentally the relaxation peak is very sharp and may be confused with a resonance; care must be taken in interpretation as the physics in the two cases is quite different.

The case of gases is the simplest to analyze. The monatomic and inert gases have only translational degrees of freedom. τ is hence very short, and there is no excess attenuation above the "classical" value. Polyatomic gases have rotational and vibrational levels that require a finite time to take up the excess energy, particularly the latter. This leads to a specific heat of the form

$$C_V = C_e + C_m\left(1 - e^{-\frac{t}{\tau}}\right) \tag{3.73}$$

which will obviously lead to a relaxation type attenuation. A classic example is CO_2, which at STP exhibits a relaxation peak centered at about 20 kHz; the maximum attenuation is of the order of 1200 times that of the classical value. Another interesting example is air. Dry air exhibits slightly greater attenuation than the classic value but for humid air below 100 kHz there is an order of magnitude increase. This is due to the reduction of τ for the vibrational mode of the oxygen molecule due to the catalytic effect of the water vapor molecules.

Similar relaxation effects have been seen in liquids and this explains the excess attenuation in nonpolar liquids such as acetone. The excess attenuation observed in water has been shown to be due to structural relaxation as explained by Hall [10]. Water is known to be a two-state liquid, part of the water in a free state, the rest in a bound state where the water molecules have a more closely packed structure. The ultrasonic wave causes transitions between the two states and the associated time delays lead to a relaxation-type phenomenon. An additional excess attenuation occurs in seawater where dissolved salts lead to a type of chemical relaxation.

Summary

Wave equation is a second-order differential equation that allows determination of the displacement $u(x, t)$ for given initial and boundary conditions.

Sound velocity in a gas is given by the general formula $V_0^2 = dP/d\rho$. For a perfect gas, $V_0 = \sqrt{\gamma RT/M}$.

Sound velocity in a liquid to a first approximation is given by $\sqrt{\gamma K_T/\rho_0}$ where K_T is the isothermal bulk modulus.

Specific acoustic impedance $Z = p/v$ carries a sign; it is positive in the forward direction and negative in the backward direction. The absolute value for plane waves is the characteristic acoustic impedance $Z_0 = \rho_0 V_0$ and is a constant for the medium.

Acoustic intensity is the average flux of acoustic energy per unit area per unit time. For a plane wave, it is given by $\overline{u_a} V_0$.

Acoustic Poynting vector is defined as $p\vec{v}$ and is the flux density of acoustic energy in a given direction.

Decibel is a log scale used to compare acoustic intensities. Acoustic attenuation in a medium is expressed in dB/m or Np/m.

Classical attenuation for liquids and gases is due to viscosity and thermal conduction.

Molecular relaxation occurs in polyatomic gases and liquids. Because of the phase lag in transferring the ultrasonic energy to the different energy levels, relaxation gives rise to an extra attenuation, which is described by the parameter $\omega\tau$. The limit $\omega\tau \ll 1$ applies to most media at ultrasonic frequencies.

Questions

1. Draw waveforms as a function of x for $u(x)$, $v(x)$, and $P(x)$ for a traveling harmonic wave. Comment on the phase relationships in the forward direction of the form in Equation 3.14.
2. Use the results of Question 1 and Equation 3.52 to calculate $I(x)$ for this wave. Calculate and sketch the graph of both instantaneous and average values of $I(x)$.
3. Using data from Table 3.1, calculate for glycerin:
   i. Viscous relaxation time τ_s
   ii. Low frequency attenuation α_s; compare this result with that of Table 3.1
4. Use an approximation for air as a perfect gas of molecular weight 29. At STP (0°C and 1 atmosphere of pressure), calculate:
   i. Mass density
   ii. Average molecular velocity
   iii. Mean free time between collisions
   iv. Mean free path between collisions
5. Justify Equation 3.7.
6. Show that the condensation s, the density change per unit density = $-S$, where S is the dilatation.
7. Find the specific acoustic impedance for a standing wave $p = p_0 \sin kx \cos \omega t$.
8. For two waves of different frequency traveling in the +x direction, show that the specific acoustic impedance is $\rho_0 V_0$.

4

Introduction to the Theory of Elasticity

The theory of elasticity is the study of the mechanics of continuous media, or in simple words, the deformation of the elements of a solid body by applied forces. In this chapter we deal with static (time independent) elasticity involving homogeneous deformations. In fact, the parameters defined here can also be used at the finite frequencies occurring in ultrasonic propagation. This is the simplest case and enables us to define concepts such as deformation, strain tensor, stress tensor, and the moduli of elasticity. We introduce tensor notation to describe the elastic parameters; it is a simple, elegant, and powerful approach that is used throughout advanced treaties in elasticity and acoustics. Complete discussions are given for tensors by Nye [11] and for elasticity by Landau and Lifshitz [12].

4.1 A Short Introduction to Tensors

Study of physics and engineering leads to categorizing measurable quantities as scalars or vectors. Scalars are physical quantities that can be represented by a simple number, e.g., temperature. Equally important, they are not associated with direction. A vector on the other hand explicitly depends on direction, for example, velocity $\vec{V}$. In 3D space, we must specify the three components V_x, V_y, and V_z to describe the velocity vector $\vec{V}$ fully.

The concept of tensor has been introduced as an extension of the idea of a vector. In anisotropic media, tensors are essential to describe the relation between two vectors. But even in isotropic media the idea of physical quantities specified by more than three components is essential, as will be seen in the theory of elasticity.

The concept of a tensor can be made concrete by a simple example, that of the electrical conductivity in a solid. For a one-dimensional system (wire), it is customary to represent the conductivity σ as the proportionality constant linking the current density J to the electric field E, $J = \sigma E$. However, for a three-dimensional medium that is anisotropic, the electric field $\vec{E}$ and the current density $\vec{J}$ will be, in general, in quite different directions. So, in general, one must write

$$
\begin{aligned}
J_1 &= \sigma_{11}E_1 + \sigma_{12}E_2 + \sigma_{13}E_3 \\
J_2 &= \sigma_{21}E_1 + \sigma_{22}E_2 + \sigma_{23}E_3 \\
J_3 &= \sigma_{31}E_1 + \sigma_{32}E_2 + \sigma_{33}E_3
\end{aligned} \tag{4.1}
$$

Thus, to specify the conductivity fully, we need to specify the nine components that are usually written in matrix form as

$$
\sigma_{ij} = \begin{bmatrix} \sigma_{11} & \sigma_{12} & \sigma_{13} \\ \sigma_{21} & \sigma_{22} & \sigma_{23} \\ \sigma_{31} & \sigma_{32} & \sigma_{33} \end{bmatrix} \tag{4.2}
$$

The notation on the left is the tensor notation; for obvious reasons, σ_{ij} is termed a tensor of the second rank.

For an isotropic system, $\vec{J}$ is always parallel to $\vec{E}$ and $|\vec{J}| = \sigma|\vec{E}|$. It follows in this case that the conductivity tensor is given by

$$
\sigma_{ij} = \begin{bmatrix} \sigma & 0 & 0 \\ 0 & \sigma & 0 \\ 0 & 0 & \sigma \end{bmatrix} \tag{4.3}
$$

A simple rule that follows from the general form of σ_{ij} is that the rank of a tensor is given by the number of indices. Thus, a scalar is a tensor of rank zero and a vector is a tensor of rank one.

At this point it should be emphasized that although a tensor can be written in matrix form, it is not just a simple matrix. A tensor represents a real physical quantity, such as conductivity, while many matrices (e.g., change of coordinates) are simple mathematical relationships. Many advanced texts show that a tensor is rigorously defined by the way that it transforms under coordinate transformation (e.g., see [11]), which will not be needed here, as all the tensors used in this book represent well-known physical properties.

From a practical standpoint, much economy of presentation and elegance can be obtained by using the Einstein convention. This convention says quite simply that when a suffix occurs twice in the same term this automatically implies summation over that suffix, which becomes a dummy index or dummy suffix.

For example, Equation 4.1 can be written

$$
\begin{aligned}
J_1 &= \sum \sigma_{1j}E_j \\
J_2 &= \sum \sigma_{2j}E_j \\
J_3 &= \sum \sigma_{3j}E_j
\end{aligned} \tag{4.4}
$$

or again

$$J_i = \sum_{j=1}^{3} \sigma_{ij} E_j$$

With the Einstein convention

$$J_i = \sigma_{ij} E_j \tag{4.5}$$

where it is understood, and never indicated explicitly, that i, j go over all available values, here 1, 2, and 3 or x, y, and z. In this relation i gives the direction of current flow.

4.2 Strain Tensor

The basic idea is that forces will be applied to solid bodies to deform them. As a starting point there is a need to describe the deformation. If a point at $\vec{r}$ from the origin is displaced to position $\vec{r}\,'$ by the force then the deformation $\vec{u} = \vec{r}_i' - \vec{r}$ is called the displacement vector. In tensor notation $u_i = x_i' - x_i$ where u_i and x_i' are functions of u_i.

Since a point is displaced during a deformation then the distance dl between two points close together is also changed. Using

$$dl^2 = dx_1^2 + dx_2^2 + dx_3^2 = dx_i^2 \quad \text{before deformation} \tag{4.6}$$

$$dl'^2 = dx_i'^2 \quad \text{after deformation} \tag{4.7}$$

Hence

$$dl'^2 = (dx_i + du_i)^2 \tag{4.8}$$

Using

$$du_i = \left(\frac{\partial u_i}{\partial x_k}\right) dx_k \tag{4.9}$$

$$dl'^2 = dl^2 + 2\frac{\partial u_i}{\partial x_k} dx_i dx_k + \frac{\partial u_i}{\partial x_k}\frac{\partial u_i}{\partial x_l} dx_k dx_l \tag{4.10}$$

This can be written as

$$dl'^2 = dl^2 + 2S_{ik} dx_i dx_k \tag{4.11}$$

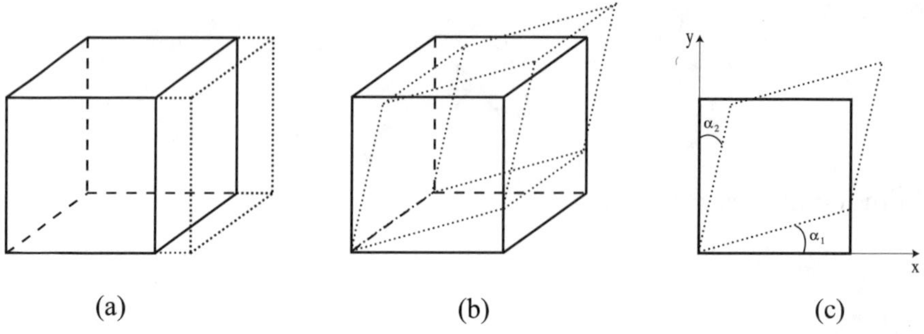

FIGURE 4.1
Strains for a unit cube. (a) Tensile strain u_{xx}. (b) Shear strain u_{xy}. (c) Definition of angles for shear strain u_{xy}.

where

$$S_{ik} = \frac{1}{2}\left(\frac{\partial u_i}{\partial x_k} + \frac{\partial u_k}{\partial x_i} + \frac{\partial u_l}{\partial x_i}\frac{\partial u_l}{\partial x_k}\right) \tag{4.12}$$

If the strains are sufficiently small, which will always be assumed to be the case in linear ultrasonics, then the quadratic terms can be ignored. The strain tensor S_{ik} is then

$$S_{ik} = \frac{1}{2}\left(\frac{\partial u_i}{\partial x_k} + \frac{\partial u_k}{\partial x_i}\right) \tag{4.13}$$

By construction the strain tensor is symmetric so that nine terms reduce to six. Clearly three of these are diagonal and three are nondiagonal. Each diagonal term ($i = k = 1, 2,$ or 3) has the simple significance shown in Figure 4.1. For example,

$$S_{11} = \frac{\partial u_1}{\partial x_1} \tag{4.14}$$

is clearly the extension per unit length in the x_1 direction. Hence, the diagonal terms correspond to compression or expansion along one of the three axes.

The off-diagonal terms can be understood with reference to Figure 4.1 for the case of a deformation of the plane perpendicular to the z axis. For small deformations

$$\tan\alpha_1 \sim \alpha_1 = \frac{\partial u_y}{\partial x}, \quad \tan\alpha_2 \sim \alpha_2 = \frac{\partial u_x}{\partial y} \tag{4.15}$$

where α_1 and α_2 are angles with the x and y axes, respectively.

Thus the change in angle between the two sides of the rectangle

$$\alpha_1 + \alpha_2 = \frac{\partial u_x}{\partial y} + \frac{\partial u_y}{\partial x} \tag{4.16}$$

is proportional to the shear strain S_{xy}.

A final property of the strain tensor can be obtained from the following mathematical results:

i. Any symmetric tensor can be diagonalized at a point by the choice of appropriate axes. If this is done then the strain tensor has diagonal components $S^{(1)}$, $S^{(2)}$, and $S^{(3)}$ and the off-diagonal terms are zero.
ii. The trace (i.e., the sum of the diagonal terms) of a symmetric tensor is invariant under change of coordinates. From (i), the trace will then be $S^{(1)} + S^{(2)} + S^{(3)}$ for the choice of any coordinate system.

Suppose that the coordinates are chosen so that S_{ik} is diagonal; then

$$\begin{aligned} dl'^2 &= (\partial_{ik} + 2S_{ik})dx_i\,dx_k \\ &= (1 + 2S^{(1)})dx_1^2 + (1 + 2S^{(2)})dx_2^2 + (1 + 2S^{(3)})dx_3^2 \end{aligned} \tag{4.17}$$

where the relative displacement along axis i is $S^{(i)}$.

Consider the volume before and after deformation of a small volume element dV. It follows that

$$\begin{aligned} dV &= dx_1\,dx_2\,dx_3 \\ dV' &= dx_1'\,dx_2'\,dx_3' \end{aligned} \tag{4.18}$$

so that

$$\begin{aligned} dV' &= dV(1 + S^{(1)})(1 + S^{(2)})(1 + S^{(3)}) \\ &\approx dV(1 + S^{(1)} + S^{(2)} + S^{(3)}) \end{aligned} \tag{4.19}$$

again neglecting quadratic terms.

In any coordinate system, the trace can be written

$$S_{ii} = S_{11} + S_{22} + S_{33} \tag{4.20}$$

This gives finally

$$dV' = dV(1 + S_{ii}) \tag{4.21}$$

so that S_{ii} gives the relative change in volume under deformation. This can be expressed as the dilatation S, which is the change in volume per unit volume, which can be expressed as

$$S = S_{ii} = S_{11} + S_{22} + S_{33} \tag{4.22}$$

4.3 Stress Tensor

We assume a body in static equilibrium under external forces such that there is no net translation or rotation. What is of concern is the effect of internal forces on a hypothetical unit cube inside the solid. These forces could arise, for example, from an ultrasonic wave impinging on the region in question. In principle there could be two types of forces acting on the cube; body forces (acting on the volume) and surface forces. Body forces such as gravity will not be considered, so that a description is needed for surface forces acting on the faces of the cube. These forces will lead to deformation of the cube, which can be described by the strain tensor treated previously. Once this description has been obtained, it will be possible to formulate a three-dimensional equivalent of Hooke's law for a relation between the forces and the deformations.

As seen in Figure 4.2, an applied force will generally be at some arbitrary angle to the unit cube. Since we are considering forces on the faces of the cube, we consider a particular face, for example, the xy face with normal along the z axis.

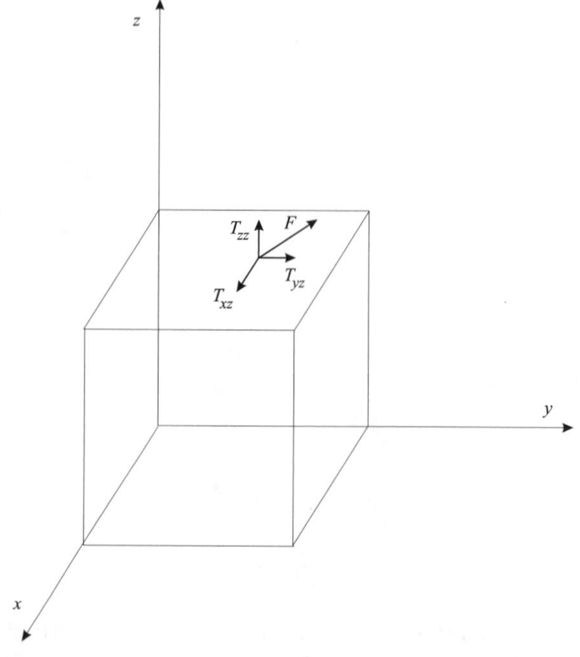

FIGURE 4.2
Definition of components of the stress tensor.

The components of the applied force can be separated into two major classes:

- Normal component to the face, which will give rise to compressive or tensile stresses.
- Tangential components, giving rise to shear stresses. For the example considered there are two of these: dF_x and dF_y.

In one dimension, the stress on a rod is defined as the force per unit area. In extending this definition to three dimensions, as above, clearly there are two vectors involved, namely the direction of the surface normal and the direction of the force. It follows that in three dimensions the stress must be described by a stress tensor of rank two.

Extending directly from one dimension

$$T_{zz} = \frac{F_z}{A_z}, \quad T_{zy} = \frac{F_z}{A_y}, \quad T_{zx} = \frac{F_z}{A_x} \tag{4.23}$$

so that all of the components are described by a stress tensor of rank two.

The condition of static equilibrium leads to symmetry of the stress tensor. The tensile stresses along any one axis must balance; otherwise the body would accelerate, so that there can only be three independent diagonal stresses. Likewise the shear stresses must balance to avoid rotation, leading to three off-diagonal stresses. An elegant demonstration of this, together with a more abstract presentation of the stress tensor is given by Landau and Lifshitz [12].

Finally, for the specific case of a liquid, the pressure is hydrostatic; it is uniform and the same in all directions. Hence, for a sphere the force in direction i on surface element dA is

$$\begin{aligned} dF_i &= -p\,dA_i = -p\,\delta_{ik}\,dA_k \\ &= T_{ik}\,dA_k \end{aligned} \tag{4.24}$$

Here δ_{ij} is the Kronecker delta, an extremely useful mathematical device. It is defined as

$$\delta_{ik} = \begin{cases} 1 & i = k \\ 0 & i \neq k \end{cases} \tag{4.25}$$

Hence, for uniform hydrostatic compression

$$\begin{aligned} T_{ik} &= -p\,\delta_{ik} \\ T_{ik} &= \begin{pmatrix} -p & 0 & 0 \\ 0 & -p & 0 \\ 0 & 0 & -p \end{pmatrix} \end{aligned} \tag{4.26}$$

The nondiagonal elements correspond to shear stress; these are zero, corresponding to the well-known fact that an inviscid liquid cannot support a shear stress.

4.4 Thermodynamics of Deformation

Assume small and slow deformations so that the latter can be assumed to be elastic (so that it returns to its original state when external forces are removed) and reversible in the thermodynamic sense.

In general, the thermodynamic identity gives

$$dU = TdS + dW \tag{4.27}$$

where

U = internal energy
T = temperature
S = entropy
W = work done on the system

For the particular case of hydrostatic compression

$$dW = -pdV = -pdS_{ii} = -p\delta_{ik}dS_{ik} = T_{ik}dS_{ik} \tag{4.28}$$

so that

$$dU = TdS + T_{ik}dS_{ik} \tag{4.29}$$

as shown elsewhere [12] this form is in fact true in the general case.

For the Helmholtz free energy $F = U - TS$, so

$$dF = -SdT + T_{ik}dS_{ik} \tag{4.30}$$

and for the Gibbs free energy $G = H - TS$

$$\begin{aligned} H &= U + pV \\ G &= U - TS - T_{ik}dS_{ik} = F - T_{ik}dS_{ik} \end{aligned} \tag{4.31}$$

From the form of a perfect differential in terms of its partial derivatives

$$T_{ik} = \left(\frac{\partial U}{\partial S_{ik}}\right)_S = \left(\frac{\partial F}{\partial S_{ik}}\right)_T \tag{4.32}$$

and

$$S_{ik} = -\left(\frac{\partial G}{\partial T_{ik}}\right)_T \tag{4.33}$$

4.5 Hooke's Law

In its simplest form Hooke's law states that for small elongations of an elastic system the stress is proportional to the strain. There are two different and equivalent approaches to Hooke's law for the isotropic solid, each important and instructive in its own way. The first [12] is based on Landau's classic expansion of the Helmholtz free energy in parameters of the system and subsequent application of statistical physics. In this case, the free energy F is expanded in terms of the strain tensor

$$F_T = F_0 + \frac{1}{2}\lambda u_{ii}^2 + \mu S_{ik}^2 \tag{4.34}$$

where λ and μ are called the Lamé coefficients. This expansion takes in account the following points:

i. For the undeformed system at constant temperature $S_{ik} = 0$ and $T_{ik} = 0$. Since $T_{ik} = (\partial F / \partial S_{ik})_T$, there is no linear term in the expansion.
ii. Since F is a scalar, every term in the expansion must be a scalar. Since the diagonal terms S_{ii}^2 and all diagonal terms S_{ik}^2 are scalars the coefficients λ and μ are also scalars.

The form of F can be rewritten to take into account the two fundamentally different forms of deformation of isotropic bodies:

i. Pure shear, corresponding to constant volume and change of shape. $S_{ii} = 0$ in this case.
ii. Pure hydrostatic compression, corresponding to change in volume at constant shape.

Any deformation can be written as the sum of two, leading to the following form

$$S_{ik} = \left(S_{ik} - \frac{1}{3}\delta_{ik}S_{ll}\right) + \frac{1}{3}\delta_{ik}S_{ll} \tag{4.35}$$

The first term is pure shear, the second hydrostatic compression.

The free energy can be rewritten to show shear and compression explicitly

$$F = \mu\left(S_{ik} - \frac{1}{3}\delta_{ik}S_{ll}\right)^2 + \frac{1}{2}KS_{ll}^2 \tag{4.36}$$

where now

$$K = \lambda + \frac{2}{3}\mu = \text{modulus of compression} \tag{4.37}$$

$$\mu = \text{modulus of rigidity}$$

This rearrangement of terms is more than a mathematical device. It will be seen that these two moduli determine the velocities of the two acoustic modes, longitudinal and shear, that can propagate in an isotropic solid.

Statistical physics tells us that the Helmholtz free energy is a minimum for a system at constant temperature in thermal equilibrium. In the absence of external forces, this minimum must occur at $S_{ik} = 0$. The two quadratic forms in Equation 4.36 must be positive, so a necessary and sufficient condition for F to be positive is that $K > 0$, $\mu > 0$.

The thermodynamic relations of the preceding section can be used to determine the relations between stress and strain, in particular Equation 4.32. Directly from Equation 4.36

$$\begin{aligned} dF &= KS_{ll}\,dS_{ll} + 2\mu\left(S_{ik} - \frac{1}{3}S_{ll}\delta_{ik}\right)d\left(S_{ik} - \frac{1}{3}S_{ll}\delta_{ik}\right) \\ &= \left[KS_{ll}\delta_{ik} + 2\mu\left(S_{ik} - \frac{1}{3}S_{ll}\delta_{ik}\right)\right]dS_{ik} \end{aligned} \tag{4.38}$$

so that finally

$$T_{ik} = \left(\frac{\partial F}{\partial S_{ik}}\right)_T = KS_{ll}\delta_{ik} + 2\mu\left(S_{ik} - \frac{1}{3}S_{ll}\delta_{ik}\right) \tag{4.39}$$

This result shows that pure compression and shear deformation give rise to stress components proportional to K and μ, respectively. It is also a manifestation of Hooke's law as in both cases stress is proportional to strain.

It is easy to find the inverse expression linking S_{ik} to T_{ik}. Directly from Equation 4.39

$$T_{ii} = 3KS_{ii} \tag{4.40}$$

Then immediately Equation 4.39 can be inverted to give

$$S_{ik} = \frac{\delta_{ik}T_{ll}}{9K} + \frac{(T_{ik} - \frac{1}{3}\delta_{ik}T_{ll})}{2\mu} \tag{4.41}$$

which again demonstrates Hooke's law. Equation 4.41 gives the important result that the diagonal components of stress and strain are uniquely connected for the case of pure hydrostatic compression. In this case, $T_{ik} = -p\delta_{ik}$ so that

$$S_{ii} = -\frac{p}{K} \tag{4.42}$$

For small variations we can write the compressibility χ as

$$\chi = \frac{1}{K} = -\frac{1}{V}\left(\frac{\partial V}{\partial p}\right)_T \tag{4.43}$$

Finally, Euler's theorem can be applied to obtain a compact form for F. Since F is quadratic in S_{ik}, Euler's theorem states that

$$S_{ik}\frac{\partial F}{\partial S_{ik}} = 2F \tag{4.44}$$

Together with $T_{ik} = (\partial F/\partial S_{ik})_T$, this gives

$$F = \frac{1}{2}T_{ik}S_{ik} \tag{4.45}$$

The second approach to Hooke's law is much more direct and will be of more practical use. T_{ij} is expanded as a Taylor's series in S_{kl}

$$T_{ij} = T_{ij}(0) + \left(\frac{\partial T_{ij}}{\partial S_{kl}}\right)_{S_{kl}=0} + \frac{1}{2}\left(\frac{\partial^2 T_{kl}}{\partial S_{ij}\partial S_{mn}}\right)_{S_{ij}=0,\,S_{mn}=0} S_{ij}S_{mn} \tag{4.46}$$

The first term $T_{ij}(0) \equiv 0$ at $S_{ij} = 0$ since stress and strain go to zero simultaneously for elastic solids. The third (nonlinear) term will be neglected here; it forms the basis of the third-order elastic constants and nonlinear acoustics. In linear elasticity, the series is truncated after the second term, leading to

$$T_{ij} = c_{ijkl}S_{kl} \tag{4.47}$$

where

$$c_{ijkl} \equiv \left(\frac{\partial T_{ij}}{\partial S_{kl}}\right)_{S_{kl}=0} \tag{4.48}$$

is known as the elastic stiffness tenor or elastic constant tensor.

A similar Taylor's series expansion of S_{ij} in terms of T_{kl} could be carried out in identical fashion, leading to the elastic compliance tensor s_{ijkl} where

$$S_{ij} = \left(\frac{\partial S_{ij}}{\partial T_{kl}}\right)_{T_{kl}=0} T_{kl} = s_{ijkl}T_{kl} \tag{4.49}$$

Each tensor can be deduced from the other by

$$s_{ijkl} = c_{ijkl}^{-1} \tag{4.50}$$

and in what follows c_{ijkl} will be used exclusively.

Since stress is proportional to strain c_{ijkl} represents Hooke's law in three dimensions and is the extension of the one-dimensional spring constant k in $F = -kx$. It is obviously a fourth-rank tensor, as it must be, as it links two second-rank tensors. Lastly, since both T_{ij} and S_{kl} are symmetric, this symmetry is reflected in c_{ijkl}, which is also itself symmetric

$$c_{ijkl} = c_{jikl} = c_{ijlk} = c_{jilk} \tag{4.51}$$

TABLE 4.1

Conversion Table from Regular Indices to Reduced Indices (Engineering Notation)

α, β	ij, kl
1	11
2	22
3	33
4	23 = 32
5	31 = 13
6	12 = 21

and

$$c_{ijkl} = c_{klij} \tag{4.52}$$

These symmetry operations reduce the number of independent constants from 81 to 36 to 21 for crystals of different symmetries. The number varies from 21 (triclinic) to 3 (cubic) as is shown in numerous advanced texts in acoustics. For isotropic solids it has already been demonstrated that there are only two independent elastic constants.

In fact it is well known [12] that for an isotropic solid

$$c_{ijkl} = \lambda\delta_{ij}\delta_{kl} + \mu(\delta_{ik}\delta_{jl} + \delta_{il}\delta_{jk}) \tag{4.53}$$

where λ and μ are the Lamé coefficients already introduced.

It is standard practice to use a reduced notation for the elastic constants, due to the symmetry of the T_{ij} and S_{kl}. Since each of the latter has six independent components, the c_{ijkl} tensor has a maximum of 36. This leads to the introduction of the so-called engineering notation where the $c_{\alpha\beta} \equiv c_{ijkl}$. Since ij and kl go in pairs, the six α and β values are as shown in Table 4.1. Again, the symmetry of c_{IJ}

$$c_{IJ} = c_{JI}$$

leads to a maximum of 21 independent constants.

Since the same symbol c is universally used for the elastic constant tensor, it is immediately obvious from the number of indices whether the full or reduced notation is being used. Thus if c_{11} is used, it can only be in reduced notation, which is, in fact, more current in the literature. Using Hooke's law and the isotropic form of c_{ijkl}, we obtain immediately

$$T_{ij} = \lambda(S_{xx} + S_{yy} + S_{zz}) + 2\mu_{ij}S_{ii} \tag{4.54}$$

for extensional stress, $i = x, y, z$ and

$$T_{ij} = 2\mu S_{ij} \tag{4.55}$$

for tangential stress with $i, j = x, y, z$ and $i \neq j$.

In reduced notation the stiffness matrix in the general case is thus

$$c_{IJ} = \begin{bmatrix} c_{11} & c_{12} & c_{13} & c_{14} & c_{15} & c_{16} \\ c_{21} & c_{22} & c_{23} & c_{24} & c_{25} & c_{26} \\ c_{31} & c_{32} & c_{33} & c_{34} & c_{35} & c_{36} \\ c_{41} & c_{42} & c_{43} & c_{44} & c_{45} & c_{46} \\ c_{51} & c_{52} & c_{53} & c_{54} & c_{55} & c_{56} \\ c_{61} & c_{62} & c_{63} & c_{64} & c_{65} & c_{66} \end{bmatrix} \tag{4.56}$$

while for the isotropic case

$$c_{IJ} = \begin{bmatrix} \lambda+2\mu & \lambda & \lambda & 0 & 0 & 0 \\ \lambda & \lambda+2\mu & \lambda & 0 & 0 & 0 \\ \lambda & \lambda & \lambda+2\mu & 0 & 0 & 0 \\ 0 & 0 & 0 & \mu & 0 & 0 \\ 0 & 0 & 0 & 0 & \mu & 0 \\ 0 & 0 & 0 & 0 & 0 & \mu \end{bmatrix} \tag{4.57}$$

where as before

$$T_J = \lambda(S_1 + S_2 + S_3) + 2\mu S_J, \quad J = 1, 2, 3 \tag{4.58}$$

for extensional stress and

$$T_J = \mu S_J, \quad J = 4, 5, 6 \tag{4.59}$$

for tangential stress.

4.6 Other Elastic Constants

Four other parameters have found practical use as they are directly related to measurements, which is not the case, for example, for the parameter λ for solids. Important mathematical relations between these parameters and values for representative materials are given in Tables 4.2 and 4.3.

i. Young's modulus E is defined as the ratio of axial stress to axial strain for a free-standing rod. E can be expressed using Equation 4.58 as follows.

TABLE 4.2
Expressions for the Elastic Constants in Terms of Different Pairs of Independent Parameters

	λ, μ	c_{11}, c_{44}	E, σ	E, μ
λ	λ	$c_{11} - 2c_{44}$	$\frac{E\sigma}{(1+\sigma)(1-2\sigma)}$	$\frac{\mu(E-2\mu)}{3\mu - E}$
μ	μ	c_{44}	$\frac{E}{2(1+\sigma)}$	μ
E	$\frac{\mu(3\lambda+2\mu)}{\lambda+\mu}$	$\frac{c_{44}(3c_{11}-4c_{44})}{c_{11}-c_{44}}$	E	E
K	$\lambda + \frac{2\mu}{3}$	$c_{11} - \frac{4c_{44}}{3}$	$\frac{E}{3(1-2\sigma)}$	$\frac{E\mu}{3(3\mu - E)}$
σ	$\frac{\lambda}{2(\lambda+\mu)}$	$\frac{c_{11}-2c_{44}}{2(c_{11}-c_{44})}$	σ	$\frac{E}{2\mu} - 1$

TABLE 4.3
Elastic Constants for Representative Isotropic Solids

Substance	Young's Modulus E 10^9 n-m^{-2}	Modulus of Compression K 10^9 n-m^{-2}	Lamé Constants λ 10^9 n-m^{-2}	Lamé Constants μ 10^9 n-m^{-2}	Poisson's Ratio σ
Epoxy	4.5	6.7	5.63	1.60	0.39
Lucite	3.9	6.5	5.60	1.39	0.4
Pyrex glass	60.3	39.6	23.4	24.21	0.25
PZT-5 A	104.1	94.0	67.4	39.6	0.32
Aluminum	67.6	78.1	61.4	25.0	0.36
Brass	104.8	140.2	114.7	38.1	0.38
Copper	128.6	209.0	178.2	46.0	0.40
Gold	80.6	169.1	150.1	28.4	0.42
Lead	34.7	98.8	90.8	12.1	0.44
Fused quartz	72.5	37.0	16.3	30.9	0.17
Steel	194.2	167.4	113.2	80.9	0.29
Beryllium	73.0	115.1	16.3	147.5	0.05
Sapphire (z)	895	298.8	201.0	145.9	0.29

Let the rod be aligned along the x axis, so that the only stress component is $T_{xx} = T_1$. Then

$$\begin{aligned} T_1 &= (\lambda+2\mu)S_1 + \lambda(S_2+S_3) \\ 0 &= (\lambda+2\mu)S_2 + \lambda(S_1+S_3) \\ 0 &= (\lambda+2\mu)S_3 + \lambda(S_1+S_2) \end{aligned} \tag{4.60}$$

Hence

$$E = \frac{T_1}{S_1} = \frac{\mu(3\lambda+2\mu)}{\lambda+\mu} \tag{4.61}$$

The usefulness of this parameter is that it is obtained in a standard laboratory measurement. Relations between E and the other elastic constants are given in Table 4.2; evidently the two independent elastic constants can be chosen to be (E,μ), (E,σ), (λ,μ), or (c_{11}, c_{44}).

ii. Poisson's ratio σ is given by the ratio of the lateral contraction to the longitudinal extension of the rod in (i).

$$\sigma = -\frac{S_3}{S_1} = -\frac{S_2}{S_1} = \frac{\lambda}{2(\lambda+\mu)} \tag{4.62}$$

σ can be measured in the same experiment as Young's modulus. It has been pointed out that by Landau and Lifshitz [12] that in principle $-1 \le \sigma \le 0.5$, although negative values of σ have never been observed. Also it can be shown that $\sigma > 0$ corresponds to $\lambda > 0$, although neither of these is thermodynamically necessary. Finally, $\sigma \sim 0.5$ corresponds to materials for which the modulus of rigidity μ is small compared to the modulus of compression K.

iii. Bulk modulus or modulus of compression

$$K \equiv -\frac{p}{S} \tag{4.63}$$

and its reciprocal, the compressibility

$$\chi \equiv -\frac{1}{V}\left(\frac{\partial V}{\partial p}\right). \tag{4.64}$$

Both parameters should be specified as being given in either adiabatic or isothermal conditions.

For a solid under uniform hydrostatic pressure

$$T_{ij} = -p\delta_{ij} \tag{4.65}$$

using

$$T_{ij} = \lambda S \delta_{ij} + 2\mu S_{ij}$$

and

$$S = S_{11} + S_{22} + S_{33}$$

this gives

$$p = -\left(\lambda + \frac{2\mu}{3}\right)S = -KS$$

Hence,

$$K = \lambda + \frac{2\mu}{3} \tag{4.66}$$

as was used earlier in Equation 4.37.

iv. Rigidity modulus μ. For a pure shear $\mu \equiv \text{shear stress}/\text{shear strain}$, for a free-standing sample. The rigidity modulus thus plays a role for shear waves analogous to that of Young's modulus for longitudinal waves in the longitudinal stretching of a free-standing rod. Since only two elastic constants are needed to describe the isotropic case fully, there are a number of possible choices. Values for each of these constants in terms of common choices for the two independent constants are given in Table 4.2. Representative values of these constants are given in Table 4.3.

Summary

Tensor of order n is a tensor requiring n indices to specify it.

Einstein notation or Einstein summation convention is a convention that repeated indices in the same term of a tensor equation are summed over all available values.

Strain tensor S_{ij} is a linearized second-order tensor describing the mechanical strain at a point. The strain tensor is symmetric.

Stress tensor T_{ij} is a second-order symmetric tensor describing the local stress. The first index gives the direction of the force, the second gives the direction of the normal to the surface on which it acts.

Lamé constants λ and μ are the constants historically chosen to describe the elastic properties of an isotropic solid.

Modulus of compression or bulk modulus K is the elastic constant corresponding to hydrostatic compression.

Compressibility is the reciprocal of the bulk modulus.

Elastic constant tensor is a fourth-order symmetric tensor giving the stress tensor as a function of the strain tensor. It is also called the elastic stiffness tensor.

Young's modulus is the elastic constant corresponding to the stretching of a free-standing bar.

Poisson's ratio is the ratio of the lateral contraction to the longitudinal extension of a free-standing bar.

Questions

1. For the case of the axial extension of a bar, what would be the implications of a negative Poisson's ratio to the deformation? What would be the consequences for the other elastic parameters?
2. In Einstein notation a spatial derivative is written using a comma, for example, $\delta u_{ij}/\delta x_j = u_{ij,j}$. Write the following differential equations

and vector algebra forms in Einstein notation:

i. grad φ

ii. curl $\vec{\Psi}$

iii. div $\vec{E}$

iv. $\nabla^2\vec{E}$

v. $\frac{\delta^2 u}{\delta t^2} = V_0^2 \frac{\delta^2 u}{\delta x^2}$

3. Write out the following equations written in Einstein notation in full Cartesian form:

   i. $u_{ij} = \frac{1}{2}(u_{i,j} + u_{j,i}) + \frac{1}{2}(u_{k,i}\, u_{k,j})$

   ii. $P_i = d_{ijk}\,\delta_{jk}$

   iii. $P_i = K_0 X_{ij} E_j$

4. Verify the results of Table 4.1.
5. Write out in full the results of Equation 4.65 to show that $K = \lambda + 2\mu/3$.

   i. A rectangular plate has length l (x direction), width w (y direction) and thickness t (z direction). A uniform stress T_{xx} is applied at the ends and a uniform stress T_{yy} on both sides, so that the width remains unchanged. Using Hooke's law, determine Poisson's ratio and Young's modulus.

   ii. Express the above results as a function of E and σ.

5

Bulk Acoustic Waves in Solids

Elasticity theory provides a complete description of the static properties of a mechanical system and in fact parameters such as the elastic moduli can also be used to describe the dynamic properties over the full ultrasonic frequency range. However, we need a dynamic theory to describe wave propagation and that is provided in the present chapter. We first generalize the one-dimensional results for fluids to the case of one-dimensional longitudinal waves in solids. We then examine the three-dimensional solid, where both longitudinal and transverse modes are present. Finally, we discuss the attenuation mechanisms in a number of important cases.

The basic results for one-dimensional propagation in fluids can be generalized to the one-dimensional propagation of a simple longitudinal mode in solids. There are of course many differences between liquids and solids regarding their acoustic properties. For our purposes some important ones are the following:

1. Compared to solids, liquids are very compressible. This is why the acoustic pressure and the compressibility are commonly used as parameters for liquids. Except for specialized applications, one never uses these parameters in solids; the stress and the elastic constants are the appropriate parameters in this case.
2. Liquids can change shape, as it were, at will, or at least to accommodate the container. Hence, a liquid cannot support a static shear stress; shear waves can only propagate in liquids at high frequencies and then only for a very short distance. However, in solids it is essential to take into account longitudinal and transverse waves to give a full description. Thus, the scalar theory is insufficient to describe the three-dimensional behavior of solids.
3. In liquids the pressure is a scalar and acts uniformly on a volume element, so that the modulus of compression (bulk modulus) is the appropriate modulus for longitudinal wave propagation. In solids, however, one can have a unidirectional compression or tension so that the appropriate modulus for longitudinal waves is not the bulk modulus.

In this chapter, we summarize the one-dimensional results and write them in the notation for longitudinal and transverse waves in solids. This is followed by the three-dimensional theory for isotropic solids. Finally, we describe the propagation properties of ultrasonic waves and attenuation mechanisms in a number of important cases.

5.1 One-Dimensional Model of Solids

We generalize the results of Chapter 3 for fluids as appropriate for longitudinal modes in solids for propagation in the x direction with wave velocity V_L. We consider an element of length l undergoing an elongation ∂u due to an external force F in the positive x direction.

The external stress is $T \equiv F/A$, so that the net stress on the element is $\partial T = l(\partial T/\partial x)$. This leads to a net force per unit volume on the element of $\partial T/\partial x$.

The strain is

$$S = \frac{\partial u}{l} = \frac{\partial u}{\partial x} \tag{5.1}$$

Hooke's law is given by $T \equiv cS$ where c is a constant.

Writing Newton's law

$$\frac{\partial T}{\partial x} = \rho \ddot{u} \tag{5.2}$$

and combining this with Hooke's law, we immediately obtain the wave equation

$$\frac{\partial^2 u}{\partial x^2} = \frac{\rho_0}{c}\frac{\partial^2 u}{\partial t^2} \tag{5.3}$$

which can also be written for the stress and the velocity, similar to the case for fluids.

The solutions for the displacement are

$$u = A \exp j(\omega t - \beta x) + B \exp j(\omega t + \beta x)$$

As for fluids the first term corresponds to propagation in the forward direction $(+x)$ and the second to the propagation in the backward direction $(-x)$.

The propagation parameters are

- wave number $\beta = \omega / V_L$
- wave velocity $V_L = \sqrt{c/\rho_0}$

The instantaneous values of the energy density follow from the expressions for the fluid and elasticity theory of Chapter 4.

$$u_K = \frac{1}{2}\rho_0 v^2 \tag{5.4}$$

$$u_P = \frac{1}{2}TS \tag{5.5}$$

and hence the average values are

$$\bar{u}_K = \frac{1}{2}\mathrm{Re}\left[\frac{1}{2}\rho v v^*\right] = \frac{1}{4}\mathrm{Re}\,[\rho v v^*] \tag{5.6}$$

$$\bar{u}_P = \frac{1}{2}\mathrm{Re}\left[\frac{1}{2}TS^*\right] \tag{5.7}$$

and finally

$$\bar{u}_a = \frac{1}{2}\mathrm{Re}[TS^*] \tag{5.8}$$

The acoustic intensity I can be written as

$$I = \bar{u}_a V_L \tag{5.9}$$

and the instantaneous acoustic Poynting vector

$$\mathcal{P} = -vT \tag{5.10}$$

which follows directly as a generalization of Equation 3.52.

5.2 Wave Equation in Three Dimensions

Following the case for optics, on physical grounds we expect to find three acoustic polarizations in three dimensions; indeed, it is well known that for $3N$ atoms there are $3N$ normal modes, three branches with N modes per branch. On physical grounds, one expects to find one longitudinal branch and two transverse branches with orthogonal polarization. This section shows how the existence of the longitudinal and transverse branches flows directly from the formalism developed thus far.

The wave equation in three dimensions can be obtained immediately by combining the following two equations already seen:

$$\frac{\partial T_{ij}}{\partial x_j} = \rho_0 \frac{\partial^2 u_i}{\partial t^2} \tag{5.11}$$

$$T_{ij} = c_{ijkl} S_{kl} \tag{5.12}$$

With the various possibilities of full and reduced notation and the Lamé constants, i.e., c_{ijkl}, c_{IJ}, λ, and μ, there are many possible choices for proceeding. Anticipating the result we choose c_{11} and c_{44}; also in this case the decoupling between longitudinal and transverse modes is most transparent. Thus

$$T_{ij} = (c_{11} - 2c_{44}) S \delta_{ij} + 2c_{44} S_{ij} = (c_{11} - 2c_{44}) S \delta_{ij} + c_{44}\left(\frac{\partial u_i}{\partial x_j} + \frac{\partial u_j}{\partial x_i}\right) \tag{5.13}$$

where

$$S = \text{dilatation} = S_{ii} = \text{div}\vec{u} = \frac{\partial u_i}{\partial x_i} \tag{5.14}$$

Thus the equation of motion becomes

$$\rho \frac{\partial^2 u_i}{\partial t^2} = \frac{\partial}{\partial x_i}\left[(c_{11} - 2c_{44})\frac{\partial u_i}{\partial x_i}\right] + c_{44}\frac{\partial^2 u_i}{\partial x_j^2} + c_{44}\frac{\partial}{\partial x_i}\left(\frac{\partial u_i}{\partial x_j}\right) \tag{5.15}$$

This can be written in vectorial form

$$\rho \frac{\partial^2 \vec{u}}{\partial t^2} = (c_{11} - c_{44})\vec{\nabla}(\vec{\nabla} \bullet \vec{u}) + c_{44}\Delta\vec{u} \tag{5.16}$$

where

$$\vec{\nabla} = \left(\frac{\partial}{\partial x_1}, \frac{\partial}{\partial x_2}, \frac{\partial}{\partial x_3}\right) \tag{5.17}$$

and

$$\Delta = \frac{\partial^2}{\partial x_k^2} \text{ is the Laplacian} \tag{5.18}$$

Finally,

$$\rho \frac{\partial^2 \vec{u}}{\partial t^2} = (c_{11} - c_{44})\vec{\nabla}(\vec{\nabla} \bullet \vec{u}) + c_{44}\nabla^2\vec{u} \tag{5.19}$$

For very good reasons it is traditional at this point to write that any vector can be written as the gradient of a scalar and the curl of a vector, the two new quantities being known as the scalar (ϕ) and vector ($\vec{\psi}$) potentials.

Thus

$$\vec{u} = \vec{\nabla}\phi + \vec{\nabla} \times \vec{\psi} \tag{5.20}$$

where

$$\vec{\nabla} \times (\vec{\nabla}\phi) \equiv 0 \tag{5.21}$$

$$\vec{\nabla} \bullet (\vec{\nabla} \times \vec{\psi}) \equiv 0 \tag{5.22}$$

Substituting in the equation of motion

$$\rho\frac{\partial^2 \phi}{\partial t^2} + \rho\frac{\partial^2(\vec{\nabla} \times \vec{\psi})}{\partial t^2} = (c_{11} - c_{44})\vec{\nabla}(\nabla^2\phi) + c_{44}\nabla^2(\vec{\nabla}\phi) + c_{44}\nabla^2(\vec{\nabla} \times \vec{\psi}) \tag{5.23}$$

Using the Helmholtz identity in vector analysis this becomes

$$\vec{\nabla}\left(\rho\frac{\partial^2\phi}{\partial t^2} - c_{11}\nabla^2\phi\right) + \vec{\nabla} \times \left(\rho\frac{\partial^2\vec{\psi}}{\partial t^2} - c_{44}\nabla^2\vec{\psi}\right) = 0 \tag{5.24}$$

Since the first term is purely a scalar and the second purely a vector, the two terms must be separately equal to zero:

$$\rho\frac{\partial^2\phi}{\partial t^2} = c_{11}\nabla^2\phi \tag{5.25}$$

$$\rho\frac{\partial^2\vec{\psi}}{\partial t^2} = c_{44}\nabla^2\vec{\psi} \tag{5.26}$$

Since $c_{11} = \lambda + 2\mu$ and $c_{44} = \mu$, we immediately associate the first equation with longitudinal waves and the second with transverse waves. It is thus natural that the scalar potential ϕ is associated with the propagation of the purely scalar property, the dilatation, and the vector potential with transverse waves that must have two (orthogonal) states of polarization. Most important, the use of scalar and vector potentials has allowed us to separate the equations of propagation of these two independent modes.

Writing more explicitly

$$\vec{u}_L = \nabla\phi, \qquad \vec{\nabla} \times \vec{u}_L \equiv 0 \tag{5.27}$$

$$\vec{u}_T = \vec{\nabla} \times \vec{\psi}, \qquad \vec{\nabla} \bullet \vec{u}_T \equiv 0 \tag{5.28}$$

we obtain

$$\frac{\partial^2 \vec{u}_L}{\partial t^2} = V_L^2 \nabla^2 \vec{u}_L, \qquad \frac{\partial^2 \vec{u}_T}{\partial t^2} = V_T^2 \nabla^2 \vec{u}_T \tag{5.29}$$

where

$$V_L = \sqrt{\frac{c_{11}}{\rho}} \qquad \text{and} \qquad V_T = \sqrt{\frac{c_{44}}{\rho}} \tag{5.30}$$

The vectorial properties of $\vec{u}_L$ and $\vec{u}_T$ confirm the previous conclusions. Since $\vec{\nabla} \bullet \vec{u}_T \equiv 0$, there is no change in volume associated with $\vec{u}_T$ (hence $\vec{\psi}$), which is as it must be for a transverse wave. Likewise $\vec{\nabla} \times \vec{u}_L \equiv 0$ means that there is no change in angle or rotation associated with $\vec{u}_L(\phi)$, which is characteristic of a longitudinal wave. Displacement deformations for typical longitudinal and transverse waves are shown in Figure 5.1.

The energy and acoustic power relations for both longitudinal and transverse waves can be extended directly from their one-dimensional forms. Thus the potential and kinetic energies per unit volume are

$$u_P = T_{ij} \frac{dS_{ij}}{dt} \tag{5.31}$$

and

$$u_K = \frac{1}{2} \rho \dot{u}_i^2 \tag{5.32}$$

The instantaneous Poynting vector $\mathcal{P}$, which gave a power flow $-vT$ per unit area in one dimension, becomes straightforwardly

$$\mathcal{P}_j(x_i, t) = -T_{ij} \frac{\partial u_i}{\partial t} \tag{5.33}$$

in three dimensions.

The above analysis shows that bulk waves consist of one longitudinal mode and two mutually orthogonal transverse modes. A standard terminology has been developed to identify these modes and it is used universally to describe bulk and guided modes. The plane of the paper (saggital plane) contains the x axis and the surface normal (z axis). The y axis is perpendicular to this plane. Calculations for bulk modes will then be carried out with longitudinal waves and transverse waves with polarization in the plane of the paper both having wave vectors in the plane of the paper. These may also be referred to as P (pressure) and SV (shear vertical) modes, respectively, following the original geophysical terminology. Transverse waves propagating in the saggital plane with polarization perpendicular to the paper (y axis) are called SH (shear horizontal) modes. In this language, the acoustic

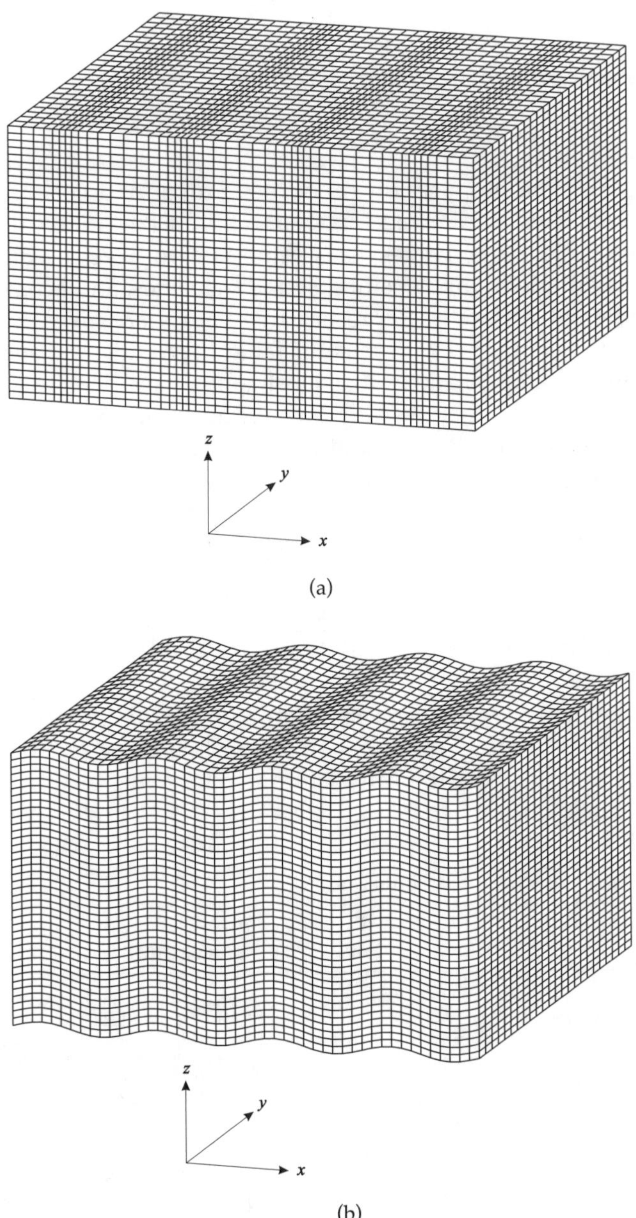

FIGURE 5.1
Grid diagrams for the deformations caused by bulk plane waves propagating along the x axis. (a) Longitudinal waves. (b) Transverse waves polarized in the z direction.

modes conveniently break up into the orthogonal, uncoupled groups of sagittal (P, SV) and SH modes.

5.3 Material Properties

We discuss first the propagation properties primarily associated with the sound velocity. This is followed by a summary of the principal sources of attenuation of ultrasonic waves. It is important to have a feeling for the orders of magnitude of the densities, sound velocities, and acoustic impedances of different materials. Representative values are given in Table 5.1, which should be compared with those of Table 3.1. A cursory glance confirms what we already know, namely that most solids have densities and sound velocities much greater than water, which are again much greater than those in air. This state of affairs is most usefully summarized in a single parameter, the acoustic impedance, given for longitudinal and transverse waves in Figures 5.2 and 5.3. It will be shown in Chapter 7 that the amplitude reflection coefficient at the interface between two media is given by

$$R = \frac{Z_2 - Z_1}{Z_2 + Z_1} \tag{5.34}$$

where the incident wave is from medium 1 and partially transmitted into medium 2. Two limiting cases are of interest. If $Z_2 = Z_1$, the reflection coefficient is zero; it is as if the wave continued traveling forward in a single

TABLE 5.1

Acoustic Properties of Various Solids

Solid	V_L (km/s)	V_S (km/s)	ρ (10^3 kg/m^3)	Z_L (MRayls)	Z_S (MRayls)
Epoxy	2.70	1.15	1.21	3.25	1.39
RTV-11 Rubber	1.05		1.18	1.24	
Lucite	2.70	1.10	1.15	3.1	1.25
Pyrex glass	5.65	3.28	2.25	13.1	7.62
Aluminum	6.42	3.04	2.70	17.33	8.21
Brass	4.70	2.10	8.64	40.6	18.15
Copper	5.01	2.27	8.93	44.6	20.2
Gold	3.24	1.20	19.7	63.8	23.6
Lead	2.16	0.7	24.6	7.83	0.44
Fused quartz	5.96	3.75	2.2		
Lithium niobate (z)	7.33		4.7	34.0	
Zinc oxide (z)	6.33		5.68	36.0	
Steel	5.9	3.2	7.90	46.0	24.9
Beryllium	12.90	8.9	1.87	24.10	16.60
Sapphire (z)	11.1	6.04	4.0	44.4	24.2

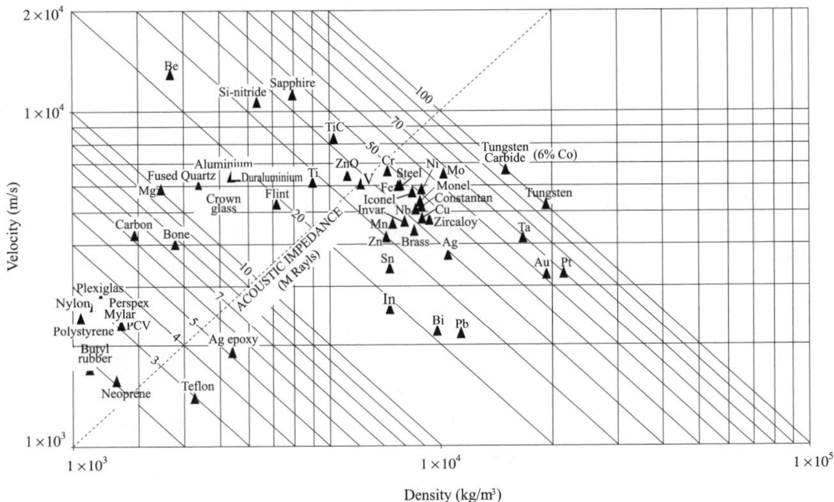

FIGURE 5.2
Density-sound velocity/longitudinal characteristic acoustic impedance plots on a log-log scale for various solids. (Based on a graph by R. C. Eggleton, described in Jipson, V. B., Acoustical Microscopy at Optical Wavelengths, Ph.D. thesis, E. L. Ginzton Laboratory, Stanford University, Stanford, CA, 1979.)

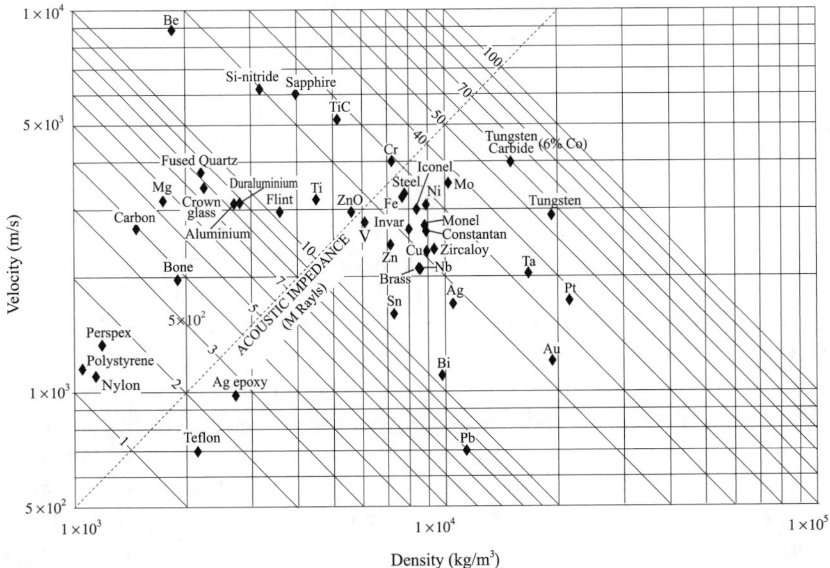

FIGURE 5.3
Density-sound velocity/transverse characteristic acoustic impedance plots on a log-log scale for various solids. (Based on a graph by R. C. Eggleton, described in Jipson, V. B., Acoustical Microscopy at Optical Wavelengths, Ph.D. thesis, E. L. Ginzton Laboratory, Stanford University, Stanford, CA, 1979.)

medium. On the other hand, if $Z_2 \gg Z_1$ then $R \sim 1$, i.e., the wave is almost totally reflected. These two limits are important because in most ultrasonic applications one is either trying to keep the wave from going into another medium (e.g., reflecting face of a delay line) or, contrariwise, maximize its transmission from one medium into another (e.g., maximum transmission from a transducer into a sample in NDE). Examples of this type come up repeatedly and in practical applications it is important to have an intuitive grasp of the magnitude of the acoustic impedances involved.

For order of magnitude purposes let us take a typical solid as having a density of 5000 $kg \cdot m^{-3}$ and a longitudinal velocity of 5000 $m \cdot s^{-1}$, giving a longitudinal acoustic impedance of 25 MRayls where the Rayl (after Lord Rayleigh) is the MKS unit of acoustic impedance. Referring to Table 5.1 it is seen that the range for typical solids is 10 to 15 MRayls, with some high-density, high-velocity materials such as tungsten going up to 100 MRayls. By comparison, plastics and rubbers are in the range 1 to 5 MRayls, water 1.5 MRayls, and air is orders of magnitude less at 400 Rayls. This is why, for off-the-cuff calculations, a solid-air or liquid-air interface can be taken to first order as totally reflecting. In some cases, the required range of sound velocities or densities of a material is fixed by other considerations (e.g., focusing properties of acoustic lenses), in which case Figures 5.2 and 5.3 are useful for showing at a glance the possible choices of common materials in a given acoustic impedance range.

The densities of materials used in ultrasonics applications are temperature independent except for very special cases. This, however, is not the case for sound velocity. From absolute zero up to room temperature, the sound velocity typically decreases by about 1%, giving a slope at room temperature $(1/V)(\delta V/\delta T) \sim 10^{-4} K^{-1}$. This is an intrinsic, thermodynamic effect that has its origin in the nonlinear acoustic properties of solids. It can be a particularly important consideration in the design and operation of acoustic surface wave devices and acoustic sensors.

Ultrasonic attenuation α in solids is a difficult parameter to specify in absolute terms, yet it is very important. In fundamental physical acoustics, a quantitative knowledge of α is often very useful for a validation of models and theories; verification of the BCS theory of superconductivity is one example, and there are many others. In applications and devices the emphasis is almost always on reducing the attenuation as much as possible to improve device performance. In some special cases (transducer backings), the opposite is desired. In either case it must be controlled, and to do this it must be understood. This is not always easy as there are many contributing factors that are difficult to control going from the state of the sample to the measuring conditions. The attenuation in many samples is almost entirely determined by the fabrication and sample preparation process. As for the measurement, to obtain an accurate value of α we require in principle a perfection exponential decay of echoes in the sample, as explained in Chapter 12. This is almost never achieved in practice even under the best laboratory conditions. Hence, accurate absolute attenuation values are never quoted and in most cases the relative attenuation is measured as a function of some parameter, such as temperature, pressure, or magnetic field. Due to these difficulties, in fundamental studies it is often more useful to measure the absolute and/or relative velocity variations, which are much less prone to experimental artifacts.

In the following, we consider mainly the principal sources of attenuation, their order of magnitude in different materials, and their variation with frequency and temperature. Only longitudinal waves will be covered unless stated otherwise. Sources of attenuation will be divided into two classes: intrinsic (thermal effects, elementary excitations) and those due to imperfections (impurities, grain boundaries, dislocations, cracks, etc. are some of the usual suspects). Detailed discussions of the physical origin of attenuation in solids are given in [7] and [13].

The intrinsic component of ultrasonic attenuation for a solid can be described from a macroscopic point of view, much as was done for liquids in Chapter 3. In the classical attenuation in a fluid, we have

$$\alpha = \frac{\omega^2}{2\rho_0 V_i^2}\left(\eta + \left(\frac{\Delta\lambda}{\lambda + 2\mu}\right)\frac{K}{C_V}\right) \tag{5.35}$$

where $\Delta\lambda$ is the difference between isothermal and adiabatic Lamé coefficients. C_V is the specific heat at constant volume per unit volume, V_i represents longitudinal or shear velocity, and the other symbols have their usual meanings. We notice immediately that since V_i^2 appears in the denominator and since on average $V_S \sim V_L/2$, the intrinsic shear attenuation is expected to dominate.

In solids it is more usual to approach the problem from a phonon point of view where the crystal lattice is represented by a gas of interacting phonons of energy $\hbar\omega$, where ω is the frequency of a lattice mode. In this picture the ultrasonic wave is composed of very many low-frequency phonons at the ultrasonic frequency. The attenuation divides into the same two components as above, namely thermoelastic loss and phonon viscosity. For simplicity we consider the case of longitudinal waves in an insulating solid where the heat is carried by the thermally excited phonons always present at temperature T, called the thermal phonons. For thermoelastic loss the regions compressed by the ultrasonic wave are heated and the excess energy is transported by thermal phonons to the rarefaction regions, which are cooler. As above, this component of attenuation can be written

$$\alpha = \frac{1}{2V}\frac{\Delta c}{c_0}\frac{\omega^2\tau_{th}}{1 + \omega^2\tau_{th}^2} \tag{5.36}$$

where $\Delta c = c_1 - c_0$ and c_0 are the relaxed and unrelaxed elastic moduli, respectively (i.e., isothermal and adiabatic). The collision time for the thermal phonons is

$$\tau_{th} = \frac{K}{C_P V^2} \tag{5.37}$$

where C_P is the heat capacity at constant pressure per unit volume. After considerable analysis this can be written in the form

$$\alpha = \frac{\gamma_G^2 C_V T}{2\rho V^3} \frac{\omega^2 \tau_{th}}{1 + \omega^2 \tau_{th}^2} \tag{5.38}$$

where γ_G is the Gruneisen constant $= 3\beta K/C_V$ and β is the linear expansion coefficient.

The viscosity component corresponds to the so-called Akhiezer loss and follows from a detailed calculation of the phonon-phonon interaction. The physical model is that application of a step function strain leads to an effective temperature change of the phonon modes, leading to a redistribution of their populations by the phonon-phonon interaction. There is a phase lag in this process and it leads to energy dissipation hence attenuation. Very detailed calculations were carried out by Bommel and Dransfeld [14], Woodruff and Ehrenreich [15], and Mason and Bateman [16]. Only the final result will be given here, which is of the form, for $\omega\tau_{th} \ll 1$ at room temperature,

$$\frac{\alpha}{f^2} = R\bar{\gamma}_G^2 \tag{5.39}$$

where $\bar{\gamma}_G$ is a modified form of the Gruneisen constant, treated as an adjustable parameter and

$$R \propto \frac{K_{\theta_D}}{M\theta_D^4 V_0^{2/3}} \tag{5.40}$$

where
- K_{θ_D} is the thermal conductivity at the Debye temperature θ_D,
- M is the average atomic mass, and
- V_0 is the atomic volume

The model predicts an attenuation that is constant and varies as f^2 at room temperature in agreement with experiment. It predicts that the attenuation will be decreased for materials with high Debye temperature and low thermal conductivity. This makes sense physically as the first condition means less thermal agitation at a given temperature while the second weakens the relaxation effect.

There are, of course, almost an infinite number of ways in which imperfections can contribute to α. Physical and chemical imperfections are usually badly characterized and theory exists only for the most simple cases. In this situation, only the simplest and most important case, that of polycrystals, will be briefly described here.

Although crystals exhibit the basic intrinsic attenuation, the same is not true of polycrystals. Polycrystals are an agglomeration of many grains, each having an orientation different from its neighbors. Zener [17] showed that the grains produce a thermal relaxation effect similar to that described previously. However, the most important effect is the scattering due to the misorientation of the grains, each of which has a different effective elastic constant in the direction of propagation. Full details have been given by Papadakis [18]. Very roughly, for scattering of an ultrasonic wave of wavelength λ by grains of a mean diameter D

$$\alpha = \beta_1 f + \beta_2 f^4, \qquad \lambda \geq 3D \tag{5.41}$$

where the first term is due to hysteresis and the second corresponds to Rayleigh scattering by the grains. Papadakis shows that this term can be written as

$$\alpha = \beta f^4 S \tag{5.42}$$

where β is the average grain volume and S is a material parameter that varies widely. In the opposite limit where $\lambda \ll D$, $\alpha \sim 1/D$ and is independent of frequency. A wealth of experimental data is reported by Papadakis [18]. Generally, Rayleigh scattering is observed in the range 1 to 10 MHz with an order of magnitude attenuation of roughly 1 dB · cm^{-1} at 10 MHz. At higher frequencies, the slope generally levels off to an f^2 variation. Average grain sizes are the order of 100 μm.

Summary

Displacement (velocity) potentials consist of a scalar (φ) and vector potential ($\vec{\Psi}$). φ governs the propagation of pure longitudinal waves and $\vec{\Psi}$ that of shear waves.

Three-dimensional wave equation for solids has solutions that are pure longitudinal and pure shear waves. The two equations are decoupled, which has the consequence that longitudinal and shear waves are independent modes of propagation in bulk solids.

Pure longitudinal bulk waves have elastic constant $\lambda + 2\mu$.

Pure shear bulk waves have elastic constant μ.

Acoustic Poynting vector in a three-dimensional isotropic solid is given by $\mathcal{P}_j = -T_{ij}(\delta u_i/\delta t)$.

Saggital modes have propagation vectors and polarization vectors in the saggital plane (plane of the paper).

SH modes have propagation vector in the saggital plane and polarization vector perpendicular to that plane.

Attenuation in isotropic solids is due to a variety of defects and elementary excitations, including impurities, grain boundaries, dislocations, cracks, phonons, electrons, magnetic excitations, etc.

Questions

1. For the one-dimensional solid derive the relation $\delta S/\delta t = \delta v/\delta z$.
2. Rederive Equations 5.25 and 5.26 in terms of λ and μ.
3. Consider a transversely isotropic solid, which is isotropic in a plane perpendicular to a principal axis. To what crystal structure is this equivalent? Enumerate the possible saggital and SH modes for the transversely isotropic solid. You should consider modes both parallel and perpendicular to the principal axis.
4. From Figures 5.2 and 5.3 and Table 5.1, determine the three solids with the lowest and highest acoustic impedance, respectively. Do the same for liquids using Table 3.1. Calculate the energy transmission coefficient at normal incidence for the case of extreme acoustic mismatch in the media chosen.
5. A plane wave of 5 MHz is incident on a steel plate. Calculate the required thickness for this wave to be retarded in phase by 90° with respect to a wave that passes through a large hole in the plate.

6

Finite Beams: Radiation, Diffraction, and Scattering

There are many advantages in using plane wave solutions; they are conceptually simple and greatly facilitate the mathematics. However, in practice, ultrasonics always involves the use of finite size transducers, hence finite beams. An immediate consequence is that diffraction effects must be considered. Furthermore, the consequences for focusing, imaging, leaky waves, and scattering from various sizes and shapes of objects, among other issues, must be addressed.

One of the many issues of this chapter will be the treatment of diffraction effects. For simplicity, we deal with scalar theory for a fluid medium although the results can be directly extended to a solid medium. We start with radiation from a point source and then extend the discussion to radiation by a circular piston. This is a classic problem in ultrasonics and the results give general guidelines for the emission of ultrasonic waves from a piezoelectric transducer. Following that, an outline is provided for the scattering of ultrasonic waves by circular and cylindrical obstacles. Finally, the main issues involved in focused ultrasonic waves, acoustic radiation pressure and the Doppler shift, are addressed.

6.1 Radiation

6.1.1 Point Source

It is convenient to write the wave equation in a totally general form for a fluid:

$$\frac{\partial^2 p}{\partial t^2} = V_0^2 \nabla^2 p \tag{6.1}$$

Since p is a scalar, we can use any coordinate system and employ the appropriate form of ∇^2.

Clearly a spherical coordinate system is best suited for problems dealing with a point source. Doing the transformation $(x, y, z) \rightarrow (r, \theta, \psi)$, we have

$$\begin{aligned} x &= r\sin\theta\cos\psi \\ y &= r\sin\theta\sin\psi \\ z &= r\cos\theta \end{aligned} \tag{6.2}$$

and

$$\Delta = \nabla^2 = \frac{\partial^2}{\partial r^2} + \frac{2}{r}\frac{\partial}{\partial r} + \frac{1}{r^2\sin\theta}\frac{\partial}{\partial\theta}\left(\sin\theta\frac{\partial}{\partial\theta}\right) + \frac{1}{r^2\sin^2\theta}\frac{\partial^2}{\partial\psi^2} \tag{6.3}$$

Substituting into Equation 6.1 and noting that for spherical waves the pressure is independent of θ and ψ:

$$\frac{\partial^2 p}{\partial t^2} = V_0^2\left(\frac{\partial^2 p}{\partial r^2} + \frac{2}{r}\frac{\partial p}{\partial r}\right) \tag{6.4}$$

Hence,

$$\frac{\partial^2 p}{\partial t^2} = V_0^2\left(\frac{1}{r}\frac{\partial^2(rp)}{\partial r^2}\right) \tag{6.5}$$

Since r and t are independent variables, this can be rewritten

$$\frac{\partial^2(rp)}{\partial t^2} = V_0^2\frac{\partial^2(rp)}{\partial r^2} \tag{6.6}$$

with solutions

$$rp = f_d(V_0 t - r) + f_c(V_0 t - r) \tag{6.7}$$

$$p = \frac{f_d(V_0 t - r)}{r} + \frac{f_c(V_0 t - r)}{r} \tag{6.8}$$

f_d, the solution for diverging waves, will be mainly useful for radiation problems, while f_c, the converging solution, will be appropriate for focused spherical waves. The diverging solution will be treated explicitly in what follows.

As for plane waves, it will useful to develop relations between displacement, particle velocity, dilatation, and pressure. From Newton's law

$$-\frac{\partial p}{\partial r} = \rho_0 \frac{\partial^2 u_r}{\partial t^2} \tag{6.9}$$

where u_r is the radial particle displacement.

This can be integrated to give

$$v_r = \frac{\partial u_r}{\partial t} = -\frac{1}{\rho_0}\int \frac{\partial p}{\partial r} dt \tag{6.10}$$

or in complex form

$$v_r = -\frac{1}{j\omega\rho_0}\frac{\partial p}{\partial r} \tag{6.11}$$

Finally, the displacement is

$$u = \int v\, dt = \frac{v}{j\omega} = \frac{1}{\rho_0 \omega^2}\frac{\partial p}{\partial r} \tag{6.12}$$

For harmonic solutions at frequency ω

$$p = \frac{A}{r}\exp j(\omega t - kr) \tag{6.13}$$

Hence,

$$S = -\frac{p}{\rho_0 V_0^2} \tag{6.14}$$

$$u_r = -\left(\frac{1}{r} + jk\right)\frac{p}{\rho_0 \omega^2} \tag{6.15}$$

$$v_r = \left(\frac{1}{r} + jk\right)\frac{p}{j\omega\rho_0} \tag{6.16}$$

Contrary to the case for plane waves, the particle velocity is in general out of phase with the pressure while the displacement always lags the pressure by $\pi/2$.

The specific acoustic impedance is given by

$$Z = \frac{p}{v} = \frac{j\omega\rho_0}{\left(\frac{1}{r} + jk\right)} = \frac{\rho_0 V_0(kr + j)}{1 + k^2r^2} \tag{6.17}$$

$$= \rho_0 V_0 \frac{k^2r^2}{1 + k^2r^2} + j\rho_0 V_0 \frac{kr}{1 + k^2r^2} \tag{6.18}$$

where the phase angle between real and imaginary parts is $\tan\theta = 1/kr$ or

$$\cos\theta = \frac{kr}{\sqrt{1 + k^2r^2}} \tag{6.19}$$

The modulus of the acoustic impedance

$$|Z| = \left|\frac{p}{v}\right| = \rho_0 V_0 \frac{kr}{\sqrt{1 + k^2r^2}} = \rho_0 V_0 \cos\theta \tag{6.20}$$

which approaches the value for plane waves for $kr \gg 1$. This is as expected because far from the source the spherical wave approximates a plane wave.

By definition $Z \equiv p/v$, so the particle velocity can be expressed in terms of the impedance as

$$v = \frac{A}{rZ} \exp j(\omega t - kr) \tag{6.21}$$

The intensity I of a spherical wave is by definition the average rate of work done per area on the surrounding medium. For a cycle of period T

$$I \equiv \frac{\int_0^T pv\,dt}{T} \tag{6.22}$$

Using the real part of p and v and the previous results for the phase angle

$$I = \frac{1}{T}\int_0^T p_0 \cos(\omega t - kr) v_0 \cos(\omega t - kr - \theta) dt$$

$$= \frac{p_0 v_0 \cos\theta}{2} \quad \text{where } p_0 = \frac{A}{r} \tag{6.23}$$

Using $v_0 \cos\theta = p_0/\rho_0 V_0$ from Equation 6.20

$$I = \frac{p_0^2}{2\rho_0 V_0} \tag{6.24}$$

It is now possible to formulate the sound field associated with spherical waves. Assuming a spherical source of radius a immersed in a fluid, the radial velocity at a point on the surface is given by

$$v = v_0 e^{j\omega t} \tag{6.25}$$

For small amplitudes, the boundary condition is continuity of the radial velocity. From the previous results

$$\frac{A}{aZ_a} \exp j(\omega t - ka) = v_0 \exp(j\omega t)$$

so that

$$A = av_0 Z_a \exp j(ka) \approx j\rho_0 V_0 ka^2 v_0, \quad \text{for } ka \ll 1 \tag{6.26}$$

This gives the desired result for a small (point) source

$$p = \frac{j\rho_0 V_0 ka^2 v_0}{r} \exp j(\omega t - kr) \tag{6.27}$$

6.1.2 Radiation from a Circular Piston

The result will, of course, be much more complex, and much more difficult to calculate, than that from a point source. The basic principle is simple; each point of the source can be treated as a pointlike source, emitting spherical waves as given by Equation 6.27 at distances far enough away from the source. Then by Huygens principle these various contributions can be summed, taking into account the amplitude and the phase from each contribution. In practice, analytical results are difficult to obtain, even for the simplest cases, leading to a choice between numerical calculation and approximate solutions. The latter approach is chosen here.

The circular piston radiator is an important example in ultrasonics as it is about the simplest approximation that can be made for radiation into an infinite medium. It is also important in audio acoustics in the theory of loudspeakers. The assumption will be made that it is mounted inside an infinite baffle, so that sound is only radiated in the forward direction. The geometry is shown in Figure 6.1, where it is assumed that the transducer is excited with uniform particle velocity across its face.

From Equation 6.27, each infinitesimal source element area $\overrightarrow{dA}$ on the piston produces a differential pressure dp at a point of observation at a distance of $\vec{r}'$ given by

$$dp = \frac{j\rho_0 V_0 k}{2\pi r'} (\vec{v} \cdot \overrightarrow{dA}) \exp j(\omega t - kr') \tag{6.28}$$

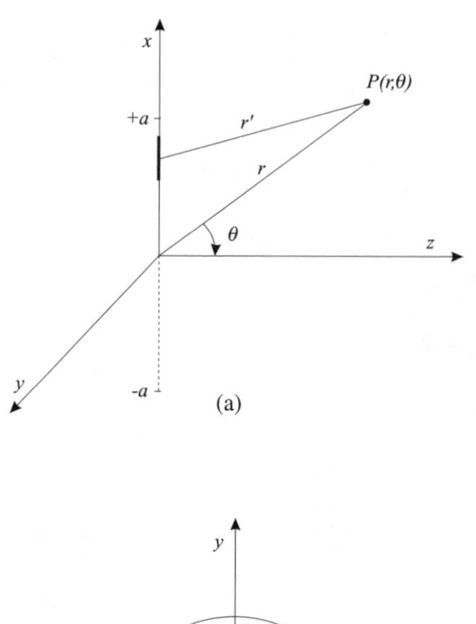

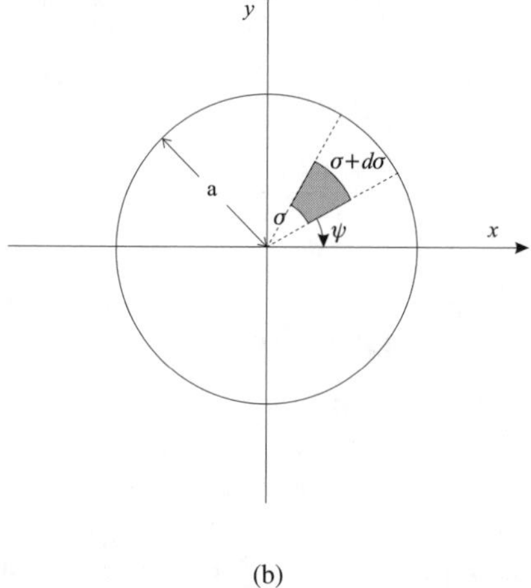

FIGURE 6.1
Geometrical variables used for the calculation of the pressure distribution for a plane piston circular radiator. (a) Axial view. (b) Radial coordinates in the plane of the radiator.

Since the motion of each element is normal to the surface, dp can be written as

$$dp = \frac{j\rho_0 V_0 k}{2\pi r'} v\, dA \exp j(\omega t - kr') \tag{6.29}$$

The total pressure p at the point (r, θ) is the integral of dp over the full radiator surface.

From elementary geometry

$$r' = (r^2 + \sigma^2 - 2r\sigma \sin\theta \cos\psi)^{\frac{1}{2}} \tag{6.30}$$

However, the resulting expression for dp is not integrable and so approximations have to be made. We first treat the far field or Fraunhofer limit, where $r \gg a$.

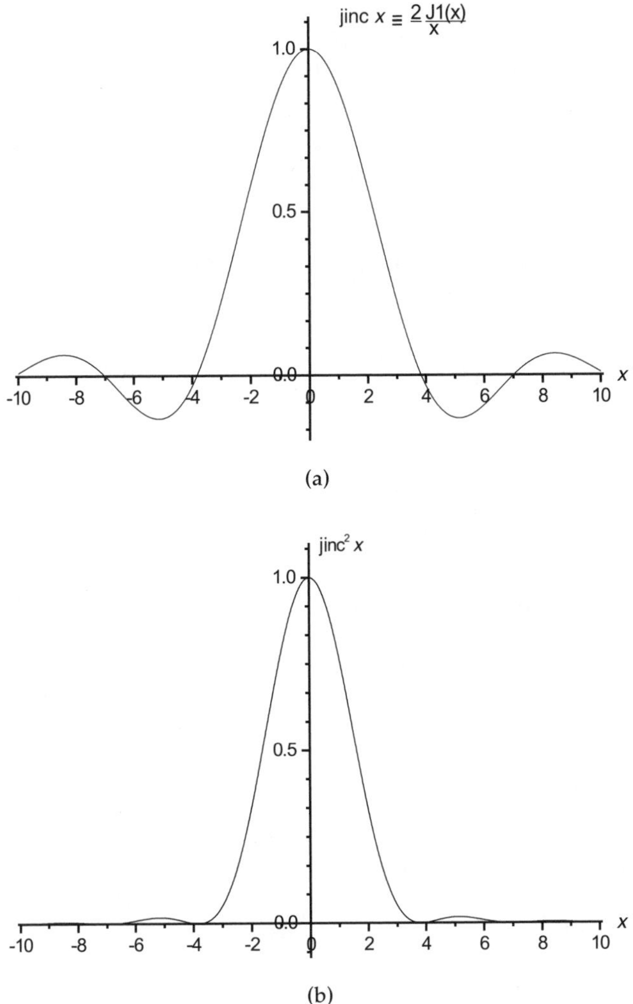

FIGURE 6.2
Directivity function jinc x for the circular radiator. (a) Pressure. (b) Intensity.

6.1.2.1 *Fraunhofer (Far Field) Region*

For $r', r \gg a$, this can be expanded in a Taylor's series with the first two terms

$$r' = r - \sigma \sin\theta \cos\psi$$

Since the distance between two neighboring points is crucial for an accurate calculation of the phase difference between pressure waves emitted from them, both of these terms must be retained for the phase. For the amplitude, $r' \approx r$ is sufficient.

We then have

$$p = \frac{j\rho_0 V_0 k}{2\pi r} v_0 e^{j(\omega t - kr)} \int_0^a \sigma d\sigma \int_0^{2\pi} e^{jk\sigma \sin\theta \cos\psi} d\psi \tag{6.31}$$

The second integral (over ψ) can be expanded as a power series and integrated to give $2\pi J_0(k\sigma \sin\theta)$ (see Appendix A). The second integral can be obtained from

$$\int x J_0(x)\, dx = x J_1(x)$$

so that

$$2\pi \int_0^a \sigma J_0(k\sigma \sin\theta)\, d\sigma = 2\pi a^2 \left[\frac{J_1(ka \sin\theta)}{ka \sin\theta}\right]$$

so that finally

$$p = \frac{j\rho_0 V_0 k a^2 v_0}{2r} e^{j(\omega t - kr)} \left[\frac{2J_1(ka \sin\theta)}{ka \sin\theta}\right] \tag{6.32}$$

The term in brackets is known as the directivity function (DF) as it gives the variation of pressure with direction. Numerical values are tabulated in Appendix A and the function is plotted in Figure 6.2.

An approximate form for the DF can be obtained by expanding $J_1(x)$, yielding

$$\frac{2J_1(x)}{x} \approx 1 - \frac{x^2}{8}$$

In particular, for points along the axis $x = 0$, then $DF = 1$ and the result for p is identical in form to that of a point-like source of area πa^2.

The first zero θ_1 of jinc$x$ occurs at $ka \sin\theta = 3.83$; hence,

$$\sin\theta_1 = \frac{3.83}{ka} = 0.61\frac{\lambda}{a} \tag{6.33}$$

which gives a measure of the angular half width of the principle lobe of the acoustic pressure. By the same token the first sidelobe is included between the angles θ_1 and θ_2 where

$$\sin\theta_2 = \frac{7.02}{ka} = 1.12\frac{\lambda}{a} \tag{6.34}$$

In this way, one can identify a whole series of lobes, on either side of the main lobe, called the sidelobes. These sidelobes are undesirable for two reasons. The main objective of an acoustic radiator is to produce a narrow collimated beam of acoustic energy to be used in some application, for example, imaging or nondestructive testing. The sidelobes represent energy lost from the main beam, which is of course undesirable. If the sidelobes are big enough, they can interfere with information obtained from the main

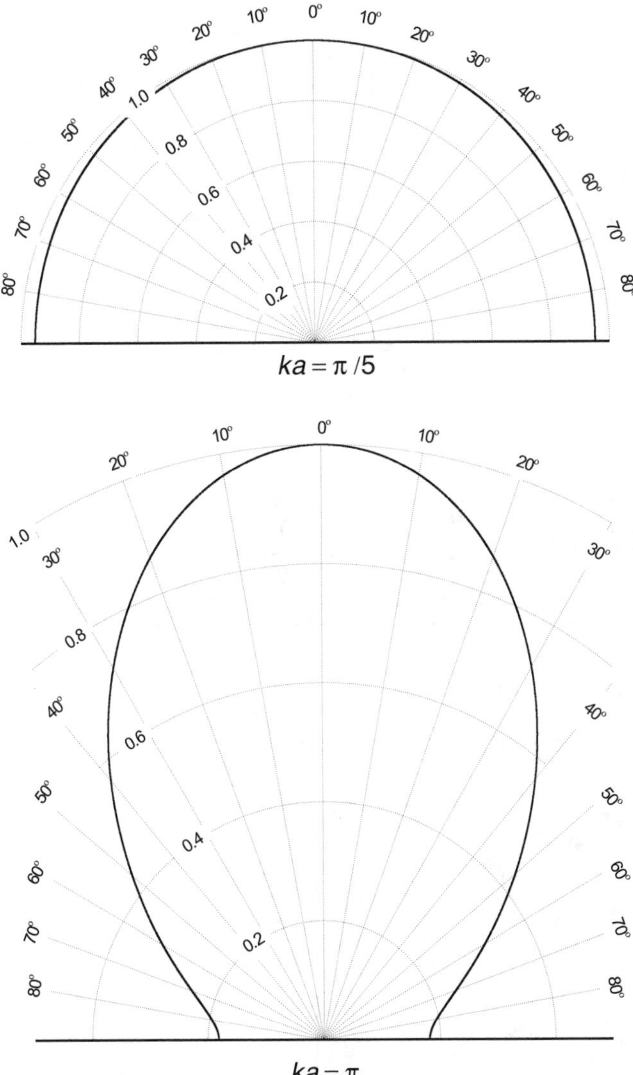

FIGURE 6.3
Polar diagram (linear scale) for circular radiators with radius/wavelength ratios of 0.1 (top) and 0.5 (bottom).

beam, which is also unwanted. Hence, an important part of the design of acoustic radiators involves sidelobe reduction.

An alternative and more efficient way to present the sidelobes is by the use of polar plots as shown in Figures 6.3 and 6.4 for several different frequencies using both dB and linear scales. It is seen that for $ka \gg 1$, there are many sidelobes. As ka decreases, the number is reduced and for $ka \ll 1$ there is really only the main lobe, for which the $DF \approx 1$. In this case, the axial intensity

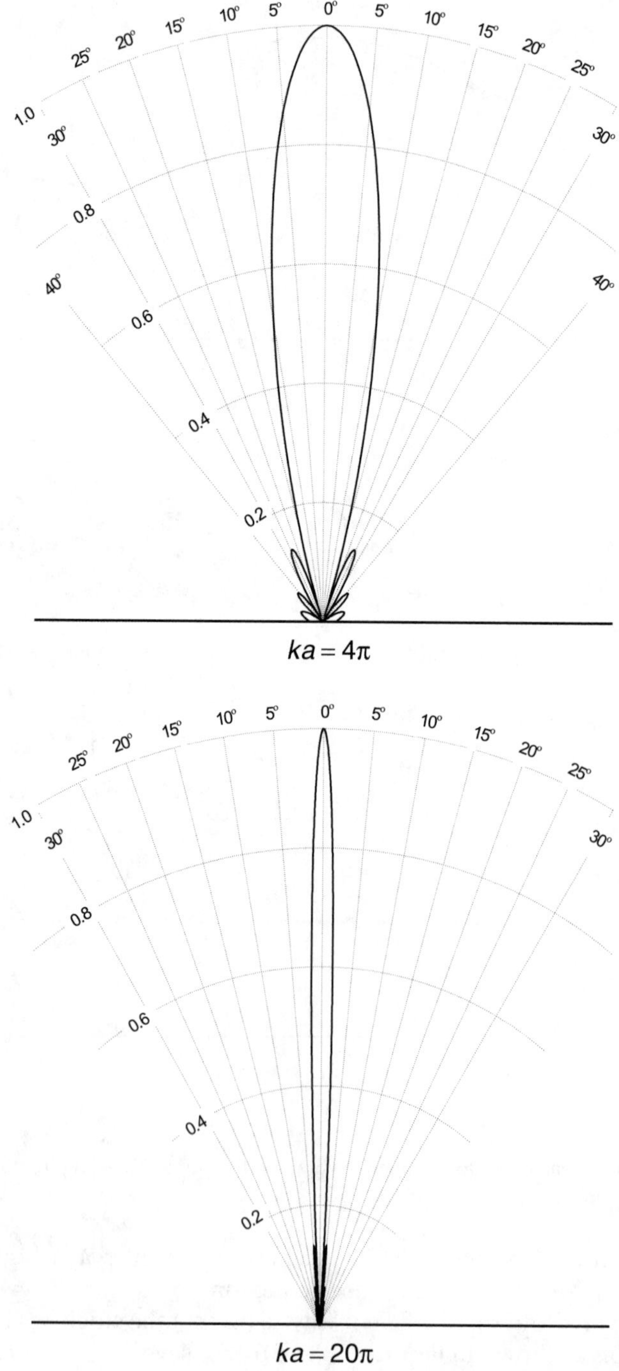

FIGURE 6.3
Polar diagram (linear scale) for circular radiators with radius/wavelength ratios of 2 (top) and 10 (bottom).

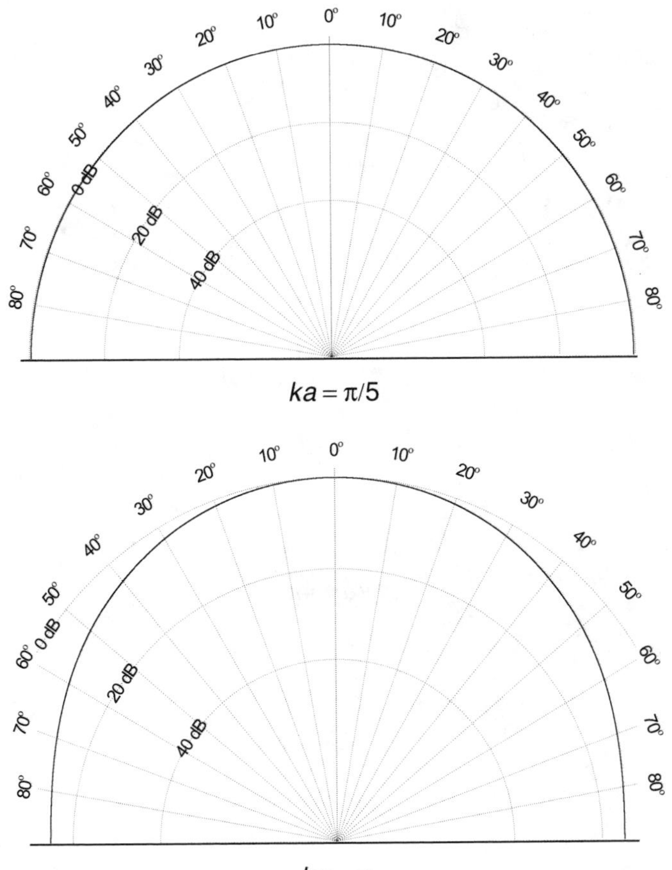

FIGURE 6.4
Polar diagrams (log scale) for circular radiators with radius/wavelength ratios of 0.1 (top) and 0.5 (bottom).

is given by

$$I_0 = \frac{\rho_0 V_0 k^2}{8\pi r^2} v_0^2 A \tag{6.35}$$

where A is the area of the piston. Deep in the far field region $I_0 \propto 1/r^2$, which is physically reasonable as far from the radiator, the latter looks like a point-like source.

6.1.2.2 Fresnel (Near-Field) Approximation

This is the opposite limiting case where the observation point is near the transducer; a quantitative criteria will be given at the end of the section. As the name implies, the situation closely resembles that of Fresnel diffraction in optics.

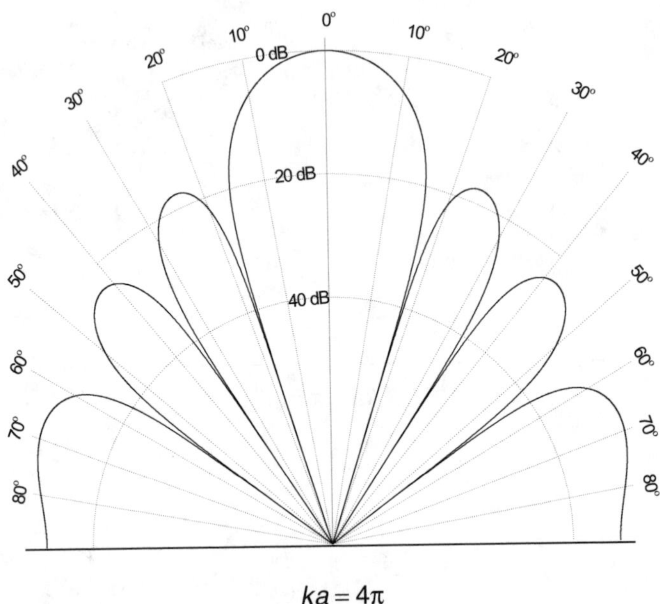

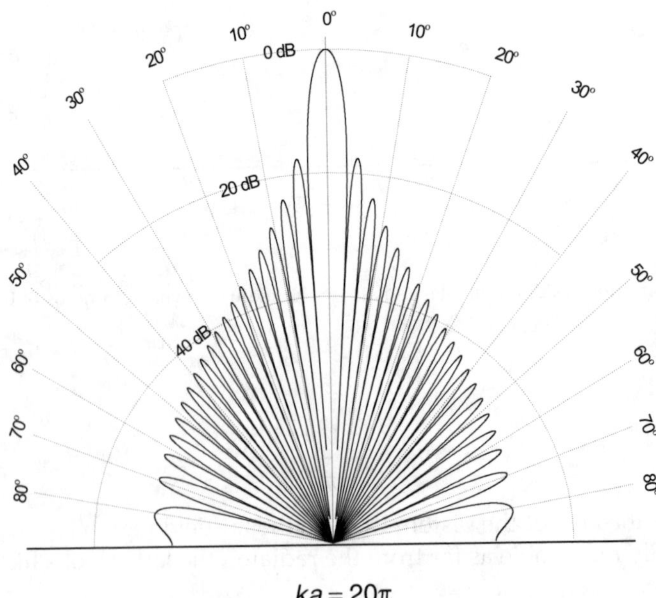

FIGURE 6.4 (Continued)
Polar diagrams (log scale) for circular radiators with radius/wavelength ratios of 2 (top) and 10 (bottom).

Analytical solutions are only available on the axis, where $r' = (r^2 + \sigma^2)^{1/2}$. Hence, from Equation 6.28

$$dp = \frac{j\rho_0 V_0 k}{2\pi} v_0 e^{j\omega t}\left(\frac{e^{-jk\sqrt{r^2+\sigma^2}}}{\sqrt{r^2+\sigma^2}}\right) dA \tag{6.36}$$

After integration

$$p = \rho_0 V_0 v_0 e^{j\omega t}\left(e^{-jk\sqrt{r^2+\sigma^2}} - e^{-jkr}\right) dA$$

with real part

$$p = \rho_0 V_0 v_0 a^2 \left[2 - 2\cos k\left(\sqrt{r^2+z^2} - z\right)\right]^{\frac{1}{2}} \tag{6.37}$$

The condition for a maximum or a minimum is given by the cosine term and is such that

$$z = \frac{4r_0^2 - n^2\lambda^2}{4n\lambda} \tag{6.38}$$

where

$n = 1, 3, 5, 7, \ldots$ for a maximum
$n = 2, 4, 6, 8, \ldots$ for a minimum

The final maximum occurs at $n = 1$ which corresponds to $z_F = a^2/\lambda$, which is called the Fresnel distance. Thus the near field corresponds to $z < z_F$ and the far field to $z > z_F$. The near field region is characterized by rapid interference maxima and minima as shown in Figure 6.5. This makes sense physically as

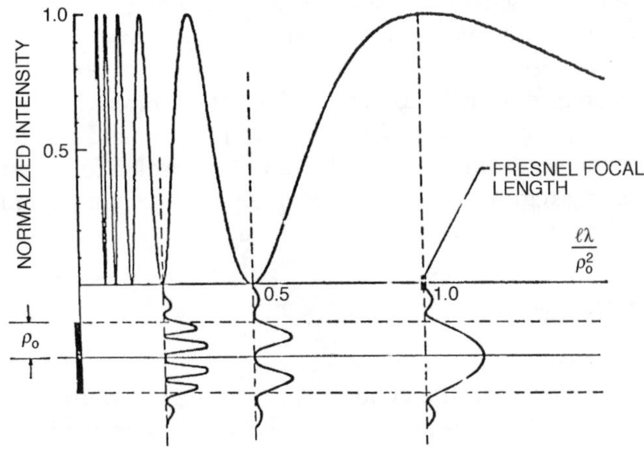

FIGURE 6.5
Axial intensity distribution produced by a circular transducer of radius ρ_0 as a function of distance ℓ from the transducer. Approximate transverse intensity distributions are plotted below this. (From Lemons, R.A., Acoustic Microscopy by Mechanical Scanning, Ph.D. thesis, E.L. Ginzton Laboratory, Stanford University, Stanford, CA. With permission.)

near the transducer a small shift along the axis leads to a relatively large shift in phase for the wavelet coming from a given surface element. This is not true in the far field where the phase shift is gradual and monotonic for all elements and the transducer acts more and more as a pointlike source.

Finally, the variation of acoustic pressure in the transverse plane is also sketched in Figure 6.5. It is seen that the beam remains well collimated up to the Fresnel distance, although there are considerable intensity variations across the beam section. Beyond the Fresnel distance, in the far field, the beam widens, as expected, due to the increasing point sourcelike behavior.

6.2 Scattering

Scattering of acoustic waves by obstacles of various sorts is, as in most branches of physics, a highly developed and mathematically very sophisticated subject. As in other areas, the main results are relatively easy to present for the case where the wavelength is either much greater or much less than the characteristic dimension of the obstacle. The problem becomes much more difficult, often intractable, when the wavelength is of the order of this dimension. In this situation, we will content ourselves with an overview of scattering by a few simple objects.

In principle, as for the case of radiation, the scattered acoustic field can be determined from Huygens principle, adding the waves emitted from secondary sources over the surface of the scattering body, taking into account their relative amplitudes and phases. For a body of arbitrary size and shape, this problem is in general intractable. For scattering by simple objects, two approaches will be used to characterize the scattering: polar diagram and total scattered intensity as a function of frequency. The polar diagram is highly useful because it gives an immediate visual clue as to the intensity of sound scattered in a given direction. The total scattered intensity is displayed as a function of ka where k is the wave number and a a characteristic dimension of the scattering center. This graph is useful for identifying the various scattering regimes mentioned above. The two main examples to be discussed will be the cylinder and the sphere.

6.2.1 The Cylinder

We suppose a plane wave incident on a rigid cylinder of radius a in a direction perpendicular to the cylinder axis. In the geometrical optics limit $ka \gg 1$, the cylinder scatters as a geometrical obstacle in the back direction and scatters as interference between the incident wave and the forward scattered wave to produce a sharply defined geometrical shadow. This limit is more common in optics than acoustics due to the length scales involved, although it is easy attainable in an ultrasonic immersion tank.

The limit $ka \gtrsim 1$ requires detailed calculation, which has been carried out by Morse [19]. The main steps of the calculation are as follows:

1. Description of an incident plane pressure wave $(p, \dot{u}_{pr})$ in cylindrical coordinates.
2. Description of the outgoing wave $(p, \dot{u}_{sr})$ in terms of the same parameters (amplitude and phase) as in the first description above.
3. Calculate the amplitudes and phases of steps 1 and 2 to satisfy $\dot{u}_{pr} + \dot{u}_{sr} \equiv 0$ at $r = a$.
4. Calculate the scattered intensity as a function of angle, βa from the solutions of step 3.

For sufficiently short wavelengths, about one half of the intensity is scattered in the forward direction and the rest is scattered approximately uniformly over the remaining solid angle. This gives rise to a cardiod-type polar plot: It becomes more and more directive in the forward direction as the wavelength decreases as shown in Figure 6.6. The total scattered intensity can also be calculated as a function of *ka*. For $ka \ll 1$, the scattered intensity rises rapidly to saturate for $ka \gg 1$.

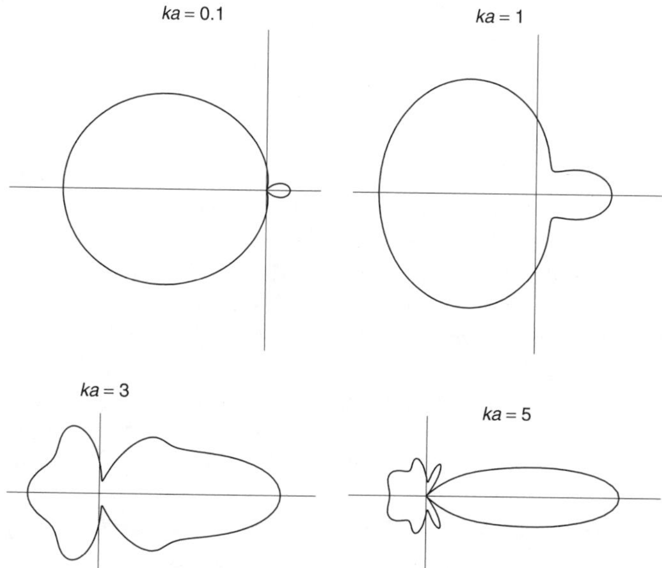

FIGURE 6.6
Polar diagrams (linear scale) for scattered radiation at wave number k from a rigid cylinder of radius a for ka = 0.1, 1, 3, and 5, respectively.

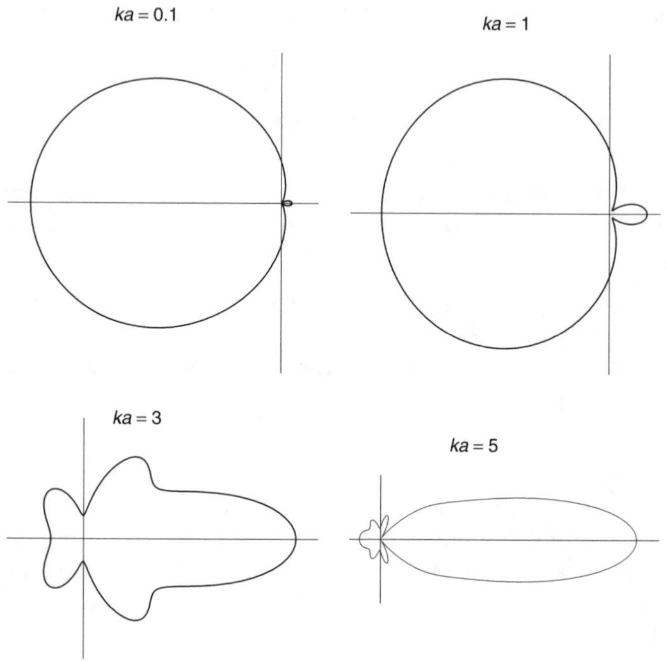

FIGURE 6.7
Polar diagrams (linear scale) for scattered radiation at wave number k from a rigid sphere of radius a for ka = 0.1, 1, 3, and 5, respectively.

6.2.2 The Sphere

The calculation follows the same lines as for the cylinder. The corresponding polar plots are shown in Figure 6.7 and the scattered intensity variation with ka in Figure 6.8. The total scattered intensity as a function of ka will be discussed. The curve can be divided into three regions. For $ka \ll 1$, the curve approaches asymptotically to the classic Rayleigh formula $I_S \sim 1/\lambda^4$ as it must in the long wavelength limit. In the opposite, short wavelength limit, $ka \gg 1$, the reflection is mainly specular and the reflected intensity saturates. In the intermediate regime the behavior is of a periodic nature due to the excitation of creeping or interface waves that travel around the curved surface of the obstacle at approximately the longitudinal sound velocity in the liquid.

The term scattering cross-section σ is commonly used to describe scattering problems; it is defined as the total scattered power divided by the incident intensity and represents the apparent area that blocks the wave. σ provides a convenient parameter to compare the scattering power of different forms of target. For example, for a sphere of radius a, $\sigma = 7/9(\pi a^2)(ka)^4$; it is seen that this form also incorporates the law for Rayleigh scattering.

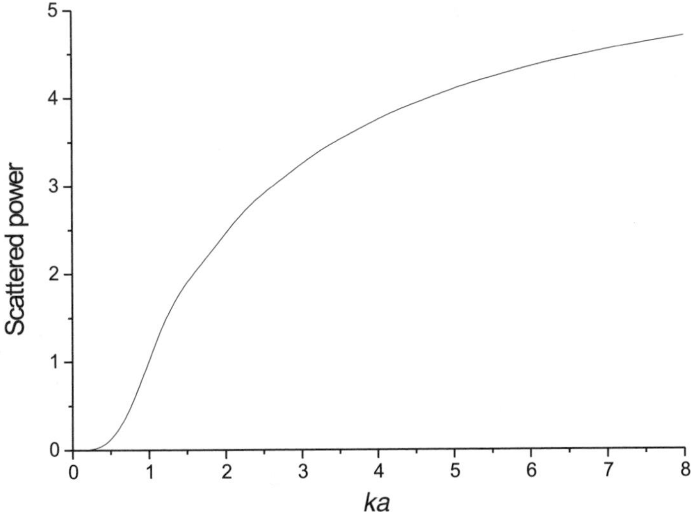

FIGURE 6.8
Scattering power of a sphere of radius *a* as a function of *ka*.

6.3 Focused Acoustic Waves

There are several levels of treatment for focused acoustic waves. The simplest, level 1, is to use geometrical optics or ray theory. For the spherically focused concave acoustic radiator to be considered in this section, level 1 immediately tells us that the acoustic energy is focused at the center of curvature. Level 2 takes into account diffraction, much in the same way that this has been handled for plane circular radiators in the previous sections of this chapter. This level demonstrates that the focal point is not an infinitesimal point but that it is spread out to the order of magnitude of the wavelength. This leads to the concept of point spread function and lateral resolution. The third level of sophistication recognizes that since the acoustic intensity is very high near the focus, nonlinear effects need to be taken into account. The main effect here is the generation of harmonics of the operating frequency in the focal region.

This book is limited to linear systems, so level 3 will not be treated here, although nonlinear effects in focusing will be discussed qualitatively in Chapter 14. Likewise, a full mathematical description of level 2 is beyond the scope of the book, and in any case has been provided in detail elsewhere by Kino [20], for example, whose general approach will be followed and summarized here. Given this we provide mainly a descriptive account of focused beams to a depth that will be sufficient to give an accurate description of acoustic lenses.

Rayleigh provided the first detailed treatment of the circular piston source described earlier, and these results will be seen to give a good first

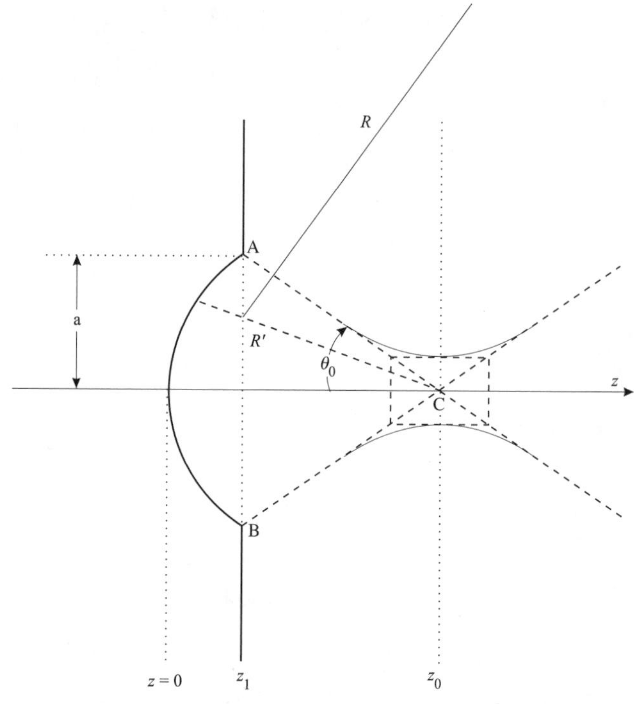

FIGURE 6.9
Focusing by a spherical radiator. The dotted cylindrical region around C gives the spatial resolution and depth of field.

approximation for circular radiators, especially when the radius of curvature is much greater than the wavelength. Early treatments were provided by Williams [21] and O'Neil [22]. Lucas and Muir [23] reduced the surface integral over the radiator to a single integral and showed that within the Fresnel approximation the boundary conditions on the curved surface could be transformed to the plane of the baffle. Recently, a numerically convergent solution consistent with all limiting cases has been provided by Chen et al. [24].

Following Kino [20] we consider the focused spherical radiator shown in Figure 6.9. Using the results of Lucas and Muir, it is possible to consider the planar element AB as an effective source by taking into account the phase difference between a point on the surface of the spherical radiator and its corresponding point on the element AB using ray theory. Kino shows that this leads to the following expression for the displacement potential in the Fresnel approximation with $a^2 \ll z^2$:

$$\phi(r,z,\psi) = -\frac{u_0}{2\pi}\iint \frac{e^{-jk(R+R_1)}\cos^2\theta}{R\cos\theta_0} r'dr'd\psi' \tag{6.39}$$

where u_0 is the amplitude of the radial displacement. Integrating over the azimuthal angle yields

$$\phi(r,z,0) = -e^{-jk(z+r^2/2z)}\frac{u_0}{z}\int_0^a J_0\left(\frac{krr'}{z}\right)e^{\frac{-jkr'^2}{2}\left(\frac{1}{z}-\frac{1}{z_0}\right)}r'\,dr' \tag{6.40}$$

where $J_0(x)$ is a Bessel function of the first kind of zero order. This result can be used to determine the displacement at the focus z_0

$$u_z(0,z_0) \sim \frac{j\pi a^2}{\lambda z_0}e^{-jkz_0}u_0 \tag{6.41}$$

for $kz_0 \gg 1$ and hence the beam intensity at the focus compared to that at the transducer.

$$\frac{I(0,z_0)}{I(0)} = \left(\frac{\pi a^2}{z_0\lambda}\right)^2 = \left(\frac{\pi}{S}\right)^2 \tag{6.42}$$

where $S = z_0\lambda/a^2$ is the Fresnel parameter. The lens will hence normally function in the regime $S < \pi$.

The lateral resolution can be determined by calculating the off-axis intensity at z_0. Equation 6.40 yields

$$\frac{I(r,z_0)}{I(0)} = \left(\frac{\pi a^2}{z_0\lambda}\right)^2 \mathrm{jinc}^2\left(\frac{ra}{\lambda z_0}\right) \tag{6.43}$$

The main result here is that the lateral intensity varies as $\mathrm{jinc}^2(ra/\lambda z_0)$, which is the same result as for a circular piston far from the source.

Equation 6.40 and its direct result, Equation 6.43, lead to quantitative criteria for the resolution.

1. Spatial resolution

   Using the Rayleigh criterion of resolution as in optics the spatial resolution is given by the position of the first zero of the $\mathrm{jinc}^2 x$ function

   $$r_0(\text{zero}) = \frac{0.61\lambda}{NA} \tag{6.44}$$

   where $NA = \sin\theta_0$ is the numerical aperture.

   The relative aperture or F number of the lens is given by

   $$F = \frac{z_0}{2a} \tag{6.45}$$

2. Sidelobes

   The sidelobes are important in radiation patterns for plane transducers as has already been seen. Likewise for focused transducers they should be reduced as much as possible to improve signal discrimination. The first sidelobe for the spherical radiator occurs at the first secondary maximum of the $\mathrm{jinc}^2 x$ function, at $kra/z_0 = 5.136$. It is 17.6 dB down in amplitude from the main lobe.

3. Depth of focus

   The axial variation of intensity can be determined from Equation 3.37, and with a suitable criterion, this can be used to determine the depth of focus. The simplest way to do this is to inscribe a cylinder in the focal region as shown in Figure 6.8. From Equation 6.44, this gives a depth of focus along the z axis

$$d_z = 1.22\lambda\left(\frac{z_0}{a}\right)^2 \quad (6.46)$$

4. Phase change of π at the focus

   It has been shown in great detail by Born and Wolf [25] that there is a π phase change at the focus of three-dimensional focusing systems. This result also follows directly from Equation 6.40. An interesting discussion on this point is given in [26]. The simple physical picture is as follows. A spherically converging wavefront at the focus comes to a point and then exits the focus as a diverging spherical front. This corresponds to a reflection with respect to the origin (rotation by π), which corresponds to the π phase change.

6.4 Radiation Pressure

Like all forms of radiation, a beam of acoustic energy will exert a force, or radiation pressure, on an object in its path. This phenomenon is important in the measurement of acoustic field and in calibration of acoustic instruments such as hydrophones. The actual effect in laboratory or in field conditions can be quite complicated and depends on the specific configuration of the system under study. In what follows we give a simple treatment of an idealized case in order to bring out the basic principles involved. A good historical and tutorial account is given by Torr [27].

Consider the case of Figure 6.10 for a perfectly absorbing target. The standard construction for the energy flux is shown; during a time Δt, the energy contained within a cylinder of length $V_0\Delta t$ will attain the wall and be absorbed. For acoustic intensity I, the energy absorbed during time Δt is $IA\Delta t$. The wall will exert a force F against the wave and during time Δt will do work equal to $FV_0\Delta t$, which must be equal to the energy absorbed. Equating the two quantities and recognizing that by Newton's third law the wave will exert an equal and opposite force on the wall $F = p_rA$, we find for the radiation pressure

$$p_r = \frac{I}{V_0} \quad \text{(absorption)} \quad (6.47)$$

For the case of a perfect reflector, the situation is similar to that for the pressure exerted by a perfect gas on the walls of the continuer. In that case, the calculation is usually made by putting the impulse, $F\Delta t$, equal to the change

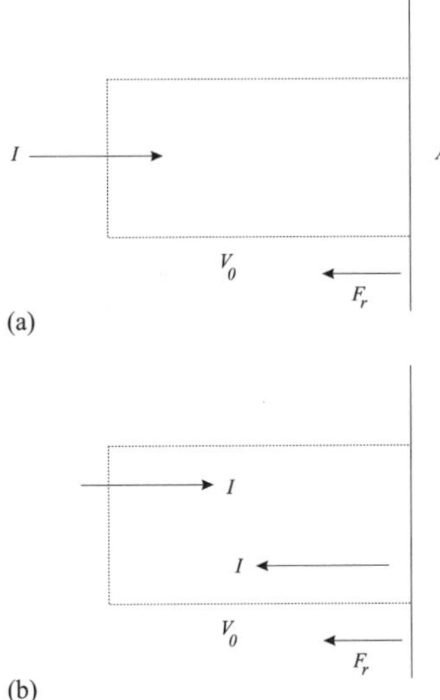

FIGURE 6.10
Geometry for acoustic radiation pressure. (a) Perfect absorber. (b) Perfect reflector.

of momentum for particles inside the cylinder of Figure 6.10. For the case of absorption, the momentum to be absorbed is simply that of the incoming wave as calculated above. However, for the reflector the direction of the momentum is reversed so that the impulse, or radiation pressure, is now determined by twice the modulus of the momentum of the incoming wave. Thus

$$p_r = \frac{2I}{V_0} \quad \text{(reflection)} \tag{6.48}$$

In general, due to partial absorption, generation of different acoustic modes in the target, partial transmission in composite targets, etc., the actual radiation pressure will have a value somewhere between that of these two limiting cases.

6.5 Doppler Effect

A classic manifestation of the Doppler effect is that experienced unconsciously by every child watching a passing train. Here the fixed observer (child) hears an apparent increase of frequency by the moving object (train) as it approaches, followed by a decrease as the train passes and then moves away. This Doppler frequency shift is important in ultrasonics, particularly for instrumentation for flowmeters, medical applications, and oceanography. In these examples, any or all of the source, medium, or receiver may be in movement.

The physical origin of the Doppler effect lies in the variation of the apparent wavelength. For the example of the moving source considered above, as the

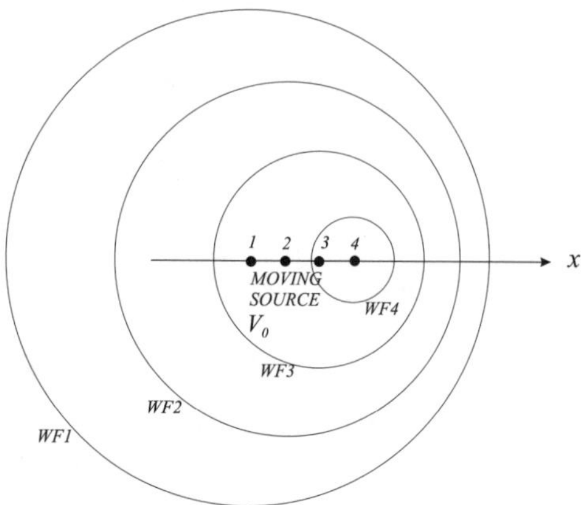

FIGURE 6.11
Crowding of wavefronts in front of a moving source leading to the Doppler shift. The situation is shown at four distinct source positions.

source emits spherical waves as it moves, the wavefronts in front of the source are scrunched together while those behind it become separated farther and farther apart as shown in Figure 6.11. The corresponding effective changes in wavelength give rise to the observed frequency changes by the fixed observer.

A quantitative estimate of the Doppler effect can be given as follows. We consider first motion along the axis for a moving source (V_S) and receiver (V_R) in a medium of sound velocity V_0, with emission by the source of a steady signal at frequency f_S. Due to the compressing of the wavefronts in front of the source, the wavefront is shortened to

$$\lambda_a = \frac{(V_0 - V_S)}{f_s} \tag{6.49}$$

If the receiver is moving away from the source it detects a frequency f_R

$$f_R = \frac{(V_0 - V_R)}{\lambda_a} \tag{6.50}$$

giving finally

$$f_R = f_S \frac{(V_0 - V_R)}{(V_0 - V_S)} \tag{6.51}$$

In a similar manner it can be shown that for a fixed source and receiver that radiates a frequency f_S toward a target with velocity V_T,

$$f_R = f_S \frac{(V_0 - V_T)}{(V_0 + V_T)} \tag{6.52}$$

More complete and rigorous demonstrations of the Doppler shift have been given in the literature, for example, Pierce [28].

Some of the applications of the Doppler shift will be mentioned here. Doppler methods for industrial flowmeters for liquids and gases are numerous and are referred to in Chapter 13. Medical applications are numerous, as this is the perfect technique for monitoring movement inside an opaque object. Bloodflow, including fetal bloodflow, is an obvious application. Others include movement of internal organic components (e.g., heart valves) and monitoring arteries for severity of athersclerosis.

Oceanography instrumentation makes widespread use of Doppler, for example, for studying ocean layer dynamics using bubbles, plankton, and detritus as scattering centers for Doppler sonar. Monitoring movement of the sea surface and navigational aid are other applications. In the nonultrasonic world, Doppler radar is also an important application.

Summary

Acoustic point source gives rise to spherical waves diverging from the point in question. The pressure amplitude of the acoustic wave varies as $1/r$.

Plane piston source is assumed to be uniformly excited across its face and to be enclosed in an infinite baffle such that acoustic energy is only radiated in the forward direction.

Fresnel distance from a circular plane piston source is given by $z_0 = a^2/\lambda$. It is the position of the last intensity maximum along the axis going out from the transducer face.

Near field is that region between the Fresnel distance and the transducer face. It is characterized by strong variations in phase and amplitude of the acoustic wave.

Far field is that region far from the source and beyond the Fresnel distance. The amplitude varies as $1/r$ and the wavefront approaches a plane wave the farther one goes from the source.

Scattering of acoustic waves can in principle be calculated from Huygens principle. The scattering amplitude is usually described by the scattering cross-section that represents the apparent area of the scattering object.

Focused acoustic radiator focuses the emitted acoustic waves at the center of curvature of the spherical radiator. The lateral intensity varies as $\text{jinc}^2(ra/\lambda z_0)$, which determines the spatial resolution at the focal point.

Radiation pressure of an acoustic wave is given by Equation 6.47 for an absorbing target and Equation 6.48 for a perfectly reflecting target.

Doppler effect is a change of observed frequency when source, target, or receiver are moving with respect to each other. The effect can be used to deduce the velocity of the target.

Questions

1. Draw the phasor diagram for the displacement, particle velocity, and pressure for the point source of Equation 6.13.
2. Draw the vector diagram for the specific acoustic resistance and reactance of the previous case.
3. Show that the specific acoustic reactance of a spherical wave is a maximum for $kr = 1$.
4. Calculate the average rate at which energy flows through a closed surface that surrounds a point source.
5. For a 5-mm radius transducer, calculate the Fresnel distance in water as a function of frequency from 1 to 1000 MHz. Graph this result. Extend this result to a family of curves for liquids with sound velocities smaller and larger than that of water.
6. Sketch the radiation patterns for transducers of radius 1 mm and 10 mm into water at frequencies of 1 MHz and 20 MHz. Explain the qualitative difference between the radiation patterns.
7. What are the implications for imaging if the side lobes of a focused beam compared to the main beam are 30 dB down and 3 dB down?
8. Reconcile Equations 5.33 and 6.24.
9. Calculate the formula for radiation pressure using the concept of momentum of a wave and Newton's second law.
10. For the case of a fixed receiver at angle θ to the motion of the source, show that Equation 6.51 becomes

$$f_R = f_S \frac{1}{1 - \frac{V_S \cos\theta}{V_0}}$$

7

Reflection and Transmission of Ultrasonic Waves at Interfaces

7.1 Introduction

Performing any operation with ultrasonic waves means transmitting them from one medium to another where the measurement or actuation is to be performed. In other cases, the objective may be to retain a wave in a given medium and prevent it from radiating out into the environment. In either case, a good understanding of the principles of reflection and transmission of ultrasonic waves is essential.

The problem is similar to that in electromagnetic and other wave phenomena. The process can be broken down into a number of simple steps:

1. Draw a diagram of the process and clearly define the interface and the coordinate system to be used.
2. Define the incident wave vector (amplitude and incidence angle) and identify all possible reflected and transmitted wave vectors.
3. Write down the velocity (displacement) potentials for each medium, and hence obtain the velocities (displacements) of each wave vector in step 2 (above). In terms of them, use the form of standard solutions of the bulk wave equation.
4. Apply the appropriate boundary conditions at the interface. Normally, the number of boundary conditions required is equal to the number of solutions to obtain.
5. Insert the solutions into the boundary conditions, thus obtaining a set of N equations for the N amplitudes to be determined.
6. Use the fact that these equations are valid for all values of the coordinate x along the interface, which invokes the principle of conservation of parallel momentum and hence Snell's law.
7. Solve the set of equations in step 5 to obtain the unknown amplitudes in terms of the incident amplitude.

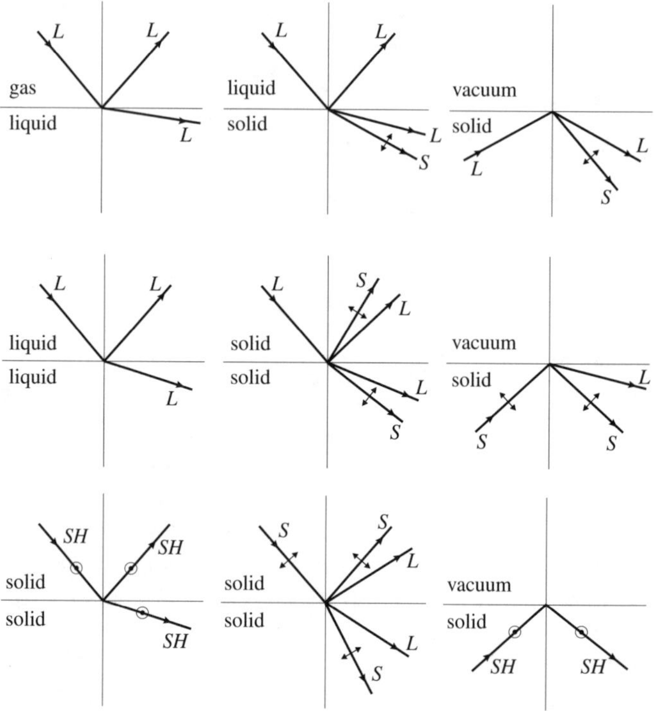

FIGURE 7.1
Typical cases of reflection and transmission of acoustic waves at interfaces between solids, liquids, and gases.

A number of typical cases are shown in Figure 7.1 in the usual convention used here in which incidence is from the upper medium. The list is not complete in the sense that the medium of incidence has been chosen arbitrarily. For example, for the solid-liquid interface, incidence from the liquid is shown, but the incident wave might be in the solid so that this case would have to be worked out separately.

The boundary conditions are easy to state superficially, but their understanding is essential to posing and solving the problem correctly. Basically, they correspond to the conditions that must be met in order to obtain a perfectly defined interface for the problem at hand. The most general case is that of the solid-solid interface. For this to be well defined, there must be no net stress on the interface or displacement of one medium with respect to the other. This leads to boundary conditions of continuity of normal and tangential components of stress and displacement, i.e., four conditions, corresponding to the four amplitudes to be determined shown in Figure 7.1. If these boundary conditions are satisfied at a given time everywhere along the interface, then the problem can be posed and solved. If, however, they are not respected locally at all times, the interface is no longer well defined and the conditions cannot be written down as valid for all values of interface coordinate x and so the problem cannot be solved straightforwardly. In fact, if the interfacial deformation is not clearly specified, or it is time dependent or irreversible, then no solution is possible. If the deformation is well defined and time independent, the problem then becomes one of nondestructive evaluation (NDE) of interfacial defects, as discussed in Chapter 15. In this and succeeding chapters we only consider perfect interfaces.

The chapter is organized as follows. In Section 7.2, we consider reflection and transmission at normal incidence for liquid-liquid interfaces. This allows us to concentrate on basic concepts such as acoustic mismatch, standing waves, and layered media in the simplest mathematical description and the most important applications area. The succeeding sections deal with oblique incidence for several important cases:

1. Fluid-fluid, the simplest case of transmission between two media.
2. Fluid-solid, which is very important in practice for sensors, NDE, acoustic microscopy, etc. It also leads into a rich case for critical angles and hence into the subject of Chapter 8, surface acoustic waves. Finally, the slowness construction is applied to the reflection or transmission problem. It has the great advantage of providing a simple, rigorous, visual demonstration of Snell's law. Subsequently, it will be fundamental to the discussion of acoustic waveguides.
3. Solid-solid, SH modes, the simplest case for transmission between two solids.
4. Solid-vacuum, the results of which will be useful for acoustic waveguides.

7.2 Reflection and Transmission at Normal Incidence

We do this case for illustrative purposes, to see the importance of impedance matching in such problems. This is the simplest case; the math is simple, and there is no mode conversion. If only longitudinal modes are considered it can be used for liquid-liquid or liquid-solid interfaces.

Consider the liquid-liquid interface shown in Figure 7.2(a), with a plane pressure wave incident from the left. Due to the difference in acoustic properties between the two media there are partial reflection and transmission at the interface. The three waves can be represented as:

$$p_i = A\exp j(\omega t - k_1 x) \tag{7.1}$$

$$p_r = AR_p \exp j(\omega t + k_1 x) \tag{7.2}$$

$$p_t = AT_p \exp j(\omega t - k_2 x) \tag{7.3}$$

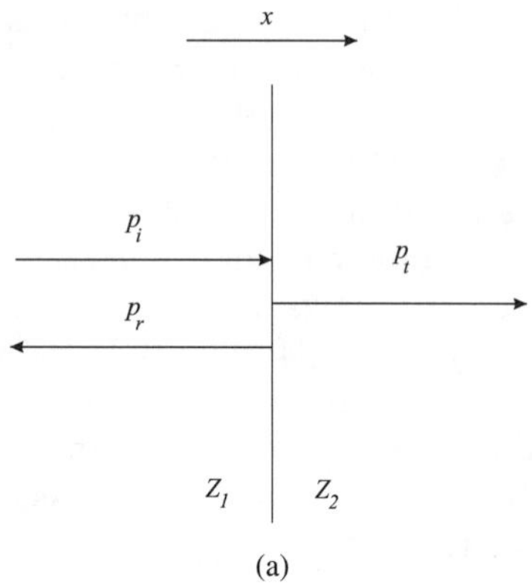

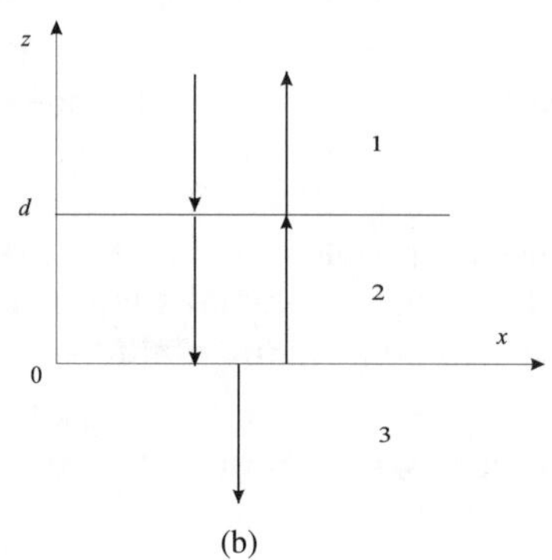

FIGURE 7.2
Configuration for reflection and transmission at normal incidence for (a) Planar interface and (b) Layer of thickness *d* between two bulk media.

Since the two media must stay in intimate contact at a perfect interface, the boundary conditions are continuity of pressure and velocity (displacement) at $x=0$; if these conditions were not met, the boundary would not be well defined. Using the definition of acoustic impedance, it follows that

$$R_p + 1 = T_p \tag{7.4}$$

$$\frac{1}{Z_1}(1 - R_p) = \frac{T_p}{Z_2} \tag{7.5}$$

where Z_1 and Z_2 are the characteristic acoustic impedances of the two media. Equations 7.4 and 7.5 can be solved to give for the pressure transmission and reflection coefficients

$$T_p = \frac{2Z_2}{Z_1 + Z_2} \tag{7.6}$$

$$R_p = \frac{Z_2 - Z_1}{Z_1 + Z_2} \tag{7.7}$$

These results give the pressure reflection coefficient ($R_p \equiv p/p_{inc}$) and the pressure transmission coefficient ($T_p \equiv p_{tr}/p_i$). Of great importance are the acoustic intensity transmission and reflection coefficient. At normal incidence these can be obtained directly from the definition of acoustic intensity $I \equiv p^2/2Z$.

Thus

$$\frac{I_t}{I_i} = \frac{Z_1}{Z_2}|T_p|^2 \tag{7.8}$$

$$\frac{I_r}{I_i} = |R_p|^2 \tag{7.9}$$

from which it can be verified that the law of conservation of energy is satisfied.

$$I_i = I_r + I_t \tag{7.10}$$

There is a lot of simple physics in this result. Let us look at the range of the modulus of R_p and T_p. If $Z_1 \equiv Z_2$, then $T_p \equiv 0$ and $R_p = 0$; it is as if there were one uniform medium so there is no reflection.

For $Z_2 \ll Z_1$, $R_p \approx -1$ and $T_p \rightarrow 0$. This is termed a free boundary, corresponding, for example, to medium 1 = water or a solid, and medium 2 = air. There is huge acoustic impedance mismatch so that nearly all of the acoustic wave is reflected. There is a phase change of π for the pressure at the interface.

The transmitted acoustic intensity for this case is given by

$$\frac{I_t}{I_i} = \frac{Z_1}{Z_2}|T_p|^2 \ll 1$$

as expected.

It is interesting to look at the numerical results for the water-air interface. For air $\rho_2 \sim 1.3\ \mathrm{kg \cdot m^{-3}}$, $V_2 \sim 330\ \mathrm{m \cdot sec^{-1}}$ and for water $\rho_1 \sim 10^3\ \mathrm{kg \cdot m^{-3}}$, $V_1 \sim 1500\ \mathrm{m \cdot sec^{-1}}$. Then

$$R_p \sim -1 + 5.8 \times 10^{-4}$$

and

$$T_p = 1 + R_p \approx 5.8 \times 10^{-4}$$

Finally,

$$\frac{I_t}{I_i} = \frac{1.5 \times 10^6}{429} \times (5.8 \times 10^{-4})^2 \approx 1.1 \times 10^{-3}$$

For the opposite case, $Z_2 \gg Z_1$, giving immediately $R_p \sim 1$ and $T \sim 2$. This case corresponds to a rigid boundary. The transmitted intensity $I_t/I_i \sim 4(Z_1/Z_2)$ is again very small as is expected as the acoustic mismatch is again very large. Numerically,

$$\frac{I_t}{I_i} \sim \frac{450 \times 10^{-6}}{1.5} \times 4 \sim 1.1 \times 10^{-3}$$

Clearly the transmitted intensity is symmetric with respect to the incident medium, i.e., the transmitted intensity is the same whether the wave is incident from air or water. This is not true for the pressure, nor the particle velocity. Symmetry considerations will be discussed later in Section 7.3.1.

7.2.1 Standing Waves

The traveling or progressive waves treated in bulk media thus far are characterized by the propagation of a disturbance (phase) and the propagation of energy. This state of affairs can be changed radically if two traveling waves, of the same frequency and mode but traveling in opposite directions, are combined. This gives rise to standing waves that form a static pattern of nodes and antinodes and for which there is no propagation of energy. Standing waves are fundamental to the operation of acoustic waveguides and resonators and as such have a central place in ultrasonics.

Standing waves can be most easily formed, and described, by the configuration of the total reflection of a plane wave treated in the previous section. Qualitatively, the situation is shown in Figure 7.3. As already shown, the reflected pressure is the negative of the incident pressure. Since the displacement is zero at the rigid boundary by the boundary conditions, the displacement in the incident wave at the boundary is also zero leading to a node. Conversely, since displacement and pressure are in quadrature there is a pressure antinode at the rigid boundary. Displacement and pressure then have a series of nodes and antinodes, the extreme values, at different times, being shown in the figure. For a free boundary, the behavior is opposite; that is, the pressure has a node at the surface and the displacement has an antinode. Again, the latter condition follows directly from the boundary conditions at a free surface. Since there are four different cases, a memory aid device is helpful. One way is to remember that the displacement is maximum (antinode) at a free surface, and that displacement-pressure and rigid-free are opposite, so that if one case is remembered the others follow automatically. This behavior is demonstrated quantitatively in what follows.

The pressure waves of the previous section (from Equations 7.1 through 7.3) lead to the following pressure field in medium 1,

$$p = p_i + p_r = \exp j(\omega t - kx) + R_p \exp j(\omega t + kx) \tag{7.11}$$

where for convenience we set the incident amplitude equal to unity.

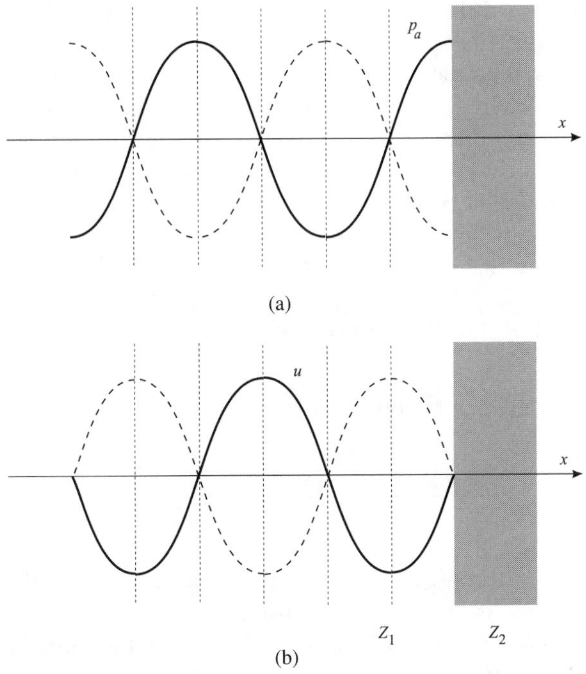

FIGURE 7.3
Standing wave pattern at a rigid boundary. (a) Incident and reflected pressure waves. (b) Incident and reflected displacements.

Hence,

$$\begin{aligned} 2p &= (1+R_p)[e^{j(\omega t-kx)}+e^{j(\omega t+kx)}]+(1-R_p)[e^{j(\omega t-kx)}-e^{j(\omega t+kx)}] \\ &= e^{j\omega t}(1+R_p)2\cos kx+(1-R_p)(-2j\sin kx)e^{j\omega t} \end{aligned} \tag{7.12}$$

The two limiting cases treated previously are of interest.

For a rigid boundary,

$$R_p = 1 \quad \text{and} \quad p = 2\cos kx e^{j\omega t} \tag{7.13}$$

For a free surface,

$$R_p = -1 \quad \text{and} \quad p = 2\sin kx e^{j\left(\omega t-\frac{\pi}{2}\right)} \tag{7.14}$$

This mathematical form gives a simple and convenient test for distinguishing between traveling and standing waves. Traveling waves correspond to propagation of a disturbance and are necessarily of the form $f(\omega t - kx) = 0$. In standing waves, the spatial and temporal variation are separated in the form $f(\omega t)g(kx) = 0$ as seen above. This provides a convenient test for categorizing an unknown waveform as either a free or a standing wave.

The ideal rigid interface or free surfaces are idealizations not always met in practice, although they are extremely good approximations, e.g., a resonator-air interface. However, very often the reflection coefficient is not unity, in which case the standing wave pattern is not complete, and in particular the amplitude at the nodes is no longer zero. The wave field can then be regarded as being part standing wave and part traveling wave. The situation is commonly described by the standing wave ratio (SWR), given by

$$\begin{aligned} \text{SWR} &= \frac{p_{antinode}}{p_{node}} \\ \text{SWR} &\equiv \frac{1+R_p}{1-R_p} \quad \text{(nonattenuating medium)} \end{aligned} \tag{7.15}$$

We can calculate the power flow for standing waves as follows. From the definition of the acoustic Poynting vector as the acoustic power per unit area transmitted across a surface and, on the other hand, the model of two reflecting surfaces to set up a standing wave, it is clear that the average acoustic intensity is zero. That is to say, there is no net propagation of acoustic energy in either the plus or the minus x direction.

This result can be seen more formally as follows. From Equation 5.10, the time-averaged acoustic Poynting vector is

$$I = \overline{p(t)\vec{v}(t)} = \frac{1}{2}\text{Re}[pv^*] \tag{7.16}$$

For a progressive wave, the acoustic pressure p and the particle velocity $\vec{v}$ are in phase and so we get a finite power flow. For standing waves the particle displacement and velocity are in phase but they are in quadrature with the pressure as shown in Figure 7.3. In this case, the time average in Equation 7.16 is equal to zero, corresponding, as we already know, to zero propagation of energy.

7.2.2 Reflection from a Layer

The input impedance of a layer sandwiched between two different media can be calculated by a direct extension of the reflection coefficient for a single interface [29]. From Equation 7.7, we have

$$R_p = \frac{Z_{in} - Z_1}{Z_{in} + Z_1} \tag{7.17}$$

where from Figure 7.2(b), Z_{in} is the input impedance presented by the layer and medium 3 at the 1-2 boundary. For simplicity, we consider normal incidence. The factor $\exp j(kx - \omega t)$ is not retained in what follows; it is common to all terms as the results are valid for all values of x.

In the layer, the pressure can be written as

$$p_2 = A \exp j(k_2 z) + B \exp j(-k_2 z) \tag{7.18}$$

Due to multiple reflections in the layer, forward and backward waves will be set up. A and B can be calculated by continuity of the impedance (since p and V_z are continuous) at the interface. The impedance associated with p_2 can be calculated using the general formula in Equation 3.28. Hence,

$$\left.\frac{-j\omega\rho_2 p_2}{\frac{\partial p_2}{\partial z}}\right|_{z=0} = Z_3 \tag{7.19}$$

which leads directly to

$$\frac{A}{B} = \frac{Z_3 - Z_2}{Z_3 + Z_2}$$

The same calculation at the 1-2 interface ($z = d$) can then be used to determine Z_{in}, which is:

$$Z_{in} = Z_2 \frac{Z_3 - jZ_2 \tan\varphi}{Z_2 - jZ_3 \tan\varphi} \tag{7.20}$$

where $\varphi = k_2 d$ is the phase change associated with the layer thickness.

A particularly important application of this result in acoustics and optics is the case where $d = \lambda_2/4$, i.e., when the thickness of the layer is one quarter wavelength. Then, by Equation 7.20,

$$Z_{in} = \frac{Z_2^2}{Z_3} \quad \text{and} \quad R_p = \frac{Z_2^2 - Z_1 Z_3}{Z_2^2 + Z_1 Z_3} \tag{7.21}$$

which gives $R_p = 0$ for $Z_2 = \sqrt{Z_1 Z_3}$. This is a very well-known and important result. It means that to obtain perfect transmission between two media of different acoustic impedance it is sufficient to provide a quarter wave layer of material between them which has an acoustic impedance equal to the geometric mean of the two end media. Of course, this result is only true at one particular frequency, that for which $d = \lambda_2/4$. Such quarter wavelength layers are used in cases where one wants to maximize the acoustic transmission between two media. The case of the single quarter wavelength layer is the one of greatest practical importance. It is, however, possible to generalize the previous result for an arbitrary number of layers, as described in [29].

7.3 Oblique Incidence: Fluid-Fluid Interface

The case of oblique incidence for the fluid-fluid interface is of some interest as it contains much of the simple physics of the fluid-solid interface but is mathematically less complicated than the latter. Moreover, certain interesting results regarding symmetry for the incident and refractive media can be determined in this case.

The situation is shown in Figure 7.1, where a wave of unit amplitude is incident on the interface at incidence angle θ to the normal. Corresponding angles are defined in the figure for the reflected and transmitted waves, which have amplitudes R and T, respectively. The velocity potentials for the three waves can be written:

$$\varphi^i = \exp j(\omega t - k_x^i x + k_z^i z) \tag{7.22}$$

$$\varphi^r = R\exp j(\omega t - k_x^r x - k_z^r z) \tag{7.23}$$

$$\varphi^t = T\exp j(\omega t - k_x^t x + k_z^t z) \tag{7.24}$$

where

$$\vec{v} = \overrightarrow{\nabla\varphi} \quad \text{and hence} \quad p = -\rho\frac{\partial\varphi}{\partial t}$$

The pressures are given by

$$p^i = -j\omega\rho_1 \exp(\omega t - k_x^i x + k_z^i z) \tag{7.25}$$

$$p^r = -j\omega\rho_1 R\exp(\omega t - k_x^r x - k_z^r z) \tag{7.26}$$

$$p^t = -j\omega\rho_2 T\exp(\omega t - k_x^t x + k_z^t z) \tag{7.27}$$

and the normal velocities by

$$v_z^i = jk_z^i\varphi^i \tag{7.28}$$

$$v_z^r = -jk_z^r\varphi^r \tag{7.29}$$

$$v_z^t = jk_z^t\varphi^t \tag{7.30}$$

At the interface $z = 0$ the boundary conditions are given by continuity of the pressure and the normal velocity. Hence,

$$p^i + p^r = p^t \tag{7.31}$$

$$v_z^i + v_z^r = v_z^t \tag{7.32}$$

These two relations will be used to determine the reflection and transmission coefficients. Before that, we can obtain the angles of reflection and transmission by noting that the boundary conditions must be valid for all values of x. It follows that

$$k_x^i = k_x^r = k_x^t \tag{7.33}$$

or

$$k^i \sin\theta_i = k^r \sin\theta_r = k^t \sin\theta_t$$

Since

$$k^r = \frac{\omega}{V_1}, \quad k^t = \frac{\omega}{V_2}$$

we have finally $\theta_i \equiv \theta_r$ and

$$\frac{\sin\theta_i}{V_1} = \frac{\sin\theta_t}{V_2} \tag{7.34}$$

which is the well-known Snell's law. Looking back at the first line, and using the quantum mechanical interpretation of $\hbar k$ as the momentum of a wave, one can say that this law corresponds to the conservation of parallel momentum, i.e., the component of momentum along the surface. This interpretation will be reinforced in the discussion of slowness curves in Section 7.4.

Putting $z = 0$ in Equations 7.31 and 7.32, we obtain

$$\rho_1(1 + R) = \rho_2 T \tag{7.35}$$

$$k_z^i(1 - R) = k_z^t T \tag{7.36}$$

which can be solved to give

$$R = \frac{\dfrac{\rho_2 V_2}{\cos\theta_2} - \dfrac{\rho_1 V_1}{\cos\theta_1}}{\dfrac{\rho_2 V_2}{\cos\theta_2} + \dfrac{\rho_1 V_1}{\cos\theta_1}} \tag{7.37}$$

$$T = \frac{\dfrac{2\rho_1 V_2}{\cos\theta_2}}{\dfrac{\rho_1 V_1}{\cos\theta_1} + \dfrac{\rho_2 V_2}{\cos\theta_2}} \tag{7.38}$$

where $\theta_i = \theta_1$ and $\theta_t = \theta_2$.

From Equations 7.37 and 7.38, we can write the reflection and transmission coefficients for the pressure as

$$R_p = \frac{p^r}{p^i} \equiv R \tag{7.39}$$

$$T_p = \frac{p^t}{p^i} = \frac{\rho_2}{\rho_1}T \tag{7.40}$$

Writing the normal acoustic impedance in standard form $Z_1 = \rho_1 V_1 / \cos\theta_1$ and $Z_2 = \rho_2 V_2 / \cos\theta_2$ we have, finally,

$$R_p = \frac{Z_2 - Z_1}{Z_2 + Z_1} \tag{7.41}$$

$$T_p = \frac{2Z_2}{Z_2 + Z_1} \tag{7.42}$$

which is the same general form as for normal incidence.

The reflection and transmission coefficients for the acoustic intensity are also of interest. Since we are concerned with transmission and reflection with respect to the boundary, only the normal component of acoustic intensity is pertinent. For a given θ, the total acoustic intensities are

$$I^i = \frac{|p^i|^2}{2Z_1}, \quad I^r = \frac{|p^r|^2}{2Z_1}, \quad \text{and} \quad I^t = \frac{|p^t|^2}{2Z_2}$$

and the normal components respect the principle of conservation of energy, as can be demonstrated from the previous results

$$I^i \cos\theta_i = I^r \cos\theta_r + I^t \cos\theta_t \tag{7.43}$$

Evidently, the acoustic intensity reflection (R_I) and transmission (T_I) coefficients are a function of incidence angle; an example will be given for the solid-liquid interface.

Let us now pause for breath to reflect on what additional information the oblique incidence treatment has given us and how to interpret the results. A first requirement is to verify the result that the velocity reflection coefficient is equal in modulus but opposite in sign to the pressure reflection coefficient that was stated in the normal incidence example given earlier. This can be obtained immediately as

$$R_v \equiv \frac{v_z^r}{v_z^i} = -\frac{jk_z^r\varphi^r}{jk_z^i\varphi^i} \tag{7.44}$$

$$R_v = -R \quad \text{at} \quad z = 0 \tag{7.45}$$

as stated previously. The same result holds evidently for the displacement.

The full consequences of Snell's law must also be explored. Let us assume that the lower medium has the higher sound velocity so that $V_1 < V_2$. The immediate consequence is that $\theta_i < \theta_t$. This means that as θ is increased, the refracted wave rapidly approaches the plane of the interface (x axis). At a critical angle θ_c, $\theta_t = \pi/2$ such that

$$\frac{\sin\theta_c}{V_1} = \frac{\sin\left(\frac{\pi}{2}\right)}{V_2} \tag{7.46}$$

so

$$\sin\theta_c = \frac{V_1}{V_2} \tag{7.47}$$

In fact, as will be developed later for the fluid-solid interface, this corresponds to the propagation of a surface wave in the plane of the interface. For angles $\theta > \theta_c$, there is total reflection and $|R_p| \equiv 1$.

It is shown in [29] that interesting conclusions can be drawn by using normalized parameters as follows. We define $n \equiv V_1/V_2$ and $m \equiv \rho_2/\rho_1$. Then we can rewrite Equations 7.41 and 7.42 as

$$R_p = \frac{m\cos\theta - \sqrt{n^2 - \sin^2\theta}}{m\cos\theta + \sqrt{n^2 + \sin^2\theta}} \tag{7.48}$$

$$T_p = \frac{2m\cos\theta}{m\cos\theta + \sqrt{n^2 - \sin^2\theta}} \tag{7.49}$$

known as the Fresnel formulae. This form facilitates the study of R and T of various material combinations for particular values of θ. Of particular interest is the region of total reflection $\theta_i > \arcsin n$. In this region

$$R_p = \exp i\varphi \tag{7.50}$$

$$\varphi = -2\arctan\frac{(\sin^2\theta_i - n^2)^{\frac{1}{2}}}{m\cos\theta} \tag{7.51}$$

In this region the modulus of the reflection coefficient is unity while the phase changes monotonically. This behavior will be of importance in the study of Rayleigh waves.

7.3.1 Symmetry Considerations

The variation of the various reflection and transmission coefficients has been treated in general in [29]. An overview of the main results is given here.

1. Angles of incidence (θ_1) and refraction (θ_2). If the direction of propagation is reversed and the refracted wave becomes the incident wave, then by Snell's law the new refracted wave is at angle θ_1. Moreover, from Equation 7.41, if the original pressure coefficient is $R_p = +V$, then reversal of propagation directions leads to a new wave with $R_p = -V$.
2. Reflection and transmission coefficients for p, v, and u. As already demonstrated at normal incidence, there are no symmetry relations for these quantities if the direction of propagation is reversed.
3. Energy transmission coefficient. The coefficient for transmission of acoustic energy normal to the interface is symmetric if the direction of propagation is reversed. As seen before,

$$T_I = \frac{I_{2z}}{I_{1z}} = \frac{\rho_1 V_1 \cos\theta_2}{\rho_2 V_2 \cos\theta_1}|T_p|^2 \tag{7.52}$$

Expressing T_p in normalized coefficients, Equation (7.49), this becomes

$$T_I = 4\frac{\cos\theta_1}{\rho_1 V_1}\frac{\cos\theta_2}{\rho_2 V_2}\left(\frac{\cos\theta_1}{\rho_1 V_1} + \frac{\cos\theta_2}{\rho_2 V_2}\right)^{-2} \tag{7.53}$$

which is symmetric with respect to interchange of the two media.

7.4 Fluid-Solid Interface

The problem is presented in Figure 7.4 where a plane wave is incident from the fluid and there is partial reflection in medium 1 and partial transmission of longitudinal and shear waves into the solid (medium 2). We wish to calculate the reflection and transmission coefficients for the stress and the

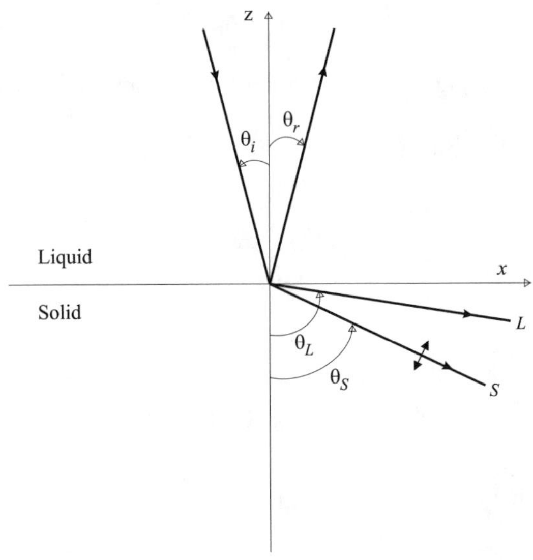

FIGURE 7.4
Coordinate system for reflection and transmission at a liquid-solid interface with incidence from the liquid.

acoustic intensity. The approach is similar to that presented by Brekhovskikh [30] and Ristic [31].

The velocity potentials can be written in the liquid and solid, respectively, as

$$\vec{v} = \overrightarrow{\nabla\varphi} \tag{7.54}$$

$$\vec{v} = \overrightarrow{\nabla\phi} + \vec{\nabla}\times\vec{\psi} \tag{7.55}$$

and the potentials can be expressed as plane wave solutions to the wave equation

$$\varphi_i = \exp j(\omega t - k\sin\theta_i x + k\cos\theta_i z) \tag{7.56}$$

$$\varphi_r = R\exp j(\omega t - k\sin\theta_r x - k\cos\theta_r z) \tag{7.57}$$

$$\phi = T_L \exp j(\omega t - k_L\sin\theta_l x + k_L\cos\theta_l z) \tag{7.58}$$

$$\psi = T_S \exp j(\omega t - k_S\sin\theta_S x + k_S\cos\theta_S z) \tag{7.59}$$

where k and k_L are wave numbers for longitudinal waves in the liquid and solid, respectively, and k_S the wave number for shear waves in the solid. R, T_L, and T_S are the reflection and transmission coefficients to be calculated. Note that these are explicitly the velocity potential reflection and transmission coefficients. In the liquid, using $p = -T = \lambda S$, $S = \vec{\nabla} \bullet \vec{u}$, $v = j\omega u$ and $v = \overrightarrow{\nabla\varphi}$ we have

$$p = \frac{\lambda_1}{j\omega}\nabla^2\varphi \tag{7.60}$$

$$V_1^2 = \frac{\lambda_1}{\rho_1} = \frac{\omega^2}{k^2} \tag{7.61}$$

In the solid, from Equations 4.54 and 4.55,

$$j\omega T_{zz} = \lambda_2\left(\frac{\partial v_x}{\partial x} + \frac{\partial v_z}{\partial z}\right) + 2\mu_2\frac{\partial v_z}{\partial z} \tag{7.62}$$

is the normal stress and

$$j\omega T_{xz} = \mu_2\left(\frac{\partial v_x}{\partial z} + \frac{\partial v_z}{\partial x}\right) \tag{7.63}$$

is the tangential stress.

Here

$$v_x = \frac{\partial\phi}{\partial x} - \frac{\partial\psi}{\partial z} \tag{7.64}$$

$$v_z = \frac{\partial\phi}{\partial z} + \frac{\partial\psi}{\partial x} \tag{7.65}$$

and as usual for bulk waves

$$V_L^2 = \frac{\lambda_2 + 2\mu_2}{\rho_2} = \frac{\omega^2}{k_L^2} \tag{7.66}$$

$$V_S^2 = \frac{\mu_2}{\rho_2} = \frac{\omega^2}{k_S^2} \tag{7.67}$$

Substituting these results in Equations 7.62 and 7.63, the stresses are easily found to be

$$j\omega T_{zz} = \lambda_2 \nabla^2 \phi + 2\mu_2 \left(\frac{\partial^2 \phi}{\partial z^2} + \frac{\partial^2 \phi}{\partial x \partial z} \right) \tag{7.68}$$

$$j\omega T_{xz} = \mu_2 \left(2\frac{\partial^2 \phi}{\partial x \partial z} + \frac{\partial^2 \psi}{\partial x^2} - \frac{\partial^2 \psi}{\partial z^2} \right) \tag{7.69}$$

These results for the stresses and velocities will be substituted into the boundary conditions; assuming an ideal nonviscous liquid, there are three boundary conditions and three amplitudes (R, T_L, and T_S) to be determined.

1. Continuity of normal velocities

$$v_{z1} \equiv v_{z2} \tag{7.70}$$

or

$$\frac{\partial \varphi}{\partial z} = \frac{\partial \phi}{\partial z} + \frac{\partial \psi}{\partial x} \tag{7.71}$$

2. Continuity of normal stress

$$p = T_{zz} \tag{7.72}$$

$$\lambda_1 \nabla^2 \varphi = \lambda_2 \nabla^2 \phi + 2\mu_2 \left(\frac{\partial^2 \phi}{\partial z^2} + \frac{\partial^2 \phi}{\partial x \partial z} \right) \tag{7.73}$$

3. Zero tangential stress since the fluid cannot support viscous stress

$$T_{xz} = 0 \tag{7.74}$$

$$\frac{\partial^2 \psi}{\partial x^2} + 2\frac{\partial^2 \phi}{\partial x \partial z} - \frac{\partial^2 \psi}{\partial z^2} = 0 \tag{7.75}$$

Since these results are valid for all values of x along the interface, substitution of the potentials in these three equations immediately yields Snell's law

$$\frac{\sin\theta_i}{V_1} = \frac{\sin\theta_r}{V_1} = \frac{\sin\theta_l}{V_L} = \frac{\sin\theta_s}{V_S} \tag{7.76}$$

hence $\theta_i = \theta_r$. The situation is very similar to that for the liquid-liquid interface and again corresponds to the conservation of parallel momentum along the surface.

The three equations coming from the boundary conditions are

$$k\cos\theta_i R + k_L\cos\theta_l T_L - k_S\sin\theta_s T_S = k\cos\theta_i \tag{7.77}$$

$$k_L^2\sin 2\theta_l T_L + k_S^2\cos 2\theta_s T_S = 0 \tag{7.78}$$

$$\rho_1 R + \rho_2\left[2\frac{k_L^2}{k_S^2}\sin^2\theta_l - 1\right]T_L + \rho_2\sin 2\theta_s T_S = 0 \tag{7.79}$$

with solutions

$$R = \frac{Z_L\cos^2 2\theta_s + Z_S\sin^2 2\theta_s - Z_1}{Z_L\cos^2 2\theta_s + Z_S\sin^2 2\theta_s + Z_1} \tag{7.80}$$

$$T_L = \left(\frac{\rho_1}{\rho_2}\right)\frac{2Z_L\cos 2\theta_s}{Z_L\cos^2 2\theta_s + Z_S\sin^2 2\theta_s + Z_1} \tag{7.81}$$

$$T_S = -\left(\frac{\rho_1}{\rho_2}\right)\frac{2Z_S\sin 2\theta_s}{Z_L\cos^2 2\theta_s + Z_S\sin^2 2\theta_s + Z_1} \tag{7.82}$$

where

$$Z_1 = \frac{\rho_1 V_1}{\cos\theta_i},\quad Z_L = \frac{\rho_2 V_L}{\cos\theta_l},\quad Z_S = \frac{\rho_2 V_S}{\cos\theta_s} \tag{7.83}$$

These expressions are very similar to those for the fluid-fluid interface but they are more complicated as they involve longitudinal and shear impedance. This can be seen explicitly by defining an effective impedance Z_{eff}

$$Z_{eff} \equiv Z_L\cos^2 2\theta_s + Z_S\sin^2 2\theta_s \tag{7.84}$$

so that the reflectance function becomes

$$R(\theta) = \frac{Z_{eff} - Z_1}{Z_{eff} + Z_1} \tag{7.85}$$

as for the fluid-fluid interface.

It is instructive to follow the variation of the reflection coefficient $R(\theta)$ over the full range of incidence angles for the case of a water-aluminum interface shown in Figure 7.5. At normal incidence, the reflection coefficient becomes that given in Equation 7.7. Its value lies between 0 and 1 depending on the acoustic mismatch between the two media. Only the longitudinal wave is

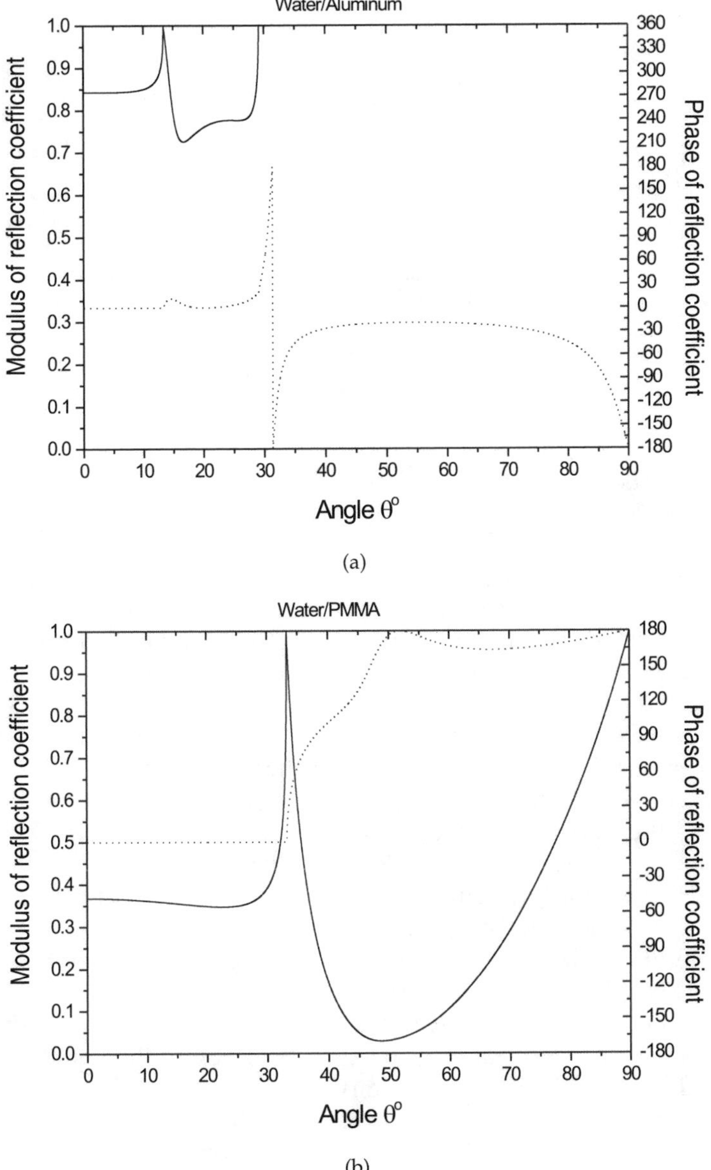

FIGURE 7.5
Reflection coefficient amplitude and phase variation with incidence angle for liquid-solid interfaces. (a) Water/aluminum. (b) Water/PMMA (small acoustic mismatch).

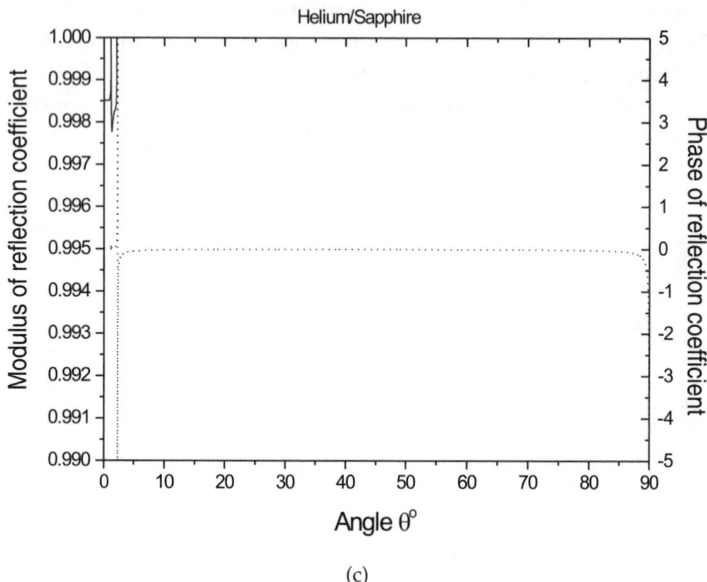

(c)

FIGURE 7.5 (Continued)
Reflection coefficient amplitude and phase variation with incidence angle for liquid-solid interfaces. (c) Liquid helium/sapphire (large acoustic mismatch).

transmitted and there is no mode conversion, i.e., no shear wave is transmitted at normal incidence. As θ increases, longitudinal and shear waves are excited in the solid. $R(\theta)$ stays more or less constant until the longitudinal critical angle, at which point it rises sharply to spike at $|R(\theta)| \equiv 1$. At this angle, the longitudinal wave propagates along the surface so no energy is propagated into the solid. The shear wave amplitude goes to zero at this angle and there is total reflection. As θ increases further, we arrive at a second critical angle θ_{cs} for shear waves, which now propagate along the surface. From θ_{cs} out to 90° there is total reflection of the incident wave, $|R(\theta)| \equiv 1$. There is also a sudden change in phase from 0 to about 2π in the region of θ_{cs}. This is due to the excitation of Rayleigh surface waves at an incidence angle $\theta_{cR} \geq \theta_{cs}$, which is the subject of Chapter 8.

Two additional limiting cases are shown in Figure 7.5. The first case, liquid helium to sapphire, corresponds to the limit of very high acoustic mismatch. $R(\theta)$ is close to unity for all θ and the values of θ_{cl} and θ_{cs} are very small, leading to a small "critical cone" of total reflection in the liquid. The other limit is that of very small acoustic mismatch, for a water-lucite interface. In this case, the sound velocity in the water is less than the longitudinal velocity in the lucite but greater than the transverse velocity. Since the acoustic impedances are relatively well matched the reflection coefficient at normal incidence is much smaller than in the other cases. There is a longitudinal critical angle but there can be no transverse critical angle, so the reflection coefficient is less than unity out to $\theta = \pi/2$.

By direct generalization of the results for the fluid-fluid interface we can write for the acoustic intensity reflection and transmission coefficients

$$\frac{I_R}{I} = |R(\theta)|^2 \tag{7.86}$$

$$\frac{I_L}{I} = \frac{\rho_2 \tan\theta}{\rho_1 \tan\theta_l}|T_L(\theta)|^2 \tag{7.87}$$

$$\frac{I_S}{I} = \frac{\rho_2 \tan\theta}{\rho_1 \tan\theta_s}|T_S(\theta)|^2 \tag{7.88}$$

These curves have been plotted for the same fluid-solid interfaces as shown in Figure 7.6. These curves show very clearly that the energy is transmitted into the solid by longitudinal waves up to θ_{cl} and by transverse waves up to θ_{ct} but not beyond.

It is useful to have a graphical method for describing reflection and refraction phenomena. This is provided by the slowness surface, which is the locus of the quantity $1/V_P$ vs. wave vector direction. Clearly, it is a surface, in $\vec{k}/\omega$ space and the radius vector from the origin to a point on the surface has length $|k|/\omega$. For a liquid, the slowness surface is a sphere and for an isotropic solid it is two concentric spheres. Clearly, a low-velocity medium such as a fluid has a large slowness surface while solids generally have smaller slowness surfaces. The slowness surface is particularly useful to determine the angles of reflection and refraction of acoustic waves at interfaces. The concept is valid for isotropic and anisotropic media.

Slowness surfaces are shown for the interface between a liquid and an isotropic solid in Figure 7.7. Since the sound velocity is generally lower in the liquid, the slowness surface is larger as shown in the figure. The solid is represented by two smaller concentric circles for the longitudinal and shear branches. The application of the slowness surface to interface problems is based on the principle of conservation of parallel wave vector which was established earlier. Since the slowness surface is drawn in wave vector space it follows that for a given incident wave, the incident reflected and refracted waves have a common k_x component as shown in the figure. Thus the reflection and refraction angles are determined by direct geometrical construction. As θ increases, θ_l and θ_s increase as the corresponding radius vectors swing up to meet the x axis. When the L ray coincides with the x axis, $\theta_l \equiv \theta_{cl}$. This is clearly the largest angle at which one can excite an L wave with a real wave vector in the solid, as for $\theta > \theta_{cl}$ the vertical line no longer intersects the L slowness circle. The same reasoning can be applied to the determination of θ_{cs}. Basically the construction corresponds to a rigorous, visual demonstration of Snell's law and the existence of critical angles. It does not, however, give any information on the transmitted and reflected amplitudes, which must be calculated directly from the boundary conditions.

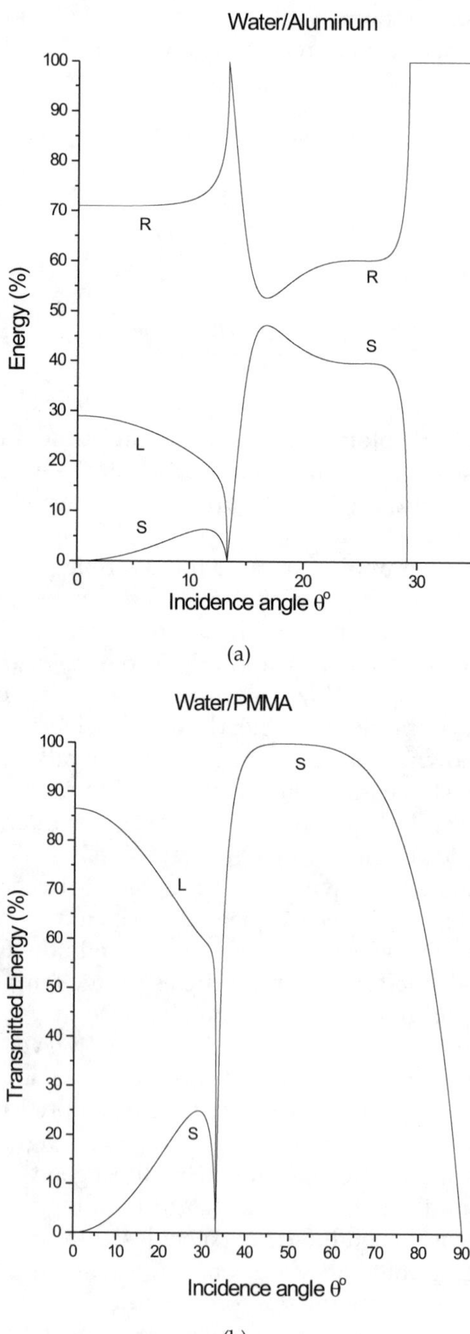

FIGURE 7.6
Energy transmission and reflection coefficients. (a) Water/aluminum. (b) Water/PMMA transmission.

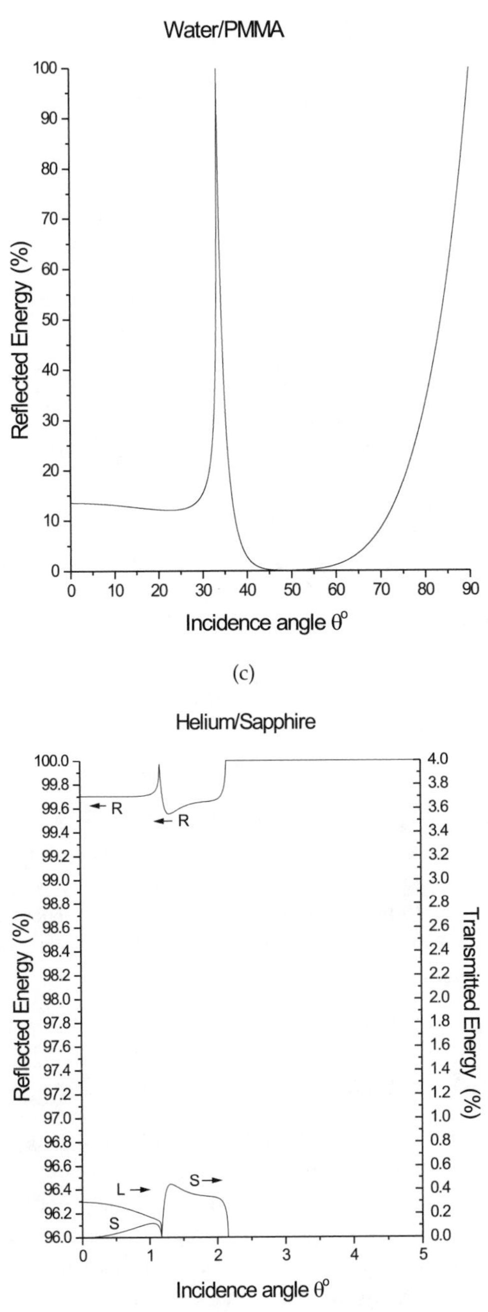

FIGURE 7.6 (Continued)
Energy transmission and reflection coefficients. (c) Water/PMMA (reflection). (d) Liquid helium/sapphire.

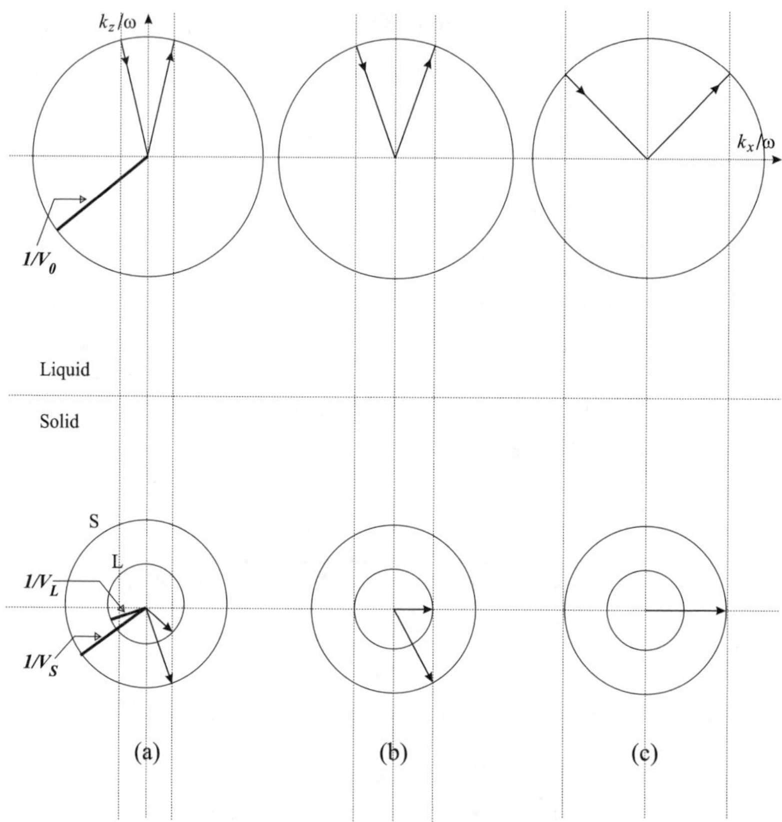

FIGURE 7.7
Slowness curves for the solid-liquid interface for increasing incidence angle. (a) Both land S waves transmitted. (b) L wave critical angle. (c) S wave critical angle.

7.5 Solid-Solid Interface

The previous examples, particularly the liquid-solid interface, demonstrate formally how the velocity potentials and reflection coefficients can be used to obtain the reflection and transmission coefficients. This formal treatment can be extended to the most general case, the solid-solid interface. For a given incident wave, whether longitudinal (P) or bulk shear (SV), there are two reflected and two transmitted waves leading to four unknown amplitudes and bringing the full set of boundary conditions into play. Several authors [29, 32] have formalized this by writing out the full set of boundary conditions for both P and SV incidence and so defining a scattering matrix. The various liquid-solid combinations that are possible can then be selected by setting the appropriate elastic constants equal to zero (e.g., $\mu = 0$ for a liquid) and using the thus simplified scattering matrix to determine the relevant reflection and transmission coefficients. In this section, we rather focus attention on several representative particular cases that are of subsequent interest for acoustic waveguides. These cases are the solid-solid interface for SH modes and the solid-vacuum interface for P and SH waves.

7.5.1 Solid-Solid Interface: SH Modes

The acoustic ray diagram is similar to that for the liquid-liquid interface as there is no coupling between SH modes and P and SV waves, but now the polarization vector for the particle velocity is in the plane of the interface. The appropriate particle velocities are v_y^i, v_y^r, and v_y^t for incident, reflected, and transmitted waves, respectively. These could be defined in terms of velocity potentials as from Equations 7.56 through 7.59, but since we know their form from the solid-fluid example, we write them directly as

$$v_y^i = A\exp j(\omega t - k\sin\theta_i x + k\cos\theta_i z) \tag{7.89}$$

$$v_y^r = B\exp j(\omega t - k\sin\theta_i x - k\cos\theta_i z) \tag{7.90}$$

$$v_y^t = C\exp j(\omega t - k\sin\theta_i x + k\cos\theta_s z) \tag{7.91}$$

The normal and tangential stress can be written, using Equations 7.62 and 7.63, as

$$T_{yx} = \frac{c_{44}}{j\omega}\left(\frac{\partial v_y}{\partial x}\right) \tag{7.92}$$

$$T_{yz} = \frac{c_{44}}{j\omega}\left(\frac{\partial v_y}{\partial z}\right) \tag{7.93}$$

using $\omega = V_S k$ and $V_S^2 = \frac{C_{44}}{\rho}$ we have for the boundary conditions at $z = 0$

$$A + B + C = 0 \tag{7.94}$$

$$-\rho_1 V_{S1}(A\cos\theta_i - B\cos\theta_i) = -\rho_2 V_{S2} C\cos\theta_s \tag{7.95}$$

which can be solved immediately to give

$$R = \frac{B}{A} = \frac{\rho_1 V_{S1}\cos\theta_i - \rho_2 V_{S2}\cos\theta_s}{\rho_1 V_{S1}\cos\theta_i + \rho_2 V_{S2}\cos\theta_s} \tag{7.96}$$

$$T = \frac{C}{A} = \frac{2\rho_1 V_{S1}\cos\theta_i}{\rho_1 V_{S1}\cos\theta_i + \rho_2 V_{S2}\cos\theta_s} \tag{7.97}$$

as reflection and transmission coefficients for the particle velocity.

7.5.2 Reflection at a Free Solid Boundary

These results are needed for the partial wave analysis used for acoustic waveguides. They follow directly from the scattering matrix [29, 32] by setting the several medium constants equal to zero. They can also be worked out directly very easily using the boundary conditions developed above and this is left as an exercise at the end of the chapter.

i. SH mode incident: free boundary

   It follows immediately from the previous treatment with $\rho_2 = 0$ and $V_{S2} = 0$ that $R_{SH} \equiv 1$ with zero phase angle. Thus an SH wave is totally reflected at a free boundary and converted into another SH wave with no mode conversion.

ii. SV mode incident: free boundary

   Using boundary conditions of zero normal and tangential stress at the boundary we obtain

$$R_{LS} = \frac{B_L}{A_S} = \frac{2\frac{V_L}{V_S}\sin 2\theta_s \cos 2\theta_s}{\sin 2\theta_s \sin 2\theta_l + \left(\frac{V_L}{V_S}\right)^2 \cos^2 2\theta_s} \tag{7.98}$$

$$R_{SS} = \frac{C_S}{A_S} = -\frac{\sin 2\theta_s \sin 2\theta_l - \left(\frac{V_L}{V_S}\right)^2 \cos^2 2\theta_s}{\sin 2\theta_s \sin 2\theta_l + \left(\frac{V_L}{V_S}\right)^2 \cos^2 2\theta_s} \tag{7.99}$$

   where A_S, B_L, and C_S are the velocity amplitudes for incident shear, reflected longitudinal, and shear waves, respectively, and

$$\frac{\sin\theta_l}{\sin\theta_s} = \frac{V_L}{V_S} \tag{7.100}$$

iii. P mode incident: free boundary

   In similar fashion for a longitudinal wave incident at a free boundary

$$R_{LL} = \frac{B_L}{A_L} = \frac{\sin 2\theta_s \sin 2\theta_l - \left(\frac{V_L}{V_S}\right)^2 \cos^2 2\theta_s}{\sin 2\theta_s \sin 2\theta_l + \left(\frac{V_L}{V_S}\right)^2 \cos^2 2\theta_s} \tag{7.101}$$

$$R_{SL} = \frac{B_S}{A_L} = \frac{2\left(\frac{V_L}{V_S}\right)\sin 2\theta_l \cos 2\theta_s}{\sin 2\theta_s \sin 2\theta_l + \left(\frac{V_L}{V_S}\right)^2 \cos^2 2\theta_s} \tag{7.102}$$

The following relations can be obtained from Equations 7.98 through 7.102

$$R_{LL} = -R_{SS} \tag{7.103}$$

$$R_{LL}^2 + R_{LS}R_{SL} = 1 \tag{7.104}$$

These results will be used in the analysis of acoustic waveguides.

Summary

Boundary conditions are the key to calculating reflection and transmission coefficients at an interface between two media. The number of boundary conditions is in general equal to the number of unknowns.

Reflection and transmission coefficients at an interface are in general different for displacement, pressure, and intensity.

Standing waves are set up by reflection at normal incidence at a perfectly reflecting interface. Such an interface may be rigid ($Z_2 \gg Z_1$) or pressure release ($Z_2 \ll Z_1$). The displacement has an antinode at a free surface and a node at a rigid surface; the opposite is true for the pressure. The reflection coefficient for the pressure is +1 at a rigid interface and –1 at a free surface; the opposite is true for the displacement.

Quarter wavelength matching layer allows perfect transmission between two media if the thickness is $\lambda/4$ and the acoustic impedance of the layer is the geometric mean of those of the two media.

Critical angles of reflection occur for incidence from low-velocity media to high-velocity media. For a solid-liquid interface, there are critical angles corresponding to transmission of longitudinal, shear, and Rayleigh waves in the solid.

Slowness surface is a surface in $\vec{k}/\omega$ space and the radius vector has modulus $1/V_P$. The slowness surface is a convenient tool for calculating the critical angles for acoustic waves at an interface.

Questions

1. Calculate $\langle I \rangle$ from Equation 7.16 for a standing wave. Sketch the result for p, v, and I.
2. Draw R and T slowness diagrams for transmission from a liquid into a solid with elliptical slowness surfaces for the case where:
   i. The major axis is parallel to the surface.
   ii. The major axis is perpendicular to the surface. Show θ_{cl} and θ_{cs} in each case.

3. State two ways in which one can obtain zero transmitted amplitude for a given mode at a liquid-solid interface.
4. Draw a figure for the boundary of a solid-solid interface for a situation where the boundary conditions for a perfect interface are not respected.
5. Draw displacement curves for standing waves corresponding to the two cases shown in Figure 7.3 for the pressure.
6. Work out in detail $R_L(\theta)$, $R_S(\theta)$, and $T(\theta)$ for a solid-liquid interface with incidence from the solid. Plot the results as a function of θ.
7. Consider a liquid-liquid interface. A source at position A will produce a certain acoustic intensity at point B in the second liquid. Now put the source at B and demonstrate the reciprocity principle, i.e., that A will receive the same acoustic intensity that B received in the first case.
8. Design a quarter wave matching layer to get perfect transmission at a sapphire-water interface at 1 GHz. Use Figure 5.2 to choose a possible material to use for this application.
9. Write down the detailed boundary conditions for each example in Figure 7.1; for each case, indicate which parameters are continuous.
10. Calculate in detail the reflection of P and SV waves at the free boundary of a solid using the notation and approach of Section 7.4.
11. Show that for a slowness curve at an interface if $\theta > \theta_c$ then k in the transmission medium cannot be real, i.e., it cannot lie on the slowness curve and must be imaginary.

8

Rayleigh Waves

8.1 Introduction

Like much of acoustics, surface acoustic waves (SAW) go back to Lord Rayleigh, and because of this, SAW and Rayleigh waves are usually used synonymously. Rayleigh's interest in the problem was brought about by his intuitive feeling that they could be a dominant acoustic signal triggered by earthquakes. His 1885 paper on the subject [33] concluded with the well-known remark "... It is not improbable that the surface waves here investigated play an important part in earthquakes, and in the collision of elastic solids. Diverging in two dimensions only, they must acquire at a great distance from the source a continually increasing preponderance." This was indeed found to be the case and Rayleigh's pioneering work stimulated a great deal of further study of other acoustic modes that could propagate in the layered structure of the earth's crust.

Rayleigh waves are now standard fare not only in seismology but also in many areas of modern technology. With the introduction of interdigital transducers (IDTs) in the 1960s, they have, as it were, been integrated into modern microelectronics in the form of filters, delay lines, and many other acousto-electronic functions. They are ubiquitous in all of the applications of ultrasonics described in this book and so it is incumbent upon us to have a good understanding of their propagation characteristics.

Rayleigh waves are the simplest cases of guided waves that we will examine. They are confined to within a wavelength or so of the surface along which they propagate. They are distinct from longitudinal and shear BAW modes, which propagate independently at different velocities. In Rayleigh waves, the longitudinal and shear motions are intimately coupled together and they travel at a common velocity. In this chapter we start with a detailed description of these waves on the surface of an isotropic solid in vacuum. In Section 8.3, the problem is generalized by placing the solid in contact with an ambient liquid. We find in this case the propagation of a perturbed Rayleigh wave, which radiates into the liquid (leaky wave). In addition there is an undamped, true interface wave at the solid-liquid interface, the Stoneley wave.

8.2 Rayleigh Wave Propagation

Consider a wave polarized in the sagittal (xz) plane with surface normal along $-\vec{z}$ and propagation in the x direction as in Figure 8.1. Hence, displacement and velocity components are in the x and z directions; there is no coupling to the transverse waves with displacement along y (SH mode), perpendicular to the sagittal plane.

As with bulk waves we define a scalar and vector potential such that

$$\vec{u} = \overrightarrow{\nabla\phi} + \vec{\nabla} \times \vec{\psi}$$

and since the displacement is in the sagittal plane the only nonzero component of $\vec{\psi}$ is in the y direction. As for bulk waves, ϕ and $\vec{\psi}$ are potentials for the longitudinal and transverse wave components, respectively, and the corresponding wave equations are given by

$$\frac{\partial^2 \phi}{\partial x^2} + \frac{\partial^2 \phi}{\partial z^2} + k_L^2 \phi = 0 \tag{8.1}$$

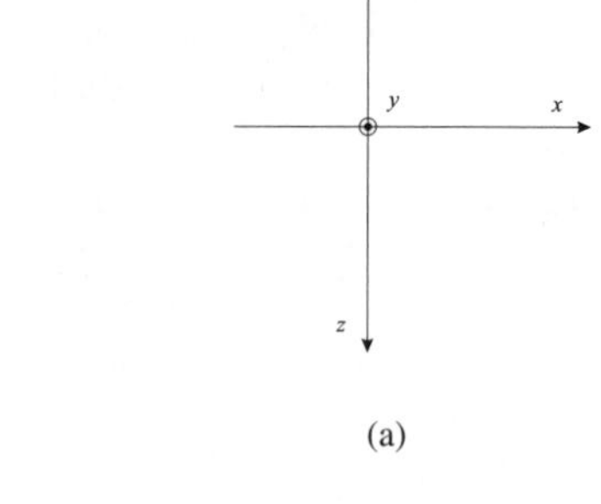

(a)

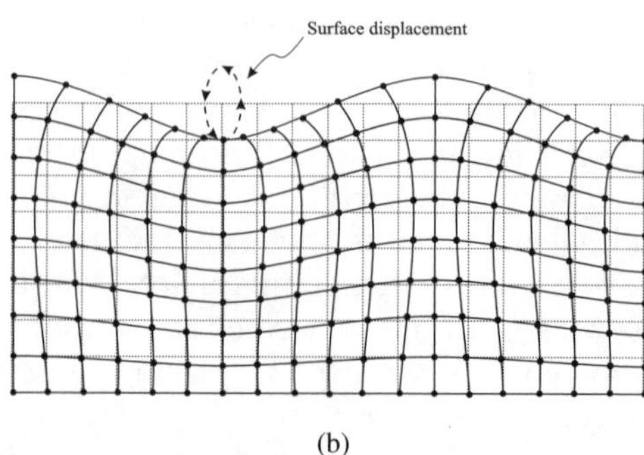

(b)

FIGURE 8.1
(a) Coordinate system for Rayleigh wave propagation. (b) Grid diagram for near-surface mechanical displacement due to Rayleigh waves.

$$\frac{\partial^2 \psi}{\partial x^2} + \frac{\partial^2 \psi}{\partial z^2} + k_S^2 \psi = 0 \tag{8.2}$$

where k_L and k_S are the usual bulk wave numbers

$$k_L = \sqrt{\frac{\rho}{\lambda + 2u}} \qquad \text{and} \qquad k_S = \sqrt{\frac{u}{\rho}}$$

Anticipating that the solutions for the surface wave equations for the two polarizations will have a common wave number, we look for solutions for ϕ and ψ propagating as harmonic waves along the x axis with wave number $\beta = k_x$ and variations in the z direction to be determined by the boundary conditions. This leads to trial solutions of the form

$$\phi = F(z)\exp j(\omega t - \beta x) \tag{8.3}$$

$$\psi = G(z)\exp j(\omega t - \beta x) \tag{8.4}$$

which give two new equations for $F(z)$ and $G(z)$ following substitution into Equations (8.1) and (8.2)

$$\frac{d^2 F}{dz^2} - (\beta^2 - k_L)^2 F = 0 \tag{8.5}$$

$$\frac{d^2 G}{dz^2} - (\beta^2 - k_S)^2 G = 0 \tag{8.6}$$

The slowness curve treatment and the known bulk wave solutions lead us to pose

$$k_L^2 < k_S^2 < \beta^2 \tag{8.7}$$

which will be confirmed *a posteriori*. Both equations have solutions of the form $\exp\pm\sqrt{\beta^2 - k_L^2}z$ and $\exp\pm\sqrt{\beta^2 - k_S^2}z$. The positive solutions are unphysical as they grow indefinitely with increasing z. We retain the negative solutions and write them in the form

$$\phi = A\exp(-\gamma_L z)\exp j(\omega t - \beta x) \tag{8.8}$$

$$\psi = B\exp(-\gamma_S z)\exp j(\omega t - \beta x) \tag{8.9}$$

where

$$\gamma_L^2 = \beta^2 - k_L^2 \tag{8.10}$$

$$\gamma_S^2 = \beta^2 - k_S^2 \tag{8.11}$$

A and B are arbitrary constants.

Unlike the problems for reflection and transmission, we are not looking for solutions for the unknown amplitudes (indeed these are arbitrary) but rather we are looking first and foremost to determining the propagation constant β and, hence, the surface wave velocity, followed by the variation of the displacements with z that are given by γ_L and γ_S.

Since we are dealing with the free surface of a semi-infinite solid the boundary conditions are particularly simple; tangential and normal stresses are zero on the surface at $z = 0$ and the displacements are undetermined. The general form of the displacements and the stress components are

$$u_x = \frac{\partial \phi}{\partial x} - \frac{\partial \psi}{\partial z} \tag{8.12}$$

$$u_z = \frac{\partial \phi}{\partial z} + \frac{\partial \psi}{\partial x} \tag{8.13}$$

$$T_{zz} = \lambda\left(\frac{\partial^2 \varphi}{\partial x^2} + \frac{\partial^2 \varphi}{\partial z^2}\right) + 2\mu\left(\frac{\partial^2 \varphi}{\partial x^2} - \frac{\partial^2 \psi}{\partial x \partial z}\right) \tag{8.14}$$

$$T_{xz} = \mu\left(\frac{\partial^2 \phi}{\partial x^2} + 2\frac{\partial^2 \varphi}{\partial x \partial z} - \frac{\partial^2 \psi}{\partial x^2}\right) \tag{8.15}$$

Putting $T_{xz} = 0$ at $z = 0$ and using the expressions for ϕ and ψ, we immediately obtain

$$\phi = A \exp j(\omega t - \beta x - \gamma_L z) \tag{8.16}$$

$$\psi = -jA \exp j(\omega t - \beta x - \gamma_S z) \tag{8.17}$$

From the characteristic equation (determinant of the coefficients equal zero) obtained from $T_{xz} = 0$ and $T_{zz} = 0$, we immediately obtain an equation for β

$$4\beta^2 \gamma_L \gamma_S - (\beta^2 + \gamma_S^2)^2 = 0 \tag{8.18}$$

This is conventionally written as a sextet equation with the definitions

$$\eta \equiv \frac{k_S}{\beta} = \frac{V}{V_S} \tag{8.19}$$

$$\xi = \frac{k_L}{k_S} = \frac{V_S}{V_L} \tag{8.20}$$

so that Equation 8.18 reduces to the Rayleigh equation

$$\eta^6 - 8\eta^4 + 8(3 - 2\xi^2)\eta^2 - 16(1 - \xi^2) = 0 \tag{8.21}$$

This equation has one real root, η_R, corresponding to the existence of a Rayleigh surface wave with the properties given by the two potential functions. Through ξ, η_R depends on Poisson's ratio σ. An approximate solution is

$$\eta_R = \frac{0.87 + 1.12\sigma}{1 + \sigma} \tag{8.22}$$

Over the allowed range of σ ($0 < \sigma < 0.5$), the Rayleigh velocity V_R thus varies from $0.87V_S$ to $0.96V_S$. This variation is shown in Figure 8.2 as a function of σ and V_S/V_L. Typical values of V_R for common materials are given in Table 8.1.

The solutions for the displacements can be obtained, knowing β and hence γ_L and γ_S, from Equations 8.12 and 8.13. The real parts of $u_x(z)$ and $u_z(z)$ are:

$$u_{xR} = A\beta_R\left(e^{-\gamma_{LR}z} - \frac{2\gamma_{LR}\gamma_{SR}}{\beta_R^2 + \gamma_{SR}^2}e^{-\gamma_{SR}z}\right)\sin(\omega t - \beta_R x) \tag{8.23}$$

$$u_{zR} = A\gamma_{LR}\left(e^{-\gamma_{LR}z} - \frac{2\beta_R^2}{\beta_R^2 + \gamma_{SR}^2}e^{-\gamma_{SR}z}\right)\cos(\omega t - \beta_R x) \tag{8.24}$$

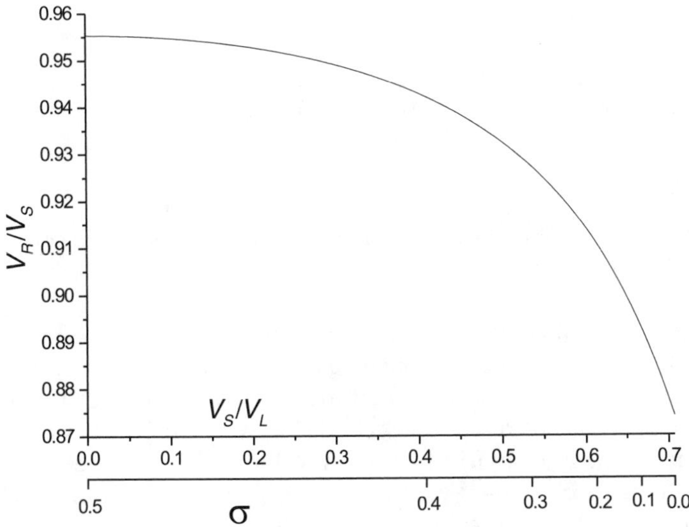

FIGURE 8.2
V_R/V_S for isotropic bodies as a function of V_S/V_L and σ, using the approximate Equation 8.22.

TABLE 8.1

Acoustic Surface Wave Parameters for Representative Piezoelectric Substrates

Material	Orientation	V_R ($m \cdot s^{-1}$)	k^2 (measured[a])	VAC (dB/μs)	AIR (dB/μs)	$\frac{1}{V_R}\frac{\partial V_R}{\partial T}$ (ppm/°C)
$LiNbO_3$	*Y*, *Z*	3488	0.045	0.88	0.19	−87
$Bi_{12}GeO_{20}$	001, 110	1681	0.015	1.45	0.19	
$LiTaO_3$	*Z*, *Y*	3329	0.0093	0.77	0.23	−52
Quartz	*Y*, *X*	3159	0.0023	2.15	0.45	38
	ST, *X*	3158	0.0016	2.62	0.47	14

Note: The total loss is given by α(dB/μs) = VAC F^2 + AIR F, where F is in GHz.

[a] Scholz, M.B., and Matsinger, J.H., *Appl. Phys. Lett.*, 20, 367, 1972.

Source: Selected data from Slobodnik, A.Z., Materials and their influence on performances, in *Acoustic Surface Waves*, Oliner, A.A., Ed., Springer-Verlag, Berlin, 1978, 300.

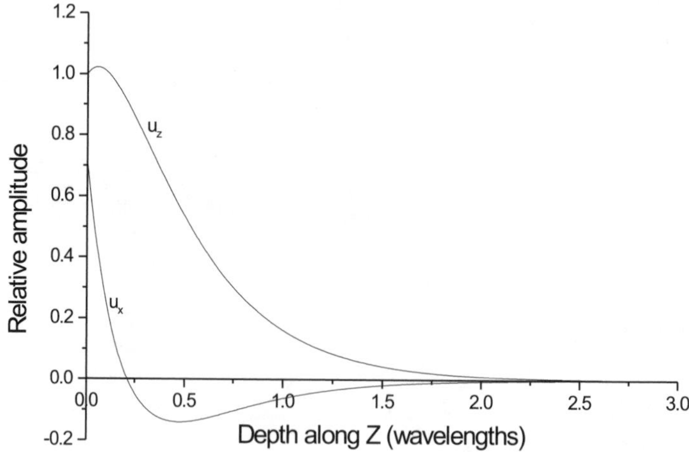

FIGURE 8.3
Relative Rayleigh wave displacements as a function of depth for fused quartz calculated from Equations 8.23 and 8.24.

The decay with depth of these solutions is shown in Figure 8.3. Several general points emerge. First, both components have a decay constant of the order of a Rayleigh wavelength, meaning that the surface disturbance is confined in a layer of thickness of order λ_R. Second, the two components are in phase quadrature so that the polarization locus is elliptical. In fact, detailed analysis shows that the displacement vector rotation is retrograde (counterclockwise) at the surface and progressive (clockwise) lower down. It should be appreciated that the actual displacements even at the surface are tiny. According to Ristic [31]: "in a device operating at 100 MHz with 10 mW average power in a beam 1 cm wide on a substrate with SAW velocity $V_R = 3$ km·s^{-1}, the wavelength is 30 μm with the peak vertical displacement on the order of 10^{-10}m."

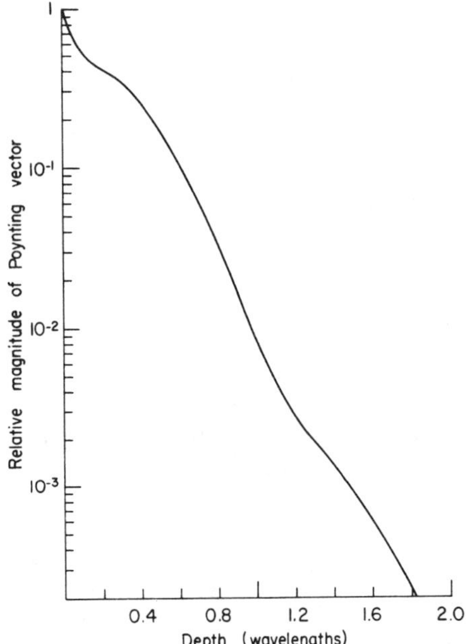

FIGURE 8.4
Relative magnitude of the Rayleigh wave Poynting vector as a function of depth for propagation along the z axis on the YZ plane of quartz. (From Farnell, G.W., Properties of elastic surface waves, in *Physical Acoustics*, IX, Mason, W.P. and Thurston, R.N., Eds., Academic Press, New York, 1972, chap. 3. With permission.)

The decrease with depth is also extremely rapid. For propagation along the z axis on the YZ plane of quartz, Farnell [34] has calculated that the magnitude of the acoustic surface wave Poynting vector decreases by four orders of magnitude in a distance of $1.8\lambda_R$, as shown in Figure 8.4.

8.3 Fluid-Loaded Surface

Waves similar to Rayleigh waves on a free surface can propagate on the surface of a fluid-loaded solid. Clearly, as the acoustic impedance of the liquid goes to zero, such waves will transform in a continuous fashion to Rayleigh waves, i.e., the fluid will act as a perturbation on the free surface wave. In fact we do not need to make this assumption, and the presence of any liquid can be taken into account by the modified boundary conditions.

Including the continuity of normal stress into the free surface boundary conditions immediately leads to a new characteristic equation [35]:

$$4\beta^2\gamma_L\gamma_S - (\beta^2 + \gamma_S)^2 = j\frac{\rho_1}{\rho_2}\frac{\gamma_L k_S^4}{\sqrt{k_L^2 - \beta^2}} \tag{8.25}$$

This equation has one real root and one complex root. The real root corresponds to a true, undamped interface wave (Stoneley wave) and will be treated in a separate section. The complex root, corresponding to a modification of the Rayleigh wave, will be treated here. For simplicity we assume that the velocity of this wave, V_R', satisfies $V_R' \geq V_R$, which will be demonstrated shortly. These surface waves modified by the presence of the fluid will be called generalized Rayleigh waves or, more commonly, leaky Rayleigh waves.

Since the velocity of the generalized Rayleigh wave is complex, it is attenuated. As the media have been assumed to be lossless, the surface wave can only be attenuated by radiating energy into the liquid. By reciprocity, a wave incident from the liquid will also generate such a wave on the surface. Generation and radiation can be simply described by a phase matching condition. As seen for reflection and transmission at the liquid-solid interface the incident wave vector component along the surface is $\beta_x = \beta \sin\theta = \omega / V_x$. For a generalized Rayleigh wave on the surface, $\beta_R' = \omega/V_R'$. As the incidence angle increases from zero, β_x increases until finally $\beta_x = \beta_R'$ at $V_x = V_R'$ at an angle θ_R such that $V_R' = V_0/\sin\theta_R$. Thus the phase velocity of the incident beam projected onto the surface for incidence at θ_R "phase matches" the velocity of the generalized Rayleigh wave, so the incident beam will amplify the latter (or generate it in the absence of an initial surface wave). This is in fact a resonance phenomenon, and the incident wave creates an extremely sharp and narrow surface wave maximum at $\theta = \theta_R$. By the same token, the Rayleigh wave radiates or "leaks" into the fluid medium at angle θ_R. In so doing, it loses acoustic energy and is attenuated, leading to the complex root for the velocity. It is for this reason that such waves are called leaky Rayleigh waves.

The phase velocity V_R' of leaky Rayleigh waves has been calculated numerically and tabulated by Viktorov [35] for different values of Poisson's ratio and density ratio. The effect is typically very small; for example, for an average interface the parameters plotted by Viktorov are $r = V_S/V_0 = 5$ and $\rho_1/\rho_2 = 0.5$, leading to $V_R'/V_R \approx 1.001$. For other values of these ratios, the value of V_R' increases monotonically. It should be noted that the numerical results by Viktorov are exact and do not make the assumption that the liquid density is very much less than that of the solid.

The attenuation factor for the leaky Rayleigh wave has also been tabulated by Viktorov. In contrast to the velocity this effect is very important, as can be verified by placing a drop of water on a SAW delay line. Even at the lowest attainable frequencies the signal disappears instantaneously. A simple estimate of the effect which clearly brings out the physics was given by Dransfeld and Saltzmann [36]. It was demonstrated earlier in the chapter that the SAW has normal and tangential components of displacement. The normal component launches compressional waves into the liquid and the efficiency of this mechanism is mediated by acoustic mismatch between the solid and the liquid. The tangential component is coupled to the fluid by viscosity and is generally much weaker. The compressional component of energy transfer can be calculated by reference to Figure 8.5 for a surface element of thickness λ and width b. Designating the normal component of the particle displacement amplitude by a

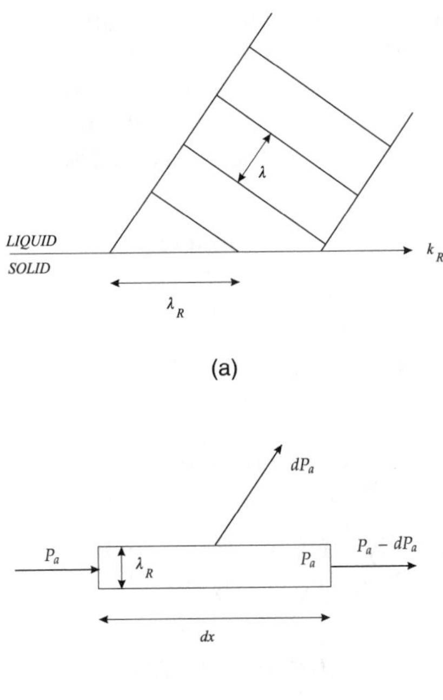

FIGURE 8.5
(a) Radiation of a Rayleigh wave from a surface element into an adjacent fluid with acoustic wavelength λ. (b) Energy balance for a surface element during time *dt* due to radiation or leaking of the Rayleigh wave into an adjacent fluid.

we have for the energy transport per second through the element [36]

$$P_a = \lambda b 2\pi^2 \rho_2 v_R^3 \left(\frac{a}{\lambda}\right)^2 \tag{8.26}$$

Since there is continuity of normal displacement at the interface, the energy emitted per second by the surface element *bdx* into the fluid is

$$dP_a = 2\pi^2 (bdx) \rho_1 V_0^3 \left(\frac{a}{\lambda_1}\right)^2 \tag{8.27}$$

so that finally the energy attenuation coefficient for the leaky Rayleigh wave is

$$\alpha_R = \frac{1}{P_a}\frac{dP_a}{dx} = \frac{\rho_1 V_0}{\rho_2 V_R \lambda} \mathrm{cm}^{-1} \tag{8.28}$$

Thus the attenuation per wavelength of the leaky wave is given by the ratio of the acoustic impedances. Viktorov gives the value $\alpha_R = 0.11$ for a typical case, so that the wave is attenuated to $1/e$ of its initial value over the distance of about ten wavelengths. This is the reason why, for nearly all practical purposes, SAW devices cannot be used in liquids.

The viscous component can also be calculated from Figure 8.5. If the width of the element shown is b, the viscous force on the element is

$$F = \eta(bdx)\frac{v}{\delta} \tag{8.29}$$

where v_0 = the particle velocity at the solid-fluid interface = ωa, where a is the particle displacement in the x direction

$$\delta = \text{viscous penetration depth} = \left(\frac{2\eta}{\rho_1\omega}\right)^{\frac{1}{2}} \tag{8.30}$$

so that $\frac{v}{\delta}$ is approximately the velocity gradient in the fluid.

The energy dissipated per second by the viscous forces is

$$dP_a' = \frac{1}{2}\left(\frac{\eta}{l}\right)v_0^2bdx \tag{8.31}$$

and using $v_0 = \omega a$, the energy flow in the Rayleigh wave is

$$P_a' = \frac{1}{2}b\rho_2V_Rv_0^2\lambda_R \tag{8.32}$$

The viscous attenuation is

$$\alpha_S = \frac{dP_a'}{P_a'dx} = \frac{\left(\rho_1\eta\frac{\omega^2}{2}\right)^{\frac{1}{2}}}{4\pi^2\rho_2V_R^2} \tag{8.33}$$

This viscous attenuation is typically a hundred times smaller than the compressional term given by Equation 8.28.

Rayleigh waves can be attenuated by many things other than ambient media: point defects, roughness, grain boundaries, electrons, phonons, and all of the defects and excitations that can attenuate bulk waves. These phenomena can best be studied per se by generating and detecting Rayleigh waves on a solid-vacuum interface. However, they do also come into play in the present context of a solid-fluid interface. On the theoretical side, we consider the reflectivity $R(\theta)$ of an infinite plane wave in the fluid incident on a perfect interface formed by a nonattenuating solid. The result is the typical theoretical $R(\theta)$ curve presented in Chapter 7 where there is total reflection for $\theta > \theta_{cs}$ where $|R(\theta)| \equiv 1$. Experimentally, spatially bounded beams must be used and these give rise to special effects discussed in the next

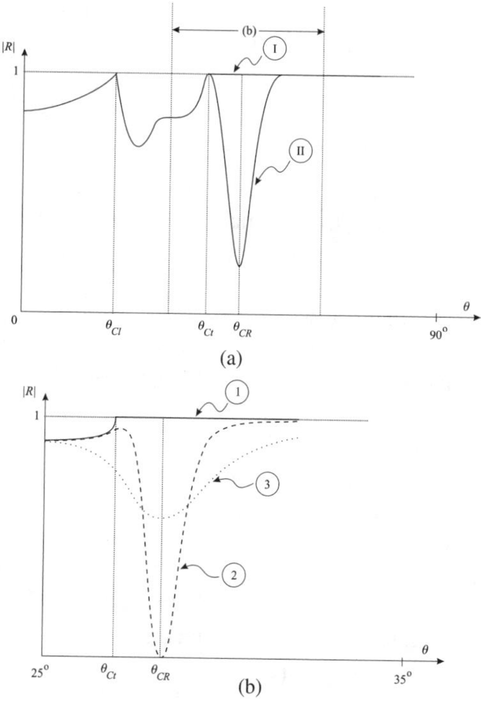

FIGURE 8.6
(a) Schematic diagram of the modulus of the reflection coefficient at a liquid-solid interface as a function of angle: I, Perfect, nonattenuating solid; II, a finite value of attenuation in the solid gives rise to the Rayleigh dip. (b) Blowup of (a) around the Rayleigh angle: 1. Zero attenuation, 2. Small but critical value of attenuation in the solid, 3. High attenuation. Increasing attenuation progressively washes out the Rayleigh dip.

section. However, making allowances for these, one still observes in reflectivity experiments on typical samples a pronounced dip at the Rayleigh angle, instead of total reflection, as shown in Figure 8.6. We call this effect the Rayleigh dip.

The existence of the Rayleigh dip can be explained in terms of attenuation of the surface wave. If there is no attenuation, the incident wave generates a Rayleigh wave, which is then re-emitted, effectively leading to total reflection. This is exactly the situation found in optics in total reflection in a prism; the evanescent wave associated with the critical angle exists, but if energy is not removed from it by dissipation, the energy in the evanescent wave is simply stored and is not propagated. Returning to the Rayleigh wave, if now an attenuation mechanism is introduced, part of the energy associated with the Rayleigh wave is absorbed. This reduces the amplitude of the re-emitted wave, leading to formation of the Rayleigh dip. In the optics analogy this corresponds to placing the face of a second prism near to the face where total reflection occurs, which taps energy stored in the evanescent wave, which in turn decreases the reflection coefficient from unity. The Rayleigh dip will be treated more fully in Section 15.2.3 on critical angle reflectivity.

8.3.1 Beam Displacement

The displacement of bounded acoustic beams at the critical angle has its counterpart in optics, which in turn has a long and venerable history going back to Newton. Newton carried out experiments with a silver plate put into contact with a glass surface at the condition of total reflection. His results were inconclusive and the question was only settled definitely in the experiments of Goos and Hänchen [37], who clearly demonstrated a lateral displacement of an optical beam that had undergone total reflection. An extensive review of the subject has been given by Lotsch [38]. Shortly after that, Schoch [39] did a complete experimental and theoretical study of the acoustic counterpart for reflectivity of a bounded ultrasonic beam at the Rayleigh angle, now called the Schoch displacement. However, Schoch's theory, modified by Brekhovskikh, lacks a physical basis and is valid only for the wide beam limit.

The first step toward a transparent physical model was made by Mott [40], followed by a complete experimental study by Neubauer [41]. The latter used Schlieren imaging to image the beam displacement and hydrophones to probe the spatial variation of the frequency dependence of the reflectivity. A Schlieren photograph by Breazeale et al. [42], Figure 8.7, shows the essential features found by Neubauer, who proposed a simple model to explain the observed structure. The standard reflectivity theory presented in Chapter 7 predicts a specularly reflected beam with a π phase reversal with respect to the incident beam which is seen on the left side of the reflected beam in Figure 8.7. In addition, at $\theta = \theta_R$ there is a Rayleigh wave in phase with the incident beam.

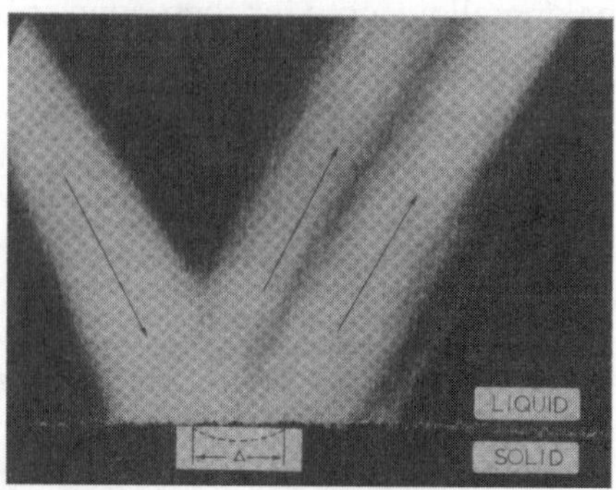

FIGURE 8.7
Schlieren photograph of an ultrasonic beam incident from the liquid at a water/aluminum interface. The specularly reflected and displaced components are clearly visible. (From Breazeale, M.A., Adler, L., and Scott, G.W., *J. Acoust. Soc. Am.*, 48, 530, 1977. With permission.)

It propagates along the surface as a leaky Rayleigh wave, radiating acoustic energy into the fluid. Initially, the specularly reflected component and the leaky Rayleigh wave are out of phase, leading to the null observed in a portion of the specular region in the left center of the Schlieren image. After that the leaky Rayleigh wave radiates into the fluid, its intensity falling off with propagation distance, as expected.

Further refinements to the model are brought into the picture using the attenuative model of Becker and Richardson [43]. At sufficiently high frequencies, that theory predicts equality of phase for the specularly reflected and leaky Rayleigh wave radiation, leading to the disappearance of the null zone at sufficiently high frequencies. This effect was also observed by Neubauer.

A rigorous theory for the beam displacement was put in place by Bertoni and Tamir [44]. A summary of the relevant parts of their work is given below; serious readers should consult the original reference. Bertoni and Tamir give a plane wave representation of the incident field particle velocity $v_{\text{inc}}(x, z)$ by the Fourier transform pair

$$v_{\text{inc}}(x,z) = \frac{1}{2\pi}\int_{-\infty}^{\infty} V(k_x)\exp[i(k_x x + k_z z)]dk_x \tag{8.34}$$

$$V(k_x) = \int_{-\infty}^{\infty} v_{\text{inc}}(x,0)\exp(-ik_x x)dx \tag{8.35}$$

where the symbol and axes have their usual meanings. The incident beam width is $2w$ so that the width projected on the surface is $2w_0$ where $w_0 = w\sec\theta_i$. Hence, the integral in Equation 8.34 is over roughly an effective width $2w_0$ and the integral over k_x in Equation 8.35 is over an interval $2\pi/w_0$, which defines the range of angles for the plane waves of amplitude $V(k_x)$. With conservation of parallel momentum $k_x = k\sin\theta_i = k_l\sin\theta_l = k_t\sin\theta_t$ as usual. In wave number space, the full reflection coefficient can be written as

$$R(k_x) = \frac{(2k_x^2 - k_s^2)^2 - 4k_x^2[(k_x^2 - k_d^2)(k_x^2 - k_s^2)]^{\frac{1}{2}} - \dfrac{ik_s^4}{Q}\left[\dfrac{(k_x^2 - k_d^2)}{(k^2 - k_x^2)}\right]^{\frac{1}{2}}}{(2k_x^2 - k_s^2)^2 - 4k_x^2[(k_x^2 - k_d^2)(k_x^2 - k_s^2)]^{\frac{1}{2}} + \dfrac{ik_s^4}{Q}\left[\dfrac{(k_x^2 - k_d^2)}{(k^2 - k_x^2)}\right]^{\frac{1}{2}}} \tag{8.36}$$

where $\rho = \rho_1/\rho_0$, and the reflected particle velocity is

$$v_{\text{refl}}(x,z) = \frac{1}{2\pi}\int_{-\infty}^{\infty} R(k_x)V(k_x)\exp[i(k_x x - k_z z)]\,dk_x \tag{8.37}$$

over the same range of wave numbers as for the incident wave. We are interested in the range that includes the Rayleigh wave number $k_R = k\sin\theta_R$.

Since k_x is in general complex, $R(k_x)$ should be considered in the complex plane, where it will exhibit poles (denominator zero) and zeros (numerator zero).

In the absence of liquid, the free surface resonant solutions for k_R can be found as zeros ($k_x = \pm k_P$) of the denominator for $\rho \to \infty$. For the free surface, k_R is real and $k_R = k\sin\theta_R$. This way of finding the solutions for k_R will be re-examined from another angle in Chapter 10 on acoustic waveguides.

If a liquid is now present, such that it is a small perturbation, then the pole of $R(k_x)$ moves from $k_x = k_R$ to $k_x = k_P$, and k_P is now complex, as can be deduced from Equation 8.36. As in the previous section we can write the solution $k_P = \beta + j\alpha = k\sin\theta_P + j\alpha$ where $\beta \approx k_R$ and α is the attenuation due to liquid loading. Hence, k_P is the wave number for a leaky wave. Taking explicit account of the poles k_P and zeros k_0 near the Rayleigh condition, the reflection coefficient for the leaky wave can be written

$$R(k_x) = \frac{k_x - k_0}{k_x - k_P} \tag{8.38}$$

For the lossless case, we already know from Section 8.3 that $|R| \equiv 1$ for $\theta > \theta_{cs}$ and the phase is π at $\theta = \theta_P$. This condition is satisfied if $k_0 \equiv k_P^*$ where $*$ is the complex conjugate.

For small losses in the solid $k_0 \neq k_P^*$ and k_0 can be calculated from Equation 8.38. In fact, $|R|$ becomes a minimum for some value of $k_x \approx k_R$, which corresponds to the frequency of minimum reflection observed experimentally.

Bertoni and Tamir carry out a calculation with a Gaussian beam to make contact with Neubauer's experimental results. Using the previous notation, the incident particle velocity can be written at the plane $z = 0$

$$v_{\text{inc}}(x,0) = \frac{\exp\left[-\left(\frac{x}{w_0}\right)^2 + ik_i x\right]}{\sqrt{\pi}\, w_0 \cos\theta_i} \tag{8.39}$$

with associated Fourier component

$$V(k_x) = \frac{\exp\left[-(k_x - k_i)^2\left(\frac{w_0}{2}\right)^2\right]}{\cos\theta_i} \tag{8.40}$$

which may be used in Equation 8.37 to find the reflected field if $R(k_x)$ is known. The key step taken by Bertoni and Tamir is to divide $R(k_x)$ in the region around the Rayleigh angle into two parts

$$R(k_x) = R_0 + R_1(k_x) \tag{8.41}$$

where

$$R_0 = R(k_i) = \frac{k_i - k_0}{k_i - k_p} \tag{8.42}$$

is the reflection coefficient for the specularly reflected (geometrical acoustics) component and

$$R_i(k_x) = \frac{k_p - k_0}{k_p - k_i} \cdot \frac{k_x - k_i}{k_x - k_p} \tag{8.43}$$

is the reflection coefficient associated with diffraction effects in re-radiation from the leaky Rayleigh wave.

Combining Equations 8.37 and 8.40, Bertoni and Tamir obtain

$$\begin{aligned} v_0(x,0) &= R_0 \frac{\exp\left[-\left(\frac{x}{w_0}\right)^2 + ik_i x\right]}{\sqrt{\pi} w_0 \cos\theta_i} \\ &= R_0 v_{\text{inc}}(x,0) \end{aligned} \tag{8.44}$$

and

$$v_1(x,0) = v_{\text{inc}}(x,0)\frac{k_p - k_0}{k_p - k_i}\left[1 + \frac{i\sqrt{\pi}\,w_0}{2}(k_p - k_i \exp(\gamma^2)\text{erfc}(\gamma))\right] \tag{8.45}$$

where erfc(γ) is the complementary error function. Comparing Equation 8.44 with Equation 8.39, we see that v_0 gives exactly the specularly reflected component. The form of $v_1(x,0)$ indicates that it is nonsymmetric, i.e., it is no longer Gaussian. The magnitude of $v_1(x,0)$ is only large near the phase matching condition $k_i = \beta$. Outside the illuminated region, i.e., $x \gg \omega_0$, Bertoni and Tamir show that

$$v_1(x,0) \approx -\frac{4}{\Delta_s} \sec\theta_i \exp\left[\left(\frac{w_0}{\Delta_s}\right)^2\right] e^{i(\beta + i\alpha)x} \tag{8.46}$$

which, from the exponential phase factor, is exactly of the form of a leaky Rayleigh wave with Schoch displacement Δ_s.

The situation is best summarized by the display of the two solutions in [44], together with their sum giving the totally reflected field. Cancellation of the specularly reflected peak and the Rayleigh peak are seen to give rise to the null, and on the right the trailing edge is clearly due to the leaky Rayleigh wave, both results as proposed by Neubauer. The origin of the displacement Δ is likewise shown in [44]. Further quantitative considerations confirm all of the other results reported by Neubauer.

8.3.2 Lateral Waves: Summary of Leaky Rayleigh Waves

A summary of the various interface waves associated with leaky Rayleigh waves has been given by Uberall [45]. The pure Rayleigh wave in contact with a vacuum has a velocity parallel to the surface (a). For the leaky wave in the limit $\rho_0/\rho_1 \ll 1$, most of the acoustic energy is in the solid but the velocity vector v'_R is now tilted toward the liquid due to leakage in that direction.

If incidence from the liquid occurs for angles of incidence other than the Rayleigh angle, then it is found that other waves exist near the interface called lateral waves. These come about directly from the theory of the reflection coefficient at the solid-liquid interface, and they are generated at the critical angles θ_{cl} and θ_{cs}. The lateral waves are effectively bulk waves that travel parallel to the surface, often called surface-skimming bulk waves. Like leaky Rayleigh waves, lateral waves also radiate into the liquid at the appropriate angle (θ_{cl}, θ_{cs}, and θ_{cR} for longitudinal and transverse lateral waves and leaky Rayleigh waves, respectively) and in the liquid these radiated waves are known as head waves. A head wave has a conical wavefront and is commonly known as a Schmidt head wave after its discoverer. All of these waves were imaged simultaneously in a classic experiment carried out by von Schmidt [46]. An electric spark in water near an aluminum surface acted as a point source, so that a whole spectrum of incident angles was emitted. Thus, L and S lateral waves and leaky Rayleigh waves were excited and propagated along the interface. These waves in turn excited conical head waves. Using Schlieren imaging, von Schmidt was able to image all of these wave fields at the same time.

8.3.3 Stoneley Waves at a Liquid-Solid Interface

Very generally, Stoneley waves are pure interface waves at the boundary between two elastic media. As will be seen later, for two solids they exist only for certain ranges of density and sound velocity ratios. However, Ewing et al. [47] have shown that they exist in all cases for the liquid-solid interface. They are pure interface waves in that they propagate without attenuation (hence, the velocity is real), and their amplitude decays exponentially on both sides of the interface. For $\rho_0/\rho_1 \ll 1$, the energy is mainly in the liquid and it decays very slowly with distance in that medium. The velocity is less than, but of the order of, the sound velocity in the liquid. On the solid side, the wave only penetrates a distance of the order of a wavelength.

It was mentioned that the characteristic equation for the general interface wave had one complex root and one real root, and it was shown in the previous section that the complex root corresponds to the leaky Rayleigh wave. The real root corresponds to the Stoneley wave, which, as stated above, propagates without attenuation in lossless media. Brekhovskikh [30] has shown that for $\rho_0/\rho_1 \ll 1$ and $V_S/V_0 \gg 1$ this root is given by

$$V_{ST} = V_0\left[1 - \frac{1}{8}\left(\frac{\rho_0 V_0^2}{\rho_1(V_S^2 - V_L^2)}\right)^2\right] \tag{8.47}$$

where subscript 0 is for the liquid and subscript 1 is for the solid and the amplitude decay into the liquid is given by

$$\exp\left[-\frac{\pi}{\lambda}\frac{\rho_0 V_0^2}{\rho_1(V_L^2 - V_S^2)}|z|\right] \tag{8.48}$$

so that the decay length is

$$\frac{\lambda}{\pi}\frac{\rho_1(V_L^2 - V_S^2)}{\rho_0 V_0^2} \tag{8.49}$$

which is very much larger than the wavelength under the stated conditions. Finally, from a practical point of view, the Stoneley wave can only be excited at glancing incidence.

Summary

Rayleigh waves are surface acoustic waves in which longitudinal and shear displacements are coupled together and travel at the same velocity. The displacements are restrained to between one and two Rayleigh wavelengths of the surface.

Rayleigh wave velocity is between 0.87 and 0.95 of the substrate transverse wave velocity.

Leaky Rayleigh waves occur for Rayleigh wave propagation at a solid-liquid interface. Acoustic energy is radiated into the liquid at the Rayleigh angle. An incoming wave from the liquid at the Rayleigh angle will likewise excite a leaky Rayleigh wave in the solid. Leaky Rayleigh waves are attenuated due to transmission of the component normal to the surface into the liquid.

Rayleigh dip is the reduction in the reflection coefficient for a solid-liquid interface at the Rayleigh angle due to attenuation of the Rayleigh wave at the solid surface.

Schoch displacement of the reflected wave due to incidence of a bounded beam from the liquid at the Rayleigh angle. The effect is analogous to the Goos-Haenchen effect in optics. It is due to phase cancellation between the directly reflected wave and the leaky Rayleigh wave.

Lateral waves are bulk waves excited near critical angles, surface-skimming bulk waves.

Head waves are leaky waves radiated into the second medium by lateral waves.

Stoneley waves at a liquid-solid interface are true interface waves. They are unattenuated and are normally localized mainly in the liquid, with a sound velocity approximately equal to the liquid sound velocity.

Questions

1. Explain how bulk fluid loading increases the generalized Rayleigh wave velocity while loading by a thin nonattenuating liquid layer usually decreases it.
2. Using

$$r = 2 - (V/V_S)^2$$

$$q^2 = 1 - (V/V_L)^2$$

$$s^2 = 1 - (V/V_S)^2$$

show that Equation 8.21 can be written as

$$r^2 - 4sq = 0$$

3. Discuss the factors that may come into play in determining the difference between BAW and SAW attenuation for a given material.
4. Show that Ψ_x and Ψ_z are not allowed for Rayleigh wave propagation.
5. Calculate and plot the Rayleigh wave polarization ellipse to scale, down to depths of $z = 3\lambda_R$.
6. Calculate α_R (Equation 8.28) and α_S (Equation 8.33) for air, water, and mercury in contact with surfaces of quartz and PMMA at 1 MHz and at 1 GHz. Explain the difference between the various cases.
7. The attenuation of Rayleigh waves on a piezoelectric substrate is given by $\alpha(Np/m) = 30f^2 + 6.5f$, where f is in GHz.
Express the attenuation constant in dB/μs if the Rayleigh wave velocity is 3200 m/s.
8. Explain in simple physical terms why the Rayleigh dip broadens out with increasing shear wave attenuation in the solid surface.
9. Early calculations of the Rayleigh wave beam displacement predicted a displacement that could be significantly greater than the beam width. In terms of the model of Section 8.3.1, is this possible? Explain.
10. Estimate the decay length in both media for Stoneley waves at a water-aluminum interface. Sketch to scale.

9

Lamb Waves

The previous chapter dealt with Rayleigh waves guided along the surface of a semi-infinite solid. This chapter deals with a similar problem, again for the case of sagittal waves, that of Lamb waves [48] propagated along a thin plate. Mathematically, the problem for Lamb waves is rather more complicated than for Rayleigh waves. We will not stress the mathematical development here but rather look at the nature of the simplest solutions, the S_0 symmetric modes and the A_0 antisymmetric modes, as well as the physical nature of the higher-order modes. The origin of the modes will be looked at from another angle, that of guided modes, in the following chapter. Very detailed and rigorous mathematical treatments of Lamb waves have been given elsewhere [26, 35, 49].

One fundamental difference between Rayleigh waves on a free surface and Lamb waves in a plate is that in the latter case there is a finite length scale, the plate thickness b. This means that for finite values of the ratio of the Lamb wavelength λ to b, the Lamb waves are dispersive. Determination of the dispersion relation and hence the variation of phase (V_P) and group (V_G) velocities with frequency is an important part of the problem. In many areas of physics, for example, lattice dynamics in solids, it is normal to describe dispersion by the $\omega(k)$ curve for the mode considered. This is also done frequently in ultrasonic waveguide problems. This particular presentation has the advantage of clearly and directly displaying the cutoff frequencies for the various modes. However, in practical ultrasonics and NDE, the curves showing phase and group velocities as a function of frequency are used much more often. This is one of the reasons that where possible we present dispersion curves as V_P as a function of fb and also V_G as a function of fb. Apart from their widespread use, these curves also have the advantage that they link experimentally observable quantities. Often for new or unusual structures there can be serious difficulties in identifying the nature of the actual acoustic modes observed experimentally. Measuring V_P and V_G over as wide a frequency range as possible and comparing directly with the theoretical curves is the best way to carry out this mode identification.

9.1 Potential Method for Lamb Waves

We follow a simple approach, developed in more detail in [26], using the coordinate system defined in Figure 9.1. The object of this section will be to obtain the dispersion equation, from which we can deduce the form of the fundamental modes in the low-frequency limit.

As in Chapter 7, the displacement can be written in terms of the scalar and vector potentials

$$u = \vec{\nabla}\phi + \vec{\nabla} \times \vec{\psi} \tag{9.1}$$

where both potentials are independent of the y coordinate. Thus

$$u_x = \frac{\partial \phi}{\partial x} + \frac{\partial \psi}{\partial z} \tag{9.2}$$

$$u_z = \frac{\partial \phi}{\partial z} - \frac{\partial \psi}{\partial x} \tag{9.3}$$

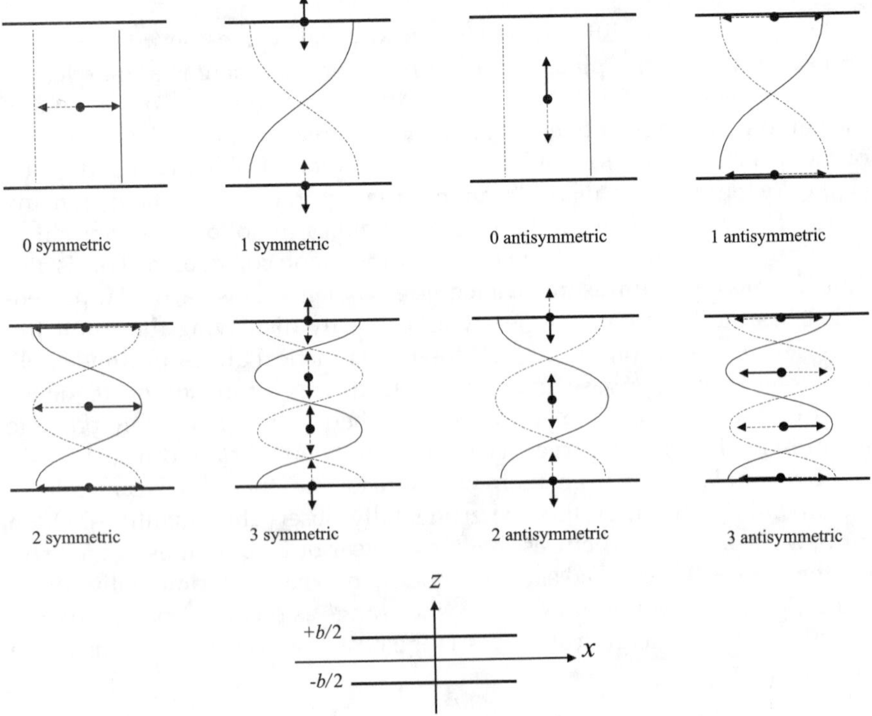

FIGURE 9.1
Coordinate system used for Lamb waves, with displacement variations for the lowest four modes in the limit $\beta S \to 0$.

The Laplacian ∇^2 can be written

$$\nabla^2 = -\beta^2 + \frac{\partial^2}{\partial z^2} \tag{9.4}$$

where wave vectors in the transverse direction, k_{tl} and k_{ts} for longitudinal and shear modes, are defined as

$$k_{tl}^2 = \frac{\omega^2}{V_L^2} - \beta^2 \tag{9.5}$$

$$k_{ts}^2 = \frac{\omega^2}{V_S^2} - \beta^2 \tag{9.6}$$

and β is the wave number in the x direction.

Following Chapter 7, we can write the normal and tangential stress for the isotropic plate as

$$\begin{aligned} T_{zz} &= c_{11}\nabla^2\phi + 2c_{44}\left(\beta^2\phi + j\beta\frac{\partial\psi}{\partial z}\right) \\ &= c_{44}\left[(\beta^2 - k_{ts}^2)\phi + 2j\beta\frac{\partial\psi}{\partial z}\right] \end{aligned} \tag{9.7}$$

for the normal stresses and

$$\begin{aligned} T_{xz} &= 2c_{44}S_{xz} = c_{44}\left(\frac{\partial u_x}{\partial z} + \frac{\partial u_z}{\partial x}\right) \\ &= c_{44}\left[(\beta^2 - k_{ts}^2)\psi - 2j\beta\frac{\partial\phi}{\partial z}\right] \end{aligned} \tag{9.8}$$

for the tangential stresses.

For stress-free boundary conditions at the free surfaces, having T_{xz} and T_{zz} equal to zero at $z = \pm b/2$ can only be satisfied simultaneously if they are either even or odd functions of z. This means in turn that the potentials ϕ and ψ must be of opposite parity, so that, omitting the factor $\exp j(\omega t - \beta x)$

$$\phi = B\cos(k_{tl}z + \alpha) \tag{9.9}$$

$$\psi = A\sin(k_{ts}z + \alpha) \tag{9.10}$$

where

$\alpha = 0$ corresponds to T_{zz} even, T_{xz} odd
$\alpha = \pi/2$ corresponds to T_{zz} odd, T_{xz} even

and the displacements become

$$u_x = -j\beta B\cos(k_{tl}z + \alpha) + k_{ts}A\cos(k_{ts}z + \alpha) \tag{9.11}$$

$$u_z = -k_{tl}B\sin(k_{tl}z + \alpha) + j\beta A\sin(k_{ts}z + \alpha) \tag{9.12}$$

These solutions divide up naturally into two groups according to whether $\alpha = 0$ or $\alpha = \pi/2$.

1. $\alpha = 0$: These are symmetric solutions with respect to z. The deformation of the plate is symmetrical with respect to the median plane $z = 0$, so that $u_z(z) = -u_z(-z)$ and $u_x(z) = u_x(-z)$.
2. $\alpha = \pi/2$: These are antisymmetric solutions with respect to z. The deformation of the plate is antisymmetric with respect to the center so that $u_z(z) = u_z(-z)$ and $u_x(z) = -u_x(-z)$.

As can be verified directly from Figure 9.1, the above considerations lead directly to the deformations of the plate displayed in Figures 9.1 and 9.2. In order to find the dispersion equation, the boundary conditions for the stress can be written explicitly as

$$(\beta^2 - k_{ts}^2)B\cos\left(k_{tl}\frac{b}{2} + \alpha\right) + 2j\beta k_{tl}A\cos\left(k_{ts}\frac{b}{2} + \alpha\right) = 0 \tag{9.13}$$

$$2j\beta k_{tl}B\sin\left(k_{tl}\frac{b}{2} + \alpha\right) + (\beta^2 - k_{ts}^2)A\sin\left(k_{ts}\frac{b}{2} + \alpha\right) = 0 \tag{9.14}$$

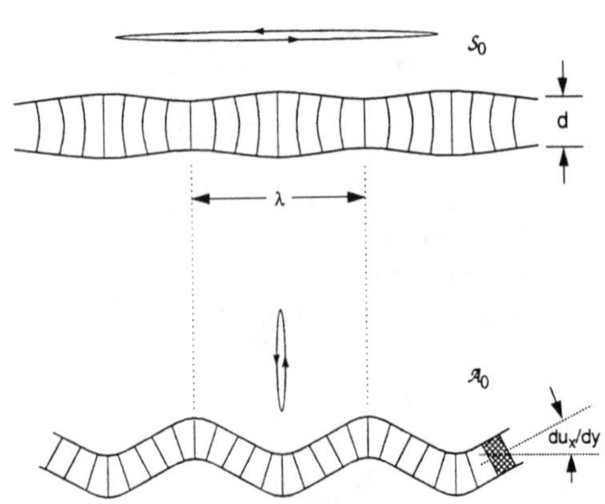

FIGURE 9.2
Mode shapes of S_0 and A_0 modes showing the deformation of particle planes and the retrograde elliptical motion at the plate surface for the case $b/\lambda = 0.03$. (From Wenzel, S.W., Applications of Ultrasonic Lamb Waves, Ph.D. thesis, University of California, Berkeley, 1992. With permission.)

leading in the usual way to the characteristic equation

$$(\beta^2 - k_{ts}^2)\cos\left(k_{tl}\frac{b}{2} + \alpha\right)\sin\left(k_{ts}\frac{b}{2} + \alpha\right)$$
$$+ 4\beta^2 k_{tl} k_{ts} \sin\left(k_{tl}\frac{b}{2} + \alpha\right)\cos\left(k_{ts}\frac{b}{2} + \alpha\right) = 0 \tag{9.15}$$

which can be rewritten as

$$\frac{\omega^4}{V_S^4} = 4\beta^2 k_{tl}^2 k_{ts}^2 \left[1 - \frac{k_{tl}}{k_{ts}} \frac{\tan\left(k_{tl}\frac{b}{2} + \alpha\right)}{\tan\left(k_{ts}\frac{b}{2} + \alpha\right)}\right] \tag{9.16}$$

where α takes on the successive values of 0 and $\frac{\pi}{2}$. The resulting two equations, together with the definitions of β, k_{tl}, and k_{ts}, can then be used to determine the dispersion relations for the two types of solutions. Numerical solutions are shown for two relatively extreme cases, brass and sapphire plates in Figures 9.3 and 9.4. The solutions clearly separate into symmetric and antisymmetric groups. For each of these in turn, we must distinguish between the fundamental modes S_0 and A_0 that extend down to zero frequency and the higher-order modes that exhibit a cutoff.

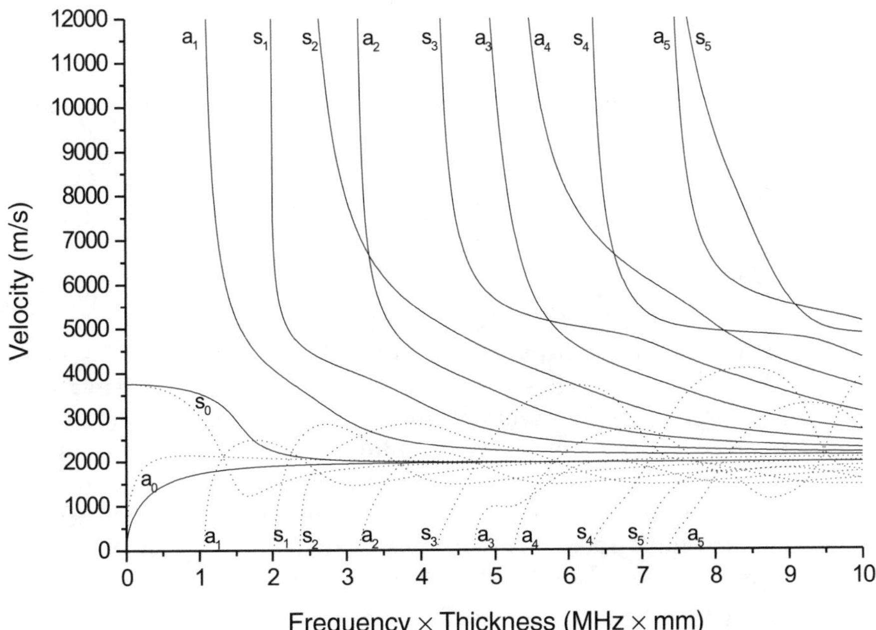

FIGURE 9.3
Phase (—) and group (···) velocities of Lamb modes in a brass plate ($V_L = 4700$ m/s, $V_S = 2100$ m/s) as a function of *fb*.

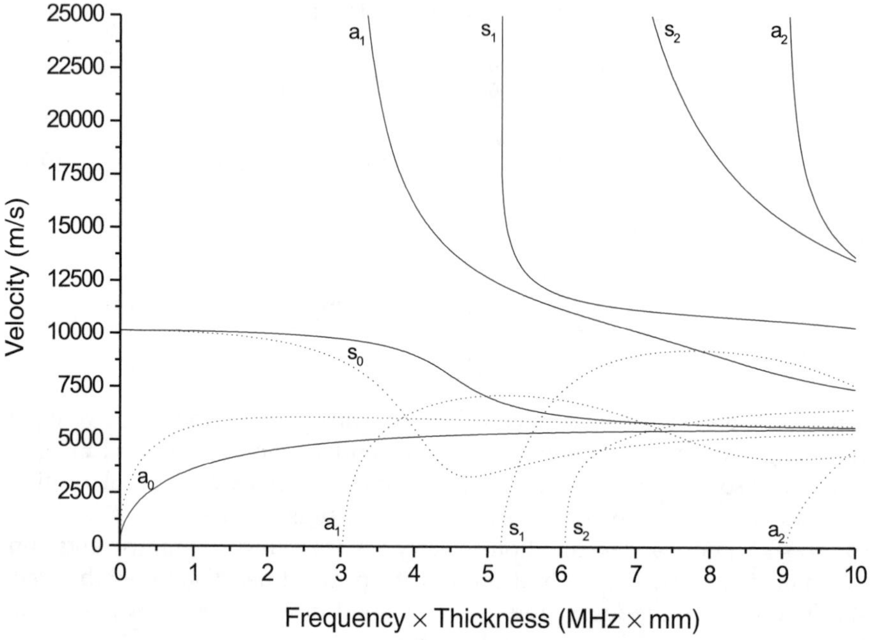

FIGURE 9.4
Phase (—) and group (···) velocities of Lamb modes in a sapphire plate (V_L = 11100 m/s, V_S = 6040 m/s) as a function of fb.

1. Symmetric mode S_0. The phase velocity tends to a constant value

$$V_P = V_{PL} = 2V_S\left(1 - \frac{V_S^2}{V_L^2}\right) \qquad (9.17)$$

$$V_S\sqrt{2} < V_{PL} < V_L \qquad (9.18)$$

Calculation of the displacement using the solutions of the dispersion Equation 9.15 shows that in this limit as $\omega \to 0$, the displacement is mainly longitudinal and constant. Thus this mode is involved in a type of Young's modulus experiment or stretching of the plate. As Rayleigh originally pointed out, this mode corresponds to stretching without bending. Thus it is physically reasonable that the displacement is almost entirely longitudinal and the limiting phase velocity is close to V_L.

2. Antisymmetric mode A_0. As $\beta \to 0$, $f \to 0$ and the phase velocity of the A_0 mode also goes to zero as

$$V_P = \frac{V_{PL}\beta b}{2\sqrt{3}} \qquad (9.19)$$

In this case, calculation of the displacement shows that there is a uniform transverse displacement across the plate, corresponding to a bending motion. Again, in Rayleigh's words, this corresponds to bending without stretching. The movement has often been described as being like that of a flag waving in the breeze; however, this analogy is not to be taken literally as the motion of the flexural modes is confined to the saggital plane.

The simple physics of the S_0 and A_0 modes at low frequency can be understood by considering the deformation of a thin sheet of paper or plastic. If the sample is gripped uniformly across the ends and stretched, this corresponds to an S_0-type deformation. There is evidently a high resistance to stretching, hence a high elastic modulus and a large value of V_{PL}. Of course, in real life the anisotropic nature of paper comes into play, but the general idea is valid for an isotropic sample.

When the paper or plastic is bent, there is almost no resistance. This corresponds to a low modulus of elasticity and a phase velocity that tends to zero as the thickness of the sheet is decreased. This is compatible with the well-known engineering result, Equation 13.54, that the bending modulus varies as b^3.

3. Higher-order modes with cutoff. As will be developed explicitly in the next chapter, cutoff corresponds to $\beta \to 0$ and hence transverse resonance of a longitudinal or transverse wave in the plate. We can obtain this condition from Equations 9.13 and 9.14 directly by putting $\beta = 0$. This gives

$$B\cos\left(\frac{\omega_c b}{2V_L} + \alpha\right) = 0 \tag{9.20}$$

$$A\sin\left(\frac{\omega_c b}{2V_S} + \alpha\right) = 0 \tag{9.21}$$

where, as before, symmetric and antisymmetric solutions correspond to $\alpha = 0$ or $\alpha = \pi/2$, respectively. Depending on the type of solution, only one of a or b is nonzero. The four possibilities that follow from Equations 9.20 and 9.21 are summarized in Table 9.1. Clearly, the behavior of symmetric and antisymmetric modes is opposite, and a series of transverse resonances of each type occurs, with alternating longitudinal and transverse displacements. The high-frequency behavior of the modes is obtained by setting $fb \to \infty$. All of the phase velocities asymptotically approach V_R of the plate, which is a physically reasonable result as it corresponds to Rayleigh waves on the surfaces of a very thick plate that do not interact with each other if the plate is sufficiently thick. The various S and A modes may cross each other although modes of a same family do not cross.

TABLE 9.1

Properties of Higher-Order Symmetric and Antisymmetric Lamb Modes

	Symmetric ($\alpha = 0$)		Antisymmetric ($\alpha = \pi/2$)	
	Even	**Odd**	**Even**	**odd**
Coefficients	$A \neq 0$ $B = 0$	$A = 0$ $B \neq 0$	$A = 0$ $B \neq 0$	$A \neq 0$ $B = 0$
Nature of displacement	Longitudinal	Transverse	Longitudinal	Transverse
Resonance equation	$\frac{\omega_c b}{2V_S} = n\pi$	$\frac{\omega_c b}{2V_L} = (2m+1)\frac{\pi}{2}$	$\frac{\omega_c b}{2V_L} = n\pi$	$\frac{\omega_c b}{2V_S} = (2m+1)\frac{\pi}{2}$
	$n = 1, 2, 3$	$m = 0, 1, 2$	$n = 1, 2, 3$	$m = 0, 1, 2$

9.2 Fluid-Loading Effects

As will be shown below, there is a direct correspondence between fluid loading of the Rayleigh wave on the surface of a semi-infinite solid and fluid loading of one side of a plate supporting Lamb waves. However, the plate case is rather more complicated as the fluid loading can be one- or two-sided with the same or different fluids. Also, a thin plate can be formed into a tube, which has the same possibilities of inside and outside loading. We describe these other cases more briefly, principally to identify the modes in question and the physical principles involved, and to provide a lead-in to active current investigations of the subject. The earlier work on fluid loading of acoustic modes on plane and curved surfaces has been summarized in an excellent review by Uberall [45].

9.2.1 Fluid-Loaded Plate: One Side

There is a one-to-one correspondence with the fluid-loaded Rayleigh wave problem. As in the latter case, there is a complex term added to the right-hand side of the dispersion relation of the Rayleigh-Lamb equation. There are now two roots, a complex root corresponding to a leaky Lamb wave (LLW) and a real root corresponding to an interface Stoneley wave. As for the case of leaky Rayleigh waves, most of the energy in the LLW resides in the plate, assuming that the liquid is a perturbation on the solid behavior. The wave leaks into the fluid at an angle $\sin\theta = V_0/V_P$ where V_P is the phase velocity of the Lamb wave considered; evidently, the same Lamb mode will be excited by a compressional wave incident from the fluid at this angle. Again, as for Rayleigh waves, there is a small change in V_P due to liquid loading, but this effect is negligible in most practical applications as the acoustic impedance of the liquid is usually much smaller than that of the solid. The applications of LLW in NDE will be discussed fully in Chapter 15. There are some specific practical points that follow from the previous results in the limit $fb \to 0$, as follows:

1. The S_0 mode is little affected by the presence of liquid as the displacement of this mode is mainly parallel to the surface.
2. At not-too-low frequencies in the sonic regime such that $V_P > V_0$, the A_0 mode is relatively highly radiative as the transverse displacements set up compressional waves in the liquid.
3. The A_0 mode in the subsonic regime, $V_P < V_0$, is trapped in the plate setting up an evanescent wave in the liquid. This makes this mode very useful for applications to liquid sensing, which will be discussed in more detail in Chapter 13.

The real root of the modified dispersion equation corresponds to a true interface wave, which is often called a Stoneley-Scholte mode (*A* mode). It propagates in the liquid parallel to the surface without attenuation. It is the direct analog of the Stoneley wave for the liquid-loaded surface. The *A* mode, however, has a phase velocity that has the same general variation with fb as the A_0 mode, as shown in Figure 9.5. As fb increases from zero, the phase velocity increases monotonically and asymptotically approaches the bulk fluid phase velocity as $fb \to \infty$.

There has been considerable recent interest in the question of mode repulsion effects between Lamb wave modes [50]. This phenomenon occurs in the present problem, as a repulsion between the *A* and the A_0 modes in the region where the phase velocity of the latter approaches the sound velocity in the liquid. This results in an upward deformation of the A_0 curve as shown in Figure 9.5. Equally important, the two modes exchange character below the interaction region, that is to say that the *A* mode now propagates predominantly in the solid and the A_0 predominantly in the fluid. The two modes propagate in both media in the interaction region, although the upper mode is very highly attenuated.

9.2.2 Fluid-Loaded Plate: Same Fluid Both Sides

This case was treated in detail in the classic paper by Osborne and Hart [51]. They found the existence of the *A* mode, described above, and in addition a new mode, analogous to the symmetric mode S_0, called the *S* mode. It was found that the *S* mode has a roughly horizontal dispersion curve, with a phase velocity V_S just slightly below that of the bulk fluid phase velocity V_0.

9.2.3 Fluid-Loaded Plate: Different Fluids

This case was considered by Bao et al. [52]. They showed that similar repulsion phenomena occur although the detailed behavior of the coupled modes is different. The *A* mode increases from zero and asymptotically approaches

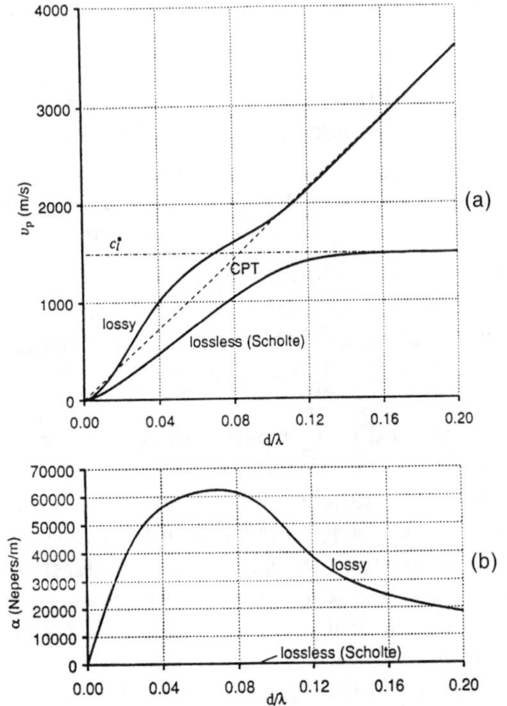

FIGURE 9.5
(a) Schematic representation of the dispersion curves for a thin plate loaded with fluid on one side. The A_0 curve is deformed from the vacuum case due to mode repulsion in the region where the phase velocity approaches the liquid sound velocity. (b) Loss of the coupled modes. (From Wenzel, S.W., Applications of Ultrasonic Lamb Waves, Ph.D. thesis, University of California, Berkeley, 1992. With permission).

the phase velocity of the lowest-velocity liquid. The *S* mode splits at low frequency and approaches the phase velocity of the highest-velocity liquid at high frequencies.

9.2.4 Fluid-Loaded Solid Cylinder

The treatment for this classic problem has been summarized by Uberall [45]. The results are analogous to those for the semi-infinite solid. A Rayleigh wave propagates around the curved surface of the cylinder and becomes leaky in the presence of the fluid. There are also higher-order Rayleigh-type modes that penetrate into the cylinder. These are called whispering gallery modes and can be represented in a ray model as multiple reflections around the inner surface of the cylinder. The analog of the Stoneley wave for a curved surface is called a Franz or creeping wave, which propagates in the liquid around the curved surface of the cylinder. There is now a difference with the plane surface, however, as the Franz waves radiate tangentially into the liquid, and hence, these modes are attenuated for geometrical reasons. This propagation path has been directly imaged by Schlieren imaging techniques [53]. This attenuation of the Franz waves is in contrast with the Stoneley waves for the plane surface, which are unattenuated.

9.2.5 Fluid-Loaded Tubes

This is a complex topic that is the subject of much current research, so we provide only a brief description. The case of the tube has all of the complexities of the plate (fluid inside, outside, etc.) as well as those provided by the curvature of the cylinder. The case of thin-walled tubes will be discussed here, where $b/a > 0.95$ with b the inner radius of the tube and a the outer. Experimental results are usually given as V_P or V_G as a function of fd ($d = a - b$ = wall thickness) although some of the theoretical results are expressed as a function of ka, where k is the mode wave number. The scale factor between the two variables is

$$fd = \frac{V_0}{2\pi}\left(1 - \frac{b}{a}\right)ka \tag{9.22}$$

and phase and group velocities are linked by

$$V_G = \left[\frac{d}{d(fd)}\frac{fd}{V_P}\right]^{-1} \tag{9.23}$$

To a first approximation the empty tube has modes very similar to those of a plate, with the exception that axial and circumferential modes are possible. Most of the oceanographic work has been carried out for the case of evacuated thin cylindrical shells immersed in a liquid. The situation is similar to that for a plate loaded on one side, except that now the A mode becomes a type of creeping (Franz) mode around the outside of the shell and radiates into the liquid as for a cylinder. Maze et al. [54] showed that the same mode repulsion and exchange of mode character between A and A_0 occurs in the region where the A_0 velocity approaches the ambient fluid sound speed. Also of interest is the case where the tube contains a filler liquid inside. This case was considered by Sessarego et al. [55] where repulsion and wave character exchange effects were found, as well as two Stoneley-type modes A and S. The group velocity of the lowest A mode has a maximum in the critical region of mode repulsion. Bao et al. [56] also found that new modes inside the tube were introduced by the fluid filling (whispering gallery type).

Again, strong coupling (repulsion) effects occur between these modes and the A_0 modes in the tube, leading to dispersion curve veering effects between the filler modes and change in wave character (fluidborne or shellborne) over the full length of the dispersion curve. The strong coupling effects were attributed to shear terms in the boundary conditions, which is a general effect as shown by Uberall et al. [50].

Summary

Lamb waves are symmetric and antisymmetric acoustic waves propagated along a thin plate. Since the wavelength is of the order of the plate thickness these waves are dispersive in nature.

A_0 and S_0 modes are the fundamental Lamb modes. The displacement is uniform across the plate thickness at low frequency. In the limit $fb \to 0$, the modes correspond to pure extensional and flexural displacements, respectively.

Stoneley-Scholte mode is a pure interface mode at the interface between a plate and a liquid. As fb increases, the phase velocity asymptotically approaches the velocity of sound in the fluid.

Franz waves or creeping waves correspond to the Stoneley-Scholte modes for a curved surface; these modes "creep" around the surface of a tube or a cylinder.

Questions

1. Explain the order of magnitude of the phase velocity for S_0 and A_0 modes in the limits $fb \to 0$ and $fb \to \infty$.
2. Describe three different experimental ways in which Lamb waves can be generated in plates.
3. Explain physically why the group velocity varies strongly with fb near a Lamb wave cutoff frequency.
4. Give a qualitative discussion on the different effects of liquid loading on the attenuation of the S_0 and A_0 modes as $fb \to 0$.
5. What are the main differences between the Lamb wave dispersion curves for a thin plastic plate compared to a thin sapphire plate?

6. Of all the fundamental acoustic modes, why is it that the phase velocity of the A_0 mode goes to zero as fb decreases to zero? How could this phenomenon be exploited in sensing applications?
7. Show that Equation 9.16 is equivalent to Equations 10.19 and 10.20.
8. Determine which Lamb modes would be excited in an aluminum plate 1 mm thick at 1 MHz and 20 MHz.
9. Compare and explain the difference between the dispersion curves for SH and Lamb waves.

10

Acoustic Waveguides

10.1 Introduction: Partial Wave Analysis

We have already described in some detail two examples of guided acoustic waves, namely Rayleigh waves on a surface and Lamb waves in a plate. Both of these problems were solved using the potential method, which can in fact be used to solve any acoustics problem in isotropic media. However, the potential method cannot be extended to anisotropic media. This is a definite shortcoming for quantitative treatment of acoustic waveguides, because while the isotropic model is simple and useful to describe the global behavior, most acoustic waveguides are in fact made from anisotropic materials. Thus it would be useful to have a formalism that works in this case, and that is provided by partial wave analysis, which will be used in this chapter.

The basic idea of the partial wave method is to consider separately the different components of the plane wave solutions involved in the particular problem at hand; these will typically be either SH or sagittal wave modes. These components, the so-called partial waves, are oriented so that they have a common wave vector β in the propagation direction along the waveguide axis. Depending on the conditions (mainly frequency of operation), the transverse components of the wave vector may be real or imaginary. The possible modes that can be set up in the waveguide are determined by transverse resonance in a manner similar to the situation for electromagnetic waveguides. This leads to low-frequency cutoff conditions and many higher-order modes as the frequency is increased. Slowness curves will prove to be very useful as a visual technique to describe the whole waveguide problem.

In this chapter, we will establish a general formalism that can be used to describe acoustic waveguide applications, based on partial wave analysis, slowness curves, and transverse resonance. In several cases these results will provide a complement to the treatments that have already been made using the potential method. As before, only isotropic media will be considered. The approach follows that adopted by Auld [32].

10.2 Waveguide Equation: SH Modes

The simplest case is provided by SH modes as there is only one direction of polarization and they are decoupled from the sagittal modes, so there is no mode conversion or reflection. The basic geometry is shown in Figure 10.1(a), where incident and reflected partial waves are shown. The local displacement (velocity) is perpendicular to the sagittal plane and the bulk shear wave velocity is relevant to the problem. The boundary conditions at the free surface lead directly to a node for the two components of stress T_{xz} and T_{zz} and an antinode for the velocity v. The principle of transverse resonance says that resonances at multiples of $\lambda/2$, which are compatible with these boundary conditions, can occur as shown in Figure 10.2(b). The fundamental mode $n = 0$ has uniform velocity down to zero frequency, so there is no cutoff. The higher modes have cutoffs at frequencies corresponding to the appropriate resonances, as shown

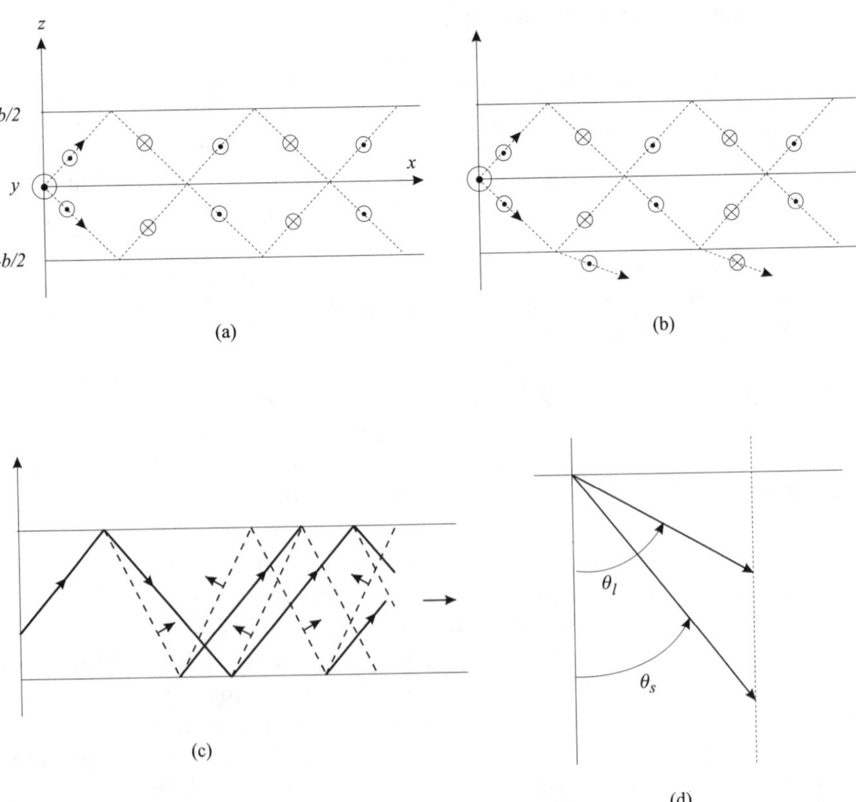

FIGURE 10.1
Partial waves used for guided wave analysis in several configurations. (a) SH modes. (b) Love waves. (c) Lamb waves. (d) Rayleigh waves.

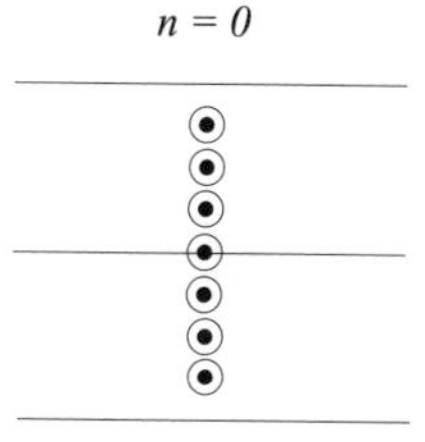

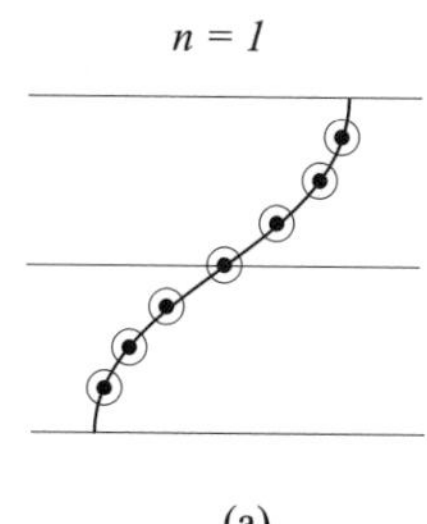

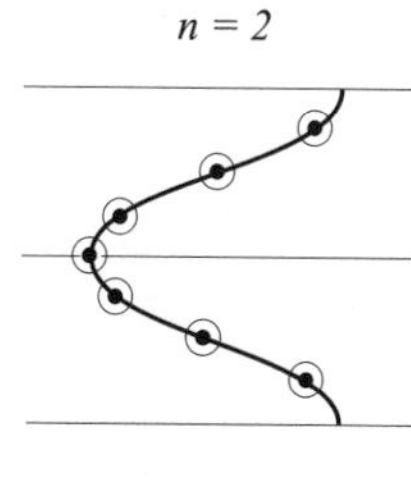

(a)

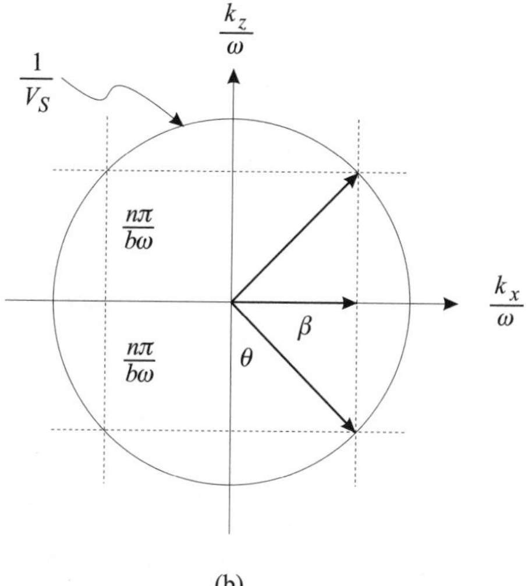

(b)

FIGURE 10.2
(a) Displacement curves for the fundamental and two lowest modes for SH waves. (b) Slowness construction for SH modes in a plate of thickness b.

for modes $n = 1$ and $n = 2$ in the figure. Thus

$$\frac{n\lambda}{2} = b \tag{10.1}$$

$$k_t = \frac{2\pi}{\lambda} = \frac{\pi n}{b} \tag{10.2}$$

where k_t is the transverse wave number.

As shown in Figure 10.1(a), the incident and reflected partial waves have a common wave number β along the propagation direction. The solution for the full wave equation is

$$k^2 = k_x^2 + k_y^2 + k_z^2 = \frac{\omega^2}{V_S^2} \tag{10.3}$$

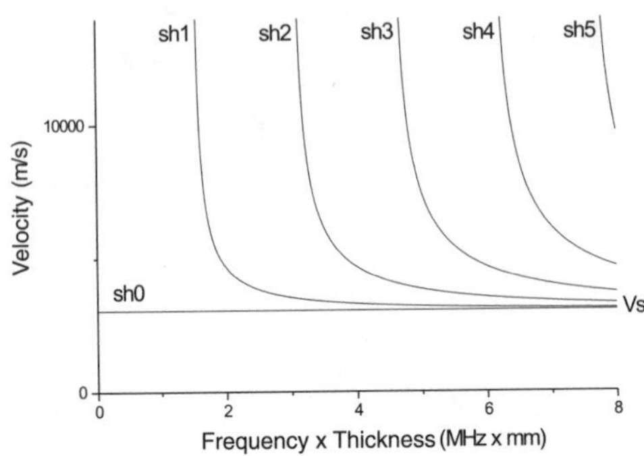

FIGURE 10.3
Dispersion curves for SH modes in an aluminum plate (V_S = 3040 m/s).

Since $k_y = 0$, and using $k_z = k_t$ from Equation 10.2 and $k_x = \beta$, we have

$$\beta^2 = \left(\frac{\omega}{V_S}\right)^2 - \left(\frac{n\pi}{b}\right)^2 \tag{10.4}$$

which we call the waveguide equation. It has a very simple geometrical interpretation in terms of the slowness curve as shown in Figure 10.2(b). The two partial waves with common wave vector are shown together with the value of k_t from Equation 10.2 and the radius of the slowness curve $1/V_S$. Thus the slowness construction corresponds exactly to the waveguide equation by Pythagoras' theorem.

It is instructive to look at the behavior for a given waveguide as the frequency is changed for a given mode number n. As the frequency is increased the transverse component $n\pi/b\omega$ decreases, θ increases, and β increases toward the boundary of the slowness curve. For $\omega \to \infty$ (very short wavelength) the transverse component goes to zero and the propagation is along x with $\beta = \omega/V_S$, corresponding to a bulk wave in this limit. As ω is decreased, θ increases until at cutoff, defined $\beta \equiv 0$, $\theta = 0$, and $n\pi/b\omega_c = k_t/\omega_c$, i.e., transverse resonance for this particular value of n. Since $\beta = 0$ there is no propagation down the guide. Furthermore, for frequencies below cutoff, $\omega < \omega_c$, the partial waves move off the slowness curve and β becomes imaginary. Thus the wave along the x becomes evanescent or nonpropagating, consistent with the notion of cutoff.

In order to obtain the full solutions for the velocity and the displacement, we need to consider the symmetry properties of the plate. For reconstruction of the partial waves, the latter must be in the same state after two reflections. This means that the amplitude must be the same, which is guaranteed by reflection at a free surface with no mode conversion, and the phase must change by a multiple of $2\pi n$. These conditions can be met in a more general way by expressing them as a symmetry principle for reflections with respect to the median (xy) plane. From the form of the transverse resonances in Figure 10.2(a), clearly for n even, there is even symmetry (symmetric mode) about the median plane and there is odd symmetry (antisymmetric mode) for

n odd. Since reflection in the central plane interchanges incident and reflected waves, then they are identical for symmetric modes and differ by a sign change for antisymmetric modes. Hence, the symmetry principle states that the amplitudes of incident and partial waves differ at most by a sign change.

The previous considerations lead to a methodology for calculations using partial waves, which is summarized for SH modes in the following steps:

1. Define the partial waves; here we have only the SH mode, so
$$v \equiv v_y$$
2. Define incident and reflected waves
$$\begin{aligned} v_i &= A \exp j(-k_{ts}z + \beta x) \\ v_r &= B \exp j(+k_{ts}z + \beta x) \end{aligned} \tag{10.5}$$
3. Apply the symmetry principle
$$B = \pm A \tag{10.6}$$
4. Use boundary conditions on reflection at $z = b/2$, $v_i = v_R$
$$\pm A \exp - j\left(\frac{k_{ts}b}{2} + \beta x\right) = A \exp - j\left(\frac{k_{ts}b}{2} + \beta x\right) \tag{10.7}$$
5. Deduce transverse resonance condition from step 4.
$$\exp jk_{ts}b = \pm 1$$
Hence
$$k_{ts} = \frac{n\pi}{b}, \quad n = 0, 1, 2, \ldots \tag{10.8}$$
6. Deduce waveguide dispersion relation and slowness description
$$\beta_n^2 = \frac{\omega^2}{V_S^2} - \left(\frac{n\pi}{b}\right)^2 = \frac{\omega^2}{V_S^2} - k_{ts}^2 \tag{10.9}$$
7. Form solutions for particle velocity from partial wave solutions. For example, for a guided wave traveling toward the right with positive velocity maximum on the upper surface
$$v_n = \cos\left[\frac{n\pi}{b}\left(z - \frac{b}{2}\right)\right] \exp -j(\beta_n)\Big] \tag{10.10}$$
8. Determine appropriate stresses from Hooke's law. For the above case
$$T_{yz} = -\frac{n\pi}{b}\frac{c_{44}}{j\omega} \sin\left[\frac{n\pi}{b}\left(z - \frac{b}{2}\right)\right] \exp -j(\beta_n) \tag{10.11}$$

The dispersion curve for the SH mode can be determined directly from the waveguide equation, Equation 10.9. The fundamental for $n = 0$ goes to the bulk shear velocity at $fd \to 0$. The higher modes have cutoff frequencies, as can be deduced directly from the waveguide equation.

10.3 Lamb Waves

The dispersion equation for Lamb waves was derived in the previous chapter using the potential method. It also provides an excellent example of the power of the partial wave method for directly solving the waveguide problem. The partial wave modes are now composed of longitudinal and transverse components in the sagittal plane as shown in Figure 10.1(c); they must obey the symmetry relations established in the previous section.

Following the methodology outlined earlier, we define the velocity fields incident and reflected partial waves as

$$v_{xi} = A_l e^{-jk_{li}\cdot\vec{r}}, \quad B_l e^{-jk_{lr}\cdot\vec{r}} \tag{10.12}$$

for the longitudinal component and

$$v_{xi} = A_s e^{-j\vec{k}_{si}\cdot\vec{r}}, \quad B_s e^{-j\vec{k}_{sr}\cdot\vec{r}} \tag{10.13}$$

for the shear component.

The symmetry conditions then require

$$\begin{aligned} B_l &= \pm A_l \\ B_s &= \pm A_s \end{aligned} \tag{10.14}$$

and the reflection condition at the surface $z = -b/2$ gives

$$\pm\begin{bmatrix} A_l e^{jk_{li}b/2} \\ A_s e^{jk_{si}b/2} \end{bmatrix} = \begin{bmatrix} R_{LL} & R_{LS} \\ R_{SL} & R_{SS} \end{bmatrix} \begin{bmatrix} A_l e^{-jk_{lr}b/2} \\ A_s e^{-jk_{sr}b/2} \end{bmatrix} \tag{10.15}$$

The determinant of this characteristic equation must vanish as a condition for nontrivial solutions, and using Equations 7.103 and 7.104, this becomes

$$\pm R_{LL} = \frac{\sin(k_{tl} + k_{ts})\frac{b}{2}}{\sin(k_{tl} - k_{ts})\frac{b}{2}} \tag{10.16}$$

expanding the sine terms, we obtain

$$\frac{\tan k_{ts}\frac{b}{2}}{\tan k_{tl}\frac{b}{2}} = -\frac{1+R_{LL}}{1-R_{LL}} \tag{10.17}$$

$$\frac{\tan k_{ts}\frac{b}{2}}{\tan k_{tl}\frac{b}{2}} = -\frac{1-R_{LL}}{1+R_{LL}} \tag{10.18}$$

for symmetric and antisymmetric modes, respectively. Expressing R_{LL} in terms of k_{tl}, k_{ts}, and β, these become finally the Rayleigh-Lamb dispersion equations

$$\frac{\tan k_{ts}\frac{b}{2}}{\tan k_{tl}\frac{b}{2}} = -\frac{4\beta^2 k_{tl}k_{ts}}{(k_{ts}^2-\beta^2)^2} \tag{10.19}$$

$$\frac{\tan k_{ts}\frac{b}{2}}{\tan k_{tl}\frac{b}{2}} = -\frac{(k_{ts}^2-\beta^2)}{4\beta^2 k_{tl}k_{ts}} \tag{10.20}$$

for antisymmetric modes.

Here the transverse wave numbers k_{tl} and k_{sl} obey the waveguide Equation 10.9. Equations 10.19 and 10.20 can be shown to be equivalent to Equation 9.16.

By putting $\beta = 0$ in Equations 10.19 and 10.20, we obtain

$$\frac{\tan k_{ts}\frac{b}{2}}{\tan k_{tl}\frac{b}{2}} = 0 \tag{10.21}$$

for symmetric modes and

$$\frac{\tan k_{ts}\frac{b}{2}}{\tan k_{tl}\frac{b}{2}} = -\infty \tag{10.22}$$

for antisymmetric modes.

For $\beta = 0$ we have $k_{ts} = \omega/V_S$ and $k_{tl} = \omega/V_L$, and Equations 10.21 and 10.22 then give the same transverse resonance conditions as described in Table 9.1.

10.4 Rayleigh Waves

It is shown by Auld [32] that in the limit $\beta b \to \infty$, the S_0 and A_0 modes become degenerate, and their displacements are tightly bound to the surface. One way to see the significance of this result is to set $b \to \infty$. For a sufficiently thick plate, the surface vibrations on the opposing surfaces become decoupled, corresponding to independent Rayleigh waves on the upper and lower surfaces. Thus the Rayleigh wave solution can be obtained by considering partial waves for one surface only. Since the two surfaces are an infinite distance apart, there will be only reflected amplitudes for the upper surface with no incident wave, as shown in Figure 10.1(d). The reflected amplitudes can be written

$$B_s = R_{SS}A_s + R_{SL}A_l \tag{10.23}$$

$$B_l = R_{LS}A_s + R_{LL}A_l \tag{10.24}$$

where the incident amplitudes A_s and A_l go to zero, and the reflection coefficients R_{ij} go to infinity. This can only be done by putting the denominator for the latter in Equations 7.101 and 7.102 equal to zero. This then gives the transverse resonance condition for Rayleigh waves as

$$\sin 2\theta_s \sin 2\theta_l + \left(\frac{V_L}{V_S}\right)\cos^2 2\theta_s = 0 \tag{10.25}$$

The reflected waves must clearly be evanescent and the transverse wave numbers can be written

$$k_{ts} = j\alpha_{ts} \tag{10.26}$$

$$k_{tl} = j\alpha_{ls} \tag{10.27}$$

and with

$$\sin\theta_s = \frac{\beta_R V_S}{\omega} \tag{10.28}$$

$$\sin\theta_l = \frac{\beta_R V_L}{\omega} \tag{10.29}$$

the dispersion equation, Equation 10.25 can be written as

$$4\beta_R^2 \alpha_{ts}\alpha_{tl} = (\alpha_{ts}^2 + \beta_R^2)^2 \tag{10.30}$$

which can easily be shown to be identical to the Rayleigh wave dispersion relation obtained by the potential method in Chapter 8.

10.5 Layered Substrates

The propagation of acoustic waves in layered half spaces developed historically in the study of seismology. In zero approximation, the earth's interior can be represented as a homogeneous half space even though it is in fact far from that approximation throughout its depth. This model accounts for the observation of bulk longitudinal (P) waves and bulk shear (SV) modes, as well as Rayleigh waves propagated along the surface. In a first approximation this half space is covered by a relatively thin crust of quite different mechanical properties. The crust can support modes analogous to those found in a free plate; in particular, modified SH plate modes or Love waves can be observed. A more detailed approach would have to account for propagation in multilayers.

Problems in seismology are of ongoing interest and would justify in their own right the study of acoustic propagation in layered systems. Modern technology has provided additional reasons for studying this subject. Microelectronics is based on varied and ingenious combinations of multilayered structures. This has favored the development of SAW technology in its planar form involving films of metallization, electrodes, and piezoelectric materials. Of more recent interest, microsensors provide another example of the application of various acoustic modes in layered systems; the layers are typically electrodes, piezoelectric films, or chemically selective films, which may be deposited on massive substrates or thin membranes.

A final important example is found in NDE. Protective layers and coatings are ubiquitous in modern manufacturing technology and their quality is an important issue. NDE techniques involve propagating ultrasonic waves in these structures and detecting echoes from defects or associated changes in acoustic properties. A thorough knowledge of the propagation of acoustic waves in such structures is obviously a prerequisite for carrying out such NDE investigations.

The previous section dealt with SH and sagittal modes in plates where they were seen to be decoupled. This is also the case for propagation in layers on substrates and there is a direct correspondence between the two cases for the simple modes. There is, however, a major difference between the two cases. For propagation in a plate in air or a vacuum, the acoustic energy is constrained to the plate, and as has been seen, the propagation can be described by incident and partial waves in the plate. If the plate is now deposited on a substrate then there is the additional possibility that there will be a wave transmitted into the substrate, i.e., the guided mode in the layer may either be trapped or may leak into the substrate. The distinction can be made in a clear and distinct manner for the case of SH modes or Love waves. A good discussion of layered substrates is given by Farnell and Adler [57].

10.5.1 Love Waves

It is a general property of isotropic media that the SH modes are separated from, and hence uncoupled to, the sagittal modes. This is obviously true in a layer on a semi-infinite substrate as shown in Figure 10.1(b) and the corresponding SH wave in the layer is known as a Love wave, discovered in 1927 by A. E. H. Love [58]. For the mode to be trapped in the layer, certain conditions have to be met. As will be demonstrated, a basic condition is that $\hat{V}_S < V_S$ where $\hat{V}_S$ is the shear velocity in the layer and V_S that in the substrate.

Following the usual procedure, we define partial waves as shown in Figure 10.1(b). In addition to the incident (i) and reflected (r) wave as for the SH plate mode, we have a transmitted (t) partial wave in the substrate. The partial waves are

$$\begin{aligned} v_{yi} &= A\exp -j(-\hat{k}_{ts}z + \beta x) \\ v_{yr} &= B\exp -j(\hat{k}_{ts}z + \beta x) \\ v_{yt} &= C\exp -j(k_{ts}z + \beta x) \end{aligned} \tag{10.31}$$

At the upper free boundary at $z = b/2$, $R_S \equiv 1$

$$R_S \equiv \left.\frac{v_{yi}}{v_{yr}}\right|_{\frac{b}{2}} = \frac{A\exp j\frac{\hat{k}_{ts}b}{2}}{B\exp j\frac{-\hat{k}_{ts}b}{2}} = 1 \tag{10.32}$$

At the lower boundary, we use the known reflection and transmission coefficients for this case, which yield

$$R_S = \frac{v_{yr}}{v_{yi}} = \frac{Z_S - \hat{Z}_S}{Z_S + \hat{Z}_S} = \frac{B\exp j\left(\frac{\hat{k}_{ts}b}{2}\right)}{A\exp j\left(-\frac{\hat{k}_{ts}b}{2}\right)} \tag{10.33}$$

$$T_S = \frac{v_{yt}}{v_{yi}} = \frac{2Z_S}{Z_S + \hat{Z}_S} = \frac{C\exp j\left(\frac{k_{ts}b}{2}\right)}{A\exp j\left(-\frac{\hat{k}_{ts}b}{2}\right)} \tag{10.34}$$

The reflection coefficients on the upper and lower surfaces must be satisfied at the same time as a condition for transverse resonance. This leads directly to

$$j\tan\hat{k}_{ts}b = \frac{c_{44}k_{ts}}{\hat{c}_{44}\hat{k}_{ts}} \tag{10.35}$$

The behavior of k_{ts} is important in this equation. From Equation 10.39, if k_{ts} is real, it corresponds to propagation of a progressive wave in the substrate, i.e., energy is leaked out of the layer. We are looking instead for solutions in which energy is trapped in the layer and, therefore, where the transmitted wave in the substrate is evanescent. This corresponds to k_{ts} being imaginary, which can be accounted for explicitly by posing $k_{ts} = -j\alpha_{ts}$ and looking for real values of α_{ts}. Combining Equation 10.35 with the usual waveguide equations for k_{ts} and α_{ts}, we obtain

$$\tan\hat{k}_{ts}b = \frac{C_{44}\alpha_{ts}}{\hat{C}_{44}\hat{k}_{ts}} \tag{10.36}$$

$$\hat{k}_{ts}^2 = \left(\frac{\omega}{\hat{V}_S}\right)^2 - \beta^2 \tag{10.37}$$

$$\alpha_{ts}^2 = \beta^2 - \left(\frac{\omega}{V_S}\right)^2 \tag{10.38}$$

The last two equations show that a necessary condition for α_{ts} to be real, hence, for trapping to occur, is for $\hat{V}_S < V_S$. This conclusion concurs with that of the slowness curve analysis of Figure 10.4.

Auld [32] solves Equations 10.36 through 10.38 graphically and hence is able to obtain threshold frequencies and dispersion relations for all modes as a function of β. In fact, Tournois and Lardat [59] have derived an implicit relation for the dispersion relation of the form

$$\tan\hat{d}\beta b = \frac{\rho V_S^2 d}{\hat{\rho}\hat{V}_S^2\hat{d}} \tag{10.39}$$

where

$$\hat{d} = \sqrt{\left(\frac{V_P}{\hat{V}_S}\right)^2 - 1}, \quad d = \sqrt{1 - \left(\frac{V_P}{V_S}\right)^2}$$

and V_P is the phase velocity of the Love wave.

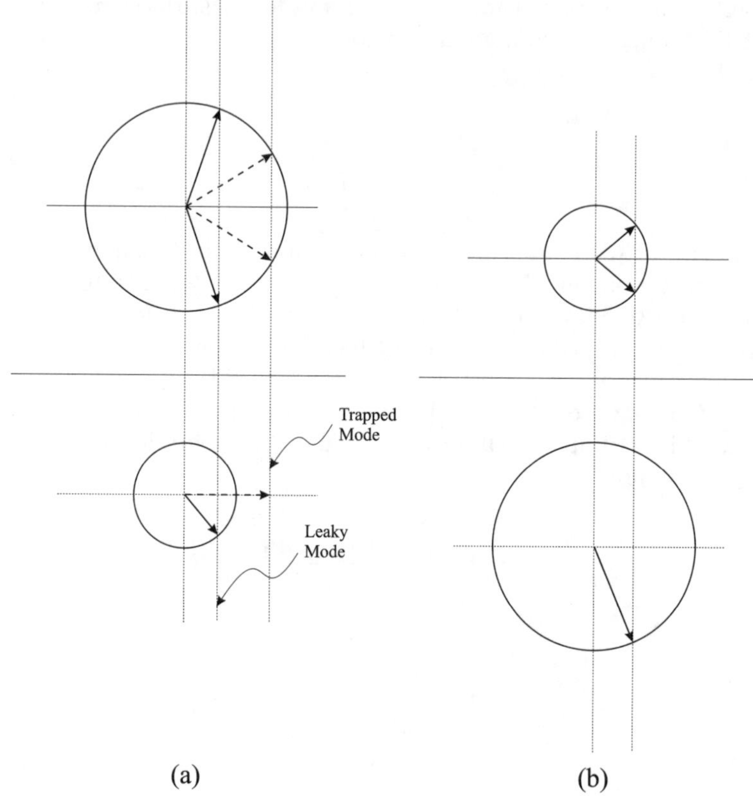

FIGURE 10.4
Slowness curves for two possible layer-substrate configurations for SH modes. (a) Love modes, showing conditions for trapped and leaky modes. (b) SH modes, showing that the SH modes are leaky under all circumstances; hence, Love waves are not possible in this case.

The phase and group velocities obtained from this relation for some of the low-order Love modes for the case of a gold film on a fused quartz substrate are shown in Figure 10.5.

The physics of the phase velocity variation with frequency can be understood by considering Figure 10.5 for a fixed layer of thickness b in the limits of very low and high frequencies. At very low frequencies the wavelength is much greater than the film thickness, so as $f \to 0$, V_P tends to the shear wave velocity in the substrate. In the opposite limit, $f \to \infty$, the wavelength is now much less than the layer thickness so that the fundamental Love mode behaves as a bulk shear wave in the layer and the phase velocity approaches the bulk shear velocity asymptotically. By the same token, the Love modes penetrate deeply into the substrate at low frequencies while they are progressively confined to the layer as the frequency increases to $\beta b \gg 1$. The fb dependence for the first Love mode for a gold layer on fused quartz is displayed in Figure 10.6.

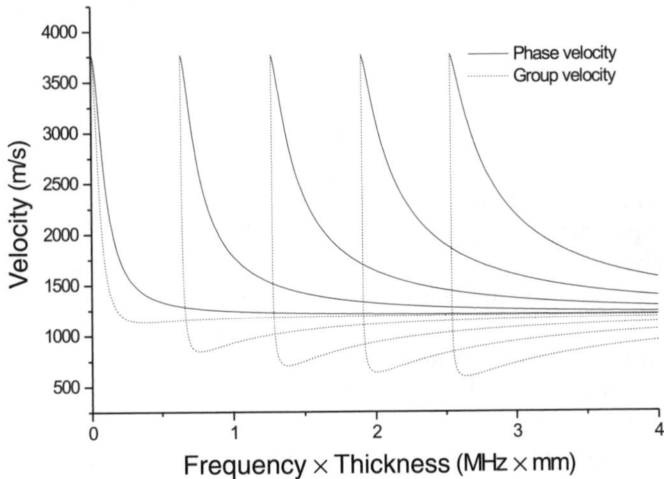

FIGURE 10.5
Fundamental and lowest-order Love modes in a gold layer (V_S = 1200 m/s) on a fused quartz substrate (V_S = 3750 m/s).

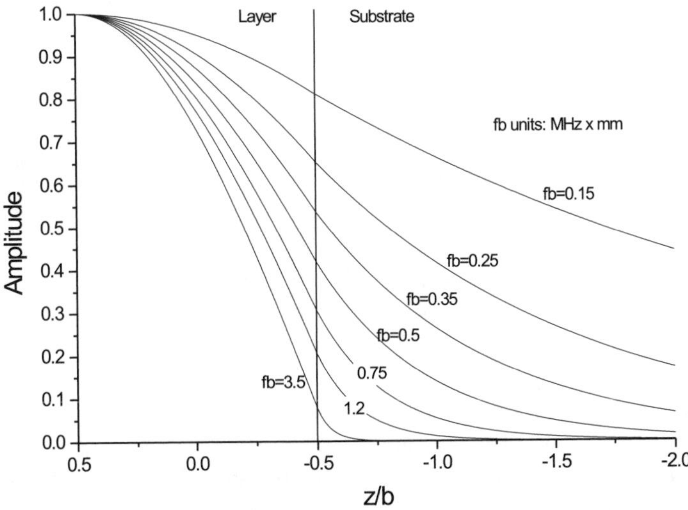

FIGURE 10.6
Displacement of the lowest-order Love mode for the case of Figure 10.5 for various values of *fb*.

10.5.2 Generalized Lamb Waves

We are concerned here with sagittal plane modes in a layer on a semi-infinite half space. As Love waves differ from the SH modes of a free plate in that they can leak into the substrate, so generalized Lamb waves share the same property with respect to Lamb waves in a free plate. For a thin layer, these sagittal plane modes can be seen as a perturbation of Rayleigh waves on a free surface so they have also been called Rayleigh-like modes.

As for Love waves one of the dominant properties of these modes is that the presence of a layer introduces a length scale (thickness) for the wavelength so that these modes are generally dispersive. Hence phase and group velocities of each mode are of importance. Further, analogous to Love waves, one can anticipate that the nature of the modes depends on the ratio of layer and substrate parameters, in particular that of the shear velocities. The problem could be solved using partial wave analysis and the waveguide equation with transverse resonance similar to the approach used for Love waves. However, the calculations become unwieldy, so we will restrict the treatment to a description of the various modes that may be excited.

The seminal work of Tiersten [60] allows a clear distinction to be made between limiting cases of layer-substrate combinations as well as providing a quantitative estimate of the phase velocity V as $\beta b \to 0$. Tiersten's approach is perturbative, which for small βb yields

$$F_0(V) + \beta b F_1(V) + (\beta b)^2 F_2(V) = 0 \tag{10.40}$$

Of particular interest is the slope of the dispersion curve at $\beta b = 0$

$$\left.\frac{dV}{d(\beta b)}\right|_{\beta b=0} = -F_1(V_R) + \left.\frac{dF_0(V)}{dV}\right|_{V_R} \tag{10.41}$$

Tiersten shows that this quantity is positive if

$$\frac{\hat{V}_S}{V_S} > \sqrt{\frac{1-\left(\frac{V_S}{V_L}\right)^2}{1-\left(\frac{\hat{V}_S}{\hat{V}_L}\right)^2}} \tag{10.42}$$

where superscript ^ is for the layer material.

The right-hand side of this relation is bounded between $\sqrt{2}$ and $1/\sqrt{2}$. The various cases to be considered are best illustrated by the normalized axes shown in Figure 10.7. For $\hat{V}_S > V_S\sqrt{2}$ the layer is said to "stiffen" the substrate, and for $\hat{V}_S < V_S/\sqrt{2}$ the layer "loads" the substrate. The intermediate region in the figure will be treated in the next section and corresponds to Stoneley waves.

For an isotropic substrate, Tiersten showed that the perturbation to the Rayleigh velocity is given explicitly by

$$\frac{\Delta V_R}{V_R} = -\frac{V_R b}{4I_R}\left[\hat{\rho}|V_{Rz}|^2 + \left(\hat{\rho} - \frac{4\hat{\mu}}{V_R^2}\cdot\frac{\hat{\lambda}+\hat{\mu}}{\hat{\lambda}+2\hat{\mu}}\right)|V_{Rx}|^2\right]_{z=0} \tag{10.43}$$

where I_R is the average unperturbed power flow per unit width along x.

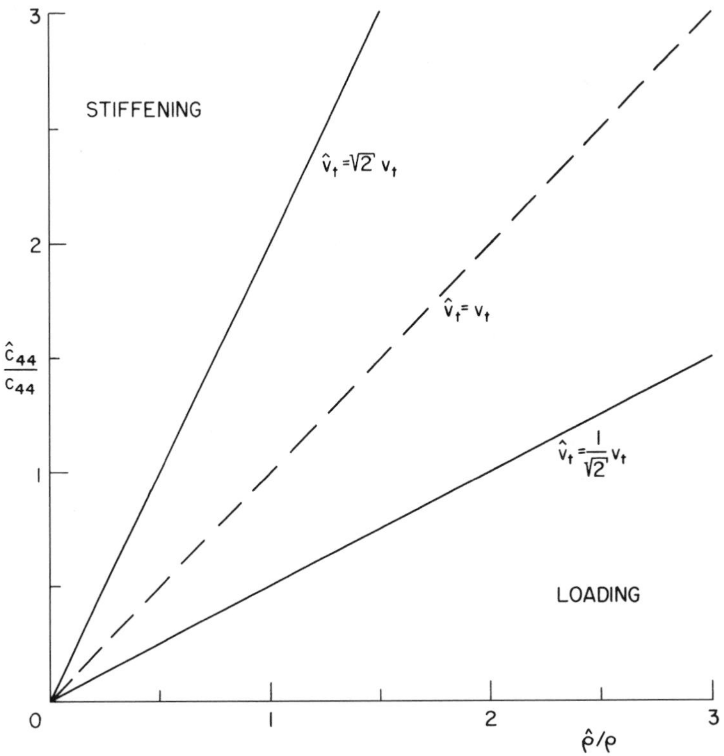

FIGURE 10.7
Sufficient conditions for stiffening and loading for isotropic material combinations. (From Farnell, G.W. and Adler, E.L., Elastic wave propagation in thin layers, in *Physical Acoustics*, IX, Mason, W.P. and Thurston, R.N., Eds., Academic Press, New York, 1972, chap. 2. With permission.)

The sign of the term in brackets is positive for stiffening and negative for loading as described above. It will be seen that the sign of $\Delta V/V$ follows naturally from the simple physics of the problem. Sufficient conditions for stiffening and loading are given in Figure 10.7. To harmonize with the notation of Figures 10.7 through 10.13 in this section we replace βb by kh.

1. Stiffening: $\hat{V}_S > V_S$

   A typical example is silicon ($\hat{V}_S$ = 5341 ms^{-1}) on a ZnO substrate (V_S = 2831 ms^{-1}) as shown in Figure 10.8(a). For vanishingly thin layer thickness ($kh \rightarrow 0$), the velocity is the Rayleigh wave velocity for the bare substrate. The high-velocity layer increases the effective surface wave velocity until it reaches the substrate shear wave velocity. For higher values of kh the partial wave leaks

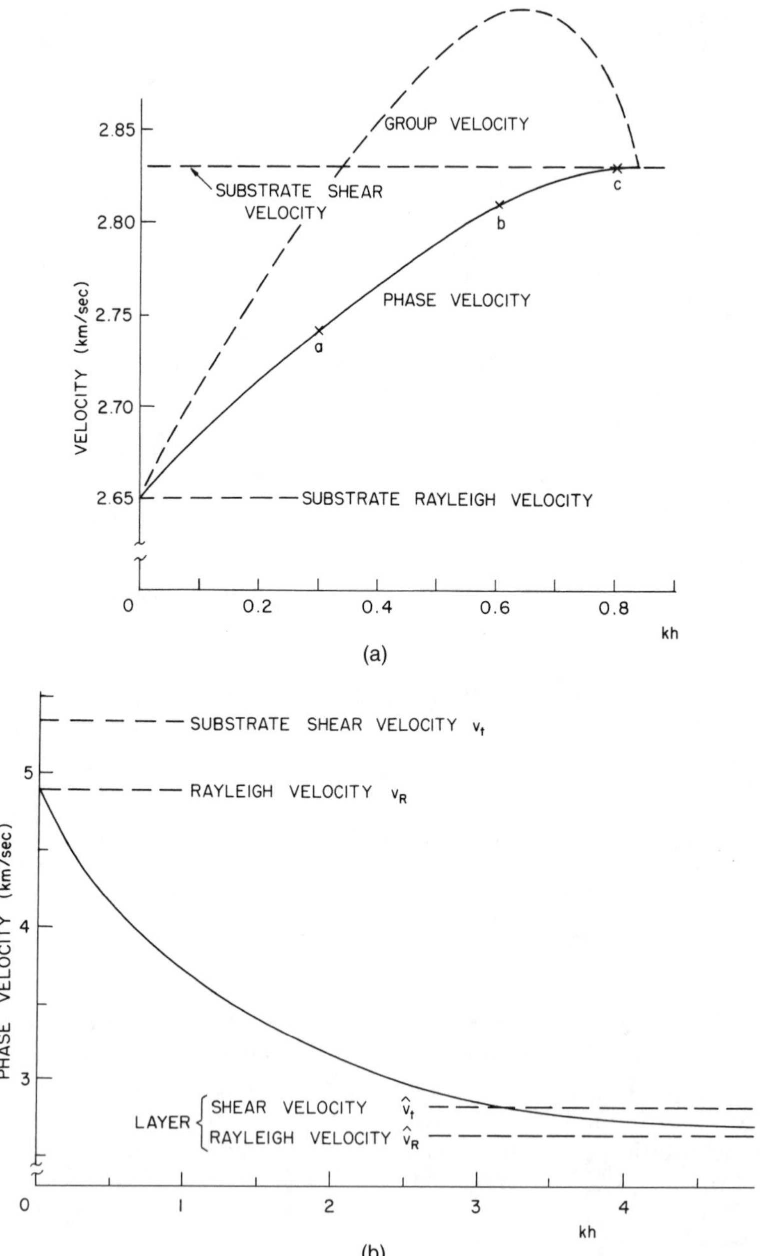

FIGURE 10.8
(a) Phase and group velocities for a silicon layer on a ZnO substrate under stiffening conditions ($\hat{V}_S > V_S$). (b) Phase velocity of the first Rayleigh mode under loading conditions ($\hat{V}_S < V_S$) for ZnO on Si. (From Farnell, G.W. and Adler, E.L., Elastic wave propagation in thin layers, in *Physical Acoustics*, IX, Mason, W.P. and Thurston, R.N., Eds., Academic Press, New York, 1972, chap. 2. With permission.)

into the substrate so that a true surface wave (evanescent decay) no longer exists, and the mode becomes a pseudo-bulk wave. Since the phase velocity reaches this condition with a horizontal slope the group velocity also goes to zero at this point. This is the only solution for the case of stiffening.

2. Loading: $\hat{V}_S < V_S$

   In this case the slope at $kh \to 0$ is negative as predicted by Equation 10.41. This can be understood very simply as follows. For $kh \to 0$, as before, the Rayleigh wave velocity approaches that of the bare substrate. As kh increases the importance of the layer increases progressively, leading to a decrease in velocity due to the effect of this low velocity material. Finally, for $kh \to \infty$ the layer dominates completely and the velocity approaches the Rayleigh wave velocity asymptotically. This explains the overall behavior of the first Rayleigh mode shown in Figure 10.8(b). As kh increases, higher-order Rayleigh modes are excited in the spirit of transverse resonance much as for Love waves. As in the latter case, each of these higher modes leaks into the substrate at a sufficiently low frequency.

   The lowest of the higher-order Rayleigh modes is important in seismology and device physics; it is the Sezawa mode, discovered by Sezawa and Kanai in 1935 [61]. The displacement components are reversed compared to the fundamental and the displacement ellipse is progressive for the Sezawa mode and regressive for the fundamental mode [57]. Displacements for the first Rayleigh mode and second Rayleigh (Sezawa) modes as a function of depth are shown in Figures 10.9 and 10.10, respectively.

10.5.3 Stoneley Waves

A Stoneley wave [62] is a sagittal interface wave between two solids that is evanescent in both media as shown for a tungsten-aluminum combination in Figure 10.11. For a solid-solid interface, these are very restrictive conditions on the existence of these modes as shown by the shaded regions in Figure 10.12. It turns out from the analysis that the Stoneley wave velocity V_{ST} lies in the range $V_R < V_{ST} < V_S$ of the dense medium and that $V_{ST} < V_S$, $\hat{V}_S$ of both media.

It is interesting to see how one can pass from the layer situation to the Stoneley case for two suitable solids by passing to the limit $kh \to \infty$. This is shown in Figure 10.13 for a tungsten-aluminum interface. Initially, if tungsten is taken as the substrate and the aluminum as the layer, the curve rises from $V_R(W)$ as this case corresponds to stiffening. As seen previously this mode exhibits in the range where it approaches asymptotically the shear wave velocity $V_S(W)$. However, in this case it rises asymptotically to V_{ST}

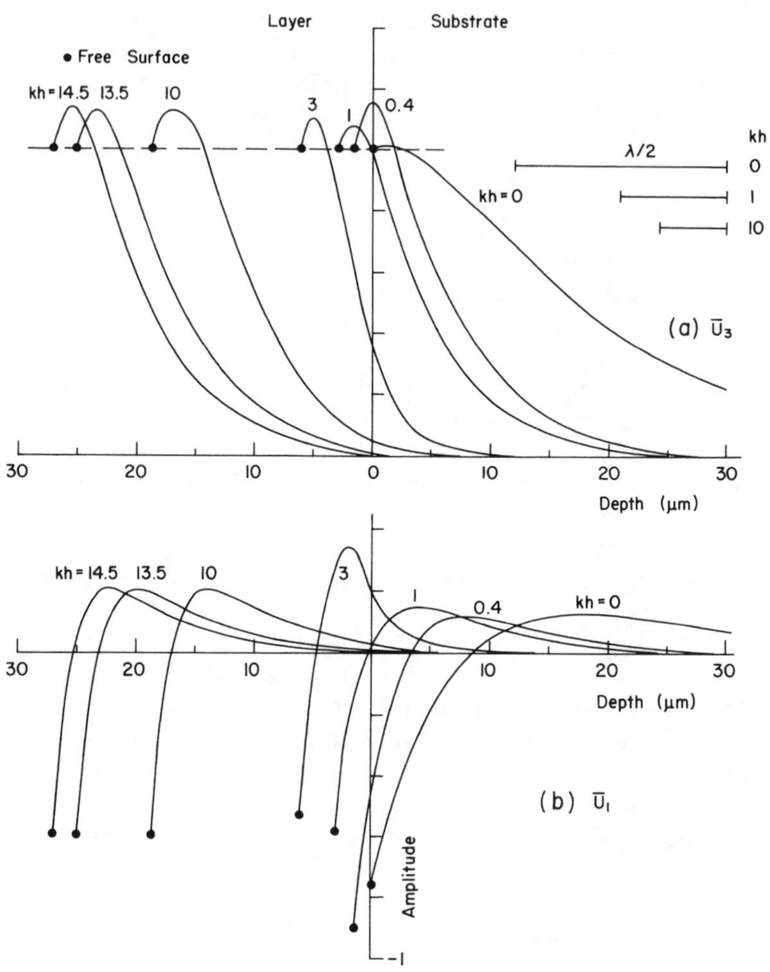

FIGURE 10.9
(a) Vertical component and (b) longitudinal component of displacement for the first Rayleigh mode at different values of *kh*. Gold on fused quartz, $f = 100$ MHz. Dots at the end of the curves indicate the position of the free surface for each *kh*. (From Farnell, G.W. and Adler, E.L., Elastic wave propagation in thin layers, in *Physical Acoustics*, IX, Mason, W.P. and Thurston, R.N., Eds., Academic Press, New York, 1972, chap. 2. With permission.)

where it becomes a Stoneley mode that exists for $h \to \infty$. If aluminum is taken as the substrate this is a loading situation and the fundamental mode velocity decreases from $V_R(Al)$ at $kh = 0$, goes through a minimum, and then approaches $V_R(W)$ for $kh \to \infty$ as seen previously. In this case, it is the first Sezawa mode whose velocity decreases from $V_S(Al)$ and approaches V_{ST} asymptotically as $kh \to \infty$ instead of becoming asymptotic to the layer shear velocity as occurred previously for the non-Stoneley cases.

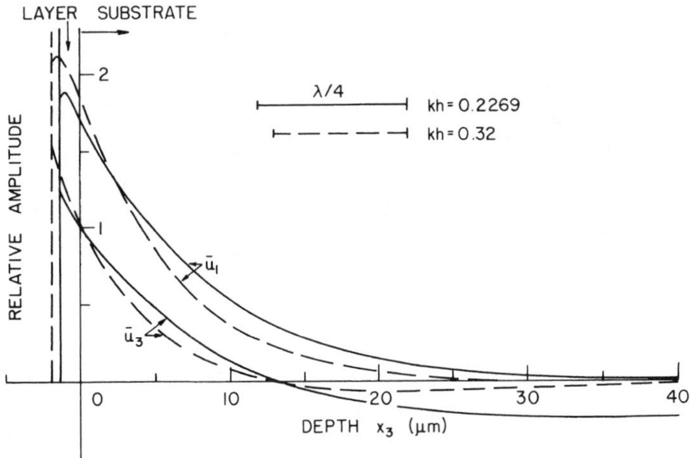

FIGURE 10.10
Displacement for the second Rayleigh (Sezawa) mode. The solid curves are for *kh* just above cutoff. F = 100 MHz. Gold on fused quartz. (From Farnell, G.W. and Adler, E.L., Elastic wave propagation in thin layers, in *Physical Acoustics*, IX, Mason, W.P. and Thurston, R.N., Eds., Academic Press, New York, 1972, chap. 2. With permission.)

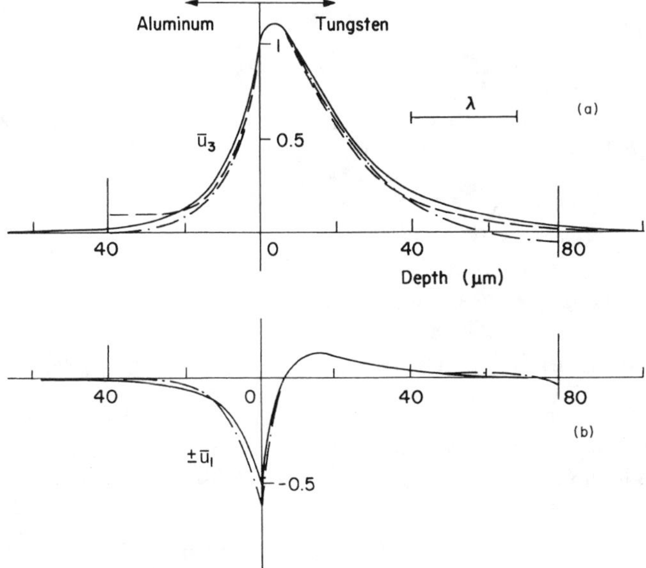

FIGURE 10.11
(a) Vertical and (b) longitudinal displacement components for aluminum-tungsten Stoneley wave (solid curves). Broken curves are for a layer of one material on a substrate of the other. F = 100 MHz. The broken curve for $\bar{u}_1$ for aluminum on tungsten is indistinguishable from Stoneley waves on this scale. (From Farnell, G.W. and Adler, E.L., Elastic wave propagation in thin layers, in *Physical Acoustics*, IX, Mason, W.P. and Thurston, R.N., Eds., Academic Press, New York, 1972, chap. 2. With permission.)

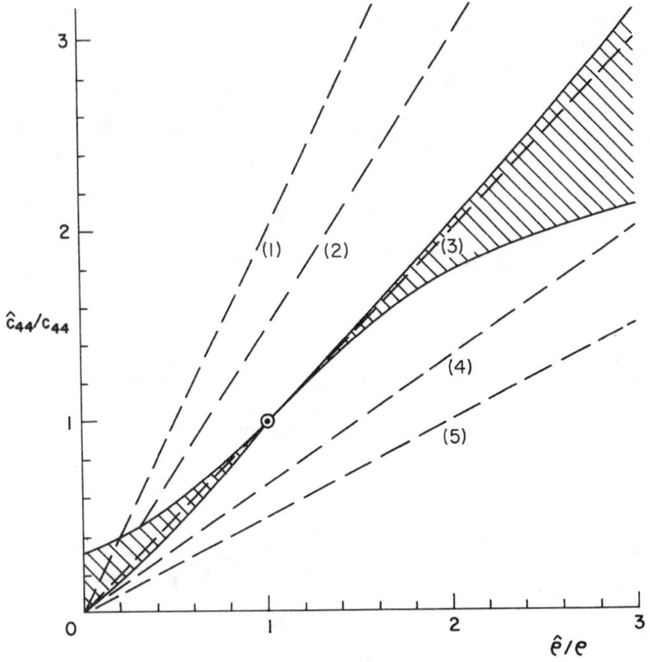

FIGURE 10.12
Region of existence of Stoneley wave (shaded region) for $V = \hat{V} = 1/\sqrt{3}$. Broken lines are lines of constant $\hat{V}_S/V_S$; material combinations above appropriate line have positive slope for dispersion curve at the origin. (1) $V = 0$, $\hat{V} = 1/\sqrt{2}$; (2) $V = 0$, $\hat{V} = 1/\sqrt{3}$; (3) $V = \hat{V}$; (4) $V = 1/\sqrt{3}$, $\hat{V} = 0$; (5) $V = 1/\sqrt{2}$, $\hat{V} = 0$. (From Farnell, G.W. and Adler, E.L., Elastic wave propagation in thin layers, in *Physical Acoustics*, IX, Mason, W.P. and Thurston, R.N., Eds., Academic Press, New York, 1972, chap. 2. With permission.)

10.6 Multilayer Structures

Although multilayer structures can act as acoustic waveguides, it is most practical to carry out a formal analysis in this direction. The extreme complexity of sagittal modes in a simple solid layer on a substrate illustrates the futility of such an exercise. Fortunately, the most important aspect of multilayer structures is the transmission or reflection of acoustic waves from them, and this can be carried out with surprising transparency. This is an important problem in many areas of ultrasonics, including NDE (adhesion, laminated materials, composites, etc.), oceanography (stratified fluids and sediments), medical ultrasonics, and instrumentation (acoustic microscopy and impedance matching of probes).

The multilayer problem has been reviewed in depth by Lowe [63] for the case of n layers between solid and media and where reference is made to studies of anisotropic layers and cylindrical layers. Achenbach et al. [64]

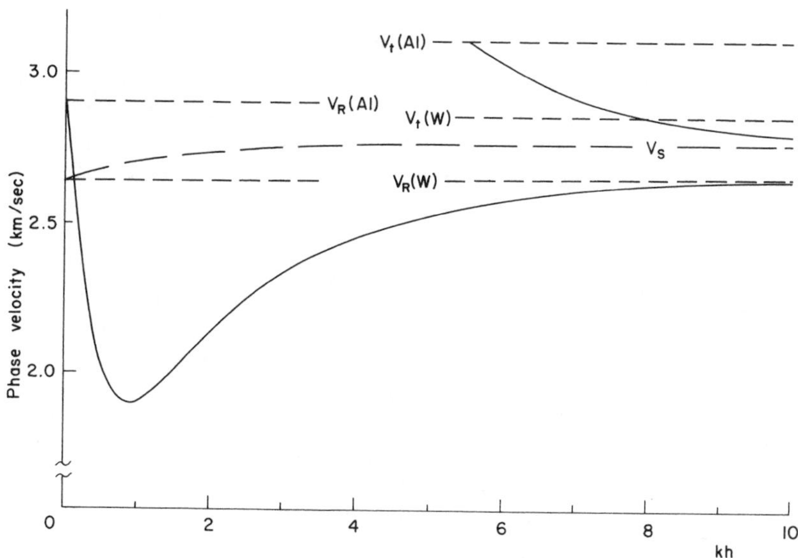

FIGURE 10.13
Dispersion curves for a tungsten layer on an aluminum substrate (solid curve) and aluminum on tungsten (broken curve). V_S is the velocity of the Stoneley wave. (From Farnell, G.W. and Adler, E.L., Elastic wave propagation in thin layers, in *Physical Acoustics*, IX, Mason, W.P. and Thurston, R.N., Eds., Academic Press, New York, 1972, chap. 2. With permission.)

provide a detailed procedure for the general analysis of anisotropic layers on an anisotropic substrate. As pointed out in [29] and [63], two main approaches have been adopted for quantitative analysis:

1. Global matrix method [65] where a single matrix is used to represent the complete system. It encompasses $4(n-1)$ equations for the n layers that correspond to the boundary conditions at each interface. The method is stable and avoids several well-known pitfalls of other approaches but it is exceedingly cumbersome to carry out. It will not be discussed further here.
2. Transfer matrix approach originally proposed by Thomson [66], corrected by Haskell [67], and formalized in the general theory by Brekhovskikh [30]. Each layer is represented by a matrix and the $n-1$ matrices are multiplied together to represent the whole system. This is a conceptually simple approach and results will be presented below.

The formalism follows [29] for the case of an isotropic substrate (medium n) supporting $(n-1)$ isotropic layers going from $(n-1)$ near the substrate to layer number 1 next to the incident fluid medium (medium 0). A longitudinal wave at angle θ is incident in the fluid and longitudinal and shear waves are emitted into the substrate.

Each layer is characterized by four waves: two longitudinal and two shear in each of forward and backward directions. Conservation of parallel wave vector is respected for all media and the usual boundary conditions of continuity of normal and tangential displacement and stress apply. For the ith layer the matrix a_{ij} links displacements and stresses at the upper and lower boundaries by

$$\begin{pmatrix} u_{ix} \\ u_{iz} \\ T_{z-} \\ T_{xz} \end{pmatrix} = \begin{pmatrix} a_{11} & a_{12} & a_{13} & a_{14} \\ a_{21} & a_{22} & a_{23} & a_{24} \\ a_{31} & a_{32} & a_{33} & a_{34} \\ a_{41} & a_{42} & a_{43} & a_{44} \end{pmatrix} \cdot \begin{pmatrix} u_{(i+1)x} \\ u_{(i+1)z} \\ T_{(i+1)zz} \\ T_{(i+1)xz} \end{pmatrix} \tag{10.44}$$

The matrix elements a_{ij} are given in [29]. For the $(n-1)$ layers, the total effect A can be obtained by multiplying the layer matrices, so that

$$A = \prod_{i=1}^{n-1} a_i \tag{10.45}$$

which means that the $(n-1)$ layers are described by

$$\begin{pmatrix} u_{1x} \\ u_{1z} \\ T_{1zz} \\ T_{1xz} \end{pmatrix} = \begin{pmatrix} A_{11} & A_{12} & A_{13} & A_{14} \\ A_{21} & A_{22} & A_{23} & A_{24} \\ A_{31} & A_{32} & A_{33} & A_{34} \\ A_{41} & A_{42} & A_{43} & A_{44} \end{pmatrix} \cdot \begin{pmatrix} u_{nx} \\ u_{nz} \\ T_{nzz} \\ T_{nxz} \end{pmatrix} \tag{10.46}$$

Finally, the reflected and transmitted waves can be calculated from the potentials

$$\varphi^{(n+1)} = \exp[-i\alpha(z-z_n)] + V\exp[i\alpha(z-z_n)], \quad z \geq z_n \tag{10.47}$$

in the fluid, and

$$\begin{aligned} \varphi^{(1)} &= W_l \exp(-i\alpha_1 z) \\ \psi^{(1)} &= W_t \exp(-i\beta_1 z) \end{aligned} \tag{10.48}$$

in the substrate.

These results give the detailed reflection and transmission factors that are presented in [29].

10.7 Free Isotropic Cylinder

In addition to providing a classic problem in acoustics for acoustic modes in simple geometries, cylindrical structures also have found some application as acoustic waveguides and delay lines in the form of thin rods [68], capillaries, tubes [69], etc. The calculation of the acoustic modes in the isotropic cylinder is best done with the potential method. Since complete accounts have been given elsewhere [26, 32, 68], we summarize the main results only briefly here.

The wave equation and potential functions are of course expressed in cylindrical coordinates (r, θ, and z) and the solutions for the potentials and the displacements are found in terms of Bessel functions. Displacement components are u_r and u_θ in the section of the cylinder and u_z along its length. As usual, the boundary conditions for the three components of stress at the free surface T_{rr}, $T_{r\theta}$, and T_{rz} are set equal to zero. The determinant of the coefficients is set equal to zero and the dispersion equation can be solved numerically. The result is that there are three families of modes, which can be described as follows:

1. Compressional modes, which are axially symmetric with displacements u_r and u_z independent of θ. The dispersion relation, known as the Pochhammer-Chree equation, qualitatively resembles Lamb waves in that the fundamental mode goes to a constant velocity $V_E = \sqrt{E/\rho}$ where E is Young's modulus and ρ is the density as $ka \to 0$, where a is the radius of the cylinder. The other modes all have cutoff frequencies and are dispersive.
2. Torsional waves. There is only one displacement component, u_θ, which is independent of θ. Again, this mode is not dispersive, with constant velocity $V_S = \sqrt{\mu/\rho}$ where μ is the shear modulus. The higher-order modes are dispersive and have cutoff frequencies. The dispersion curves for the whole family of torsional modes resemble those for SH modes in a plate.
3. Flexural waves. These are the most complicated as they involve displacement components u_r, u_θ, and u_z, which vary with θ as $\sin n\theta$ and $\cos n\theta$. These modes qualitatively resemble the antisymmetric modes of a plate; in particular, the fundamental has no cutoff and propagates down to zero frequency, with velocity $V \approx \sqrt{\omega a/2}\sqrt{V_E}$ while the higher modes have cutoffs and are dispersive.

10.8 Waveguide Configurations

Most waveguide configurations of current use are based on the SAW configuration. Especially when considered in the context of modern microelectronic technology, the standard SAW configuration has several major disadvantages. These include beam spread due to diffraction, which can lead to undesired crosstalk; large width (up to 100 wavelengths), which can lead to large areas; and single orientation, i.e., uni- or bidirectional, and inability to turn corners or go from one device layer to another.

These disadvantages can be overcome by the use of acoustic waveguides; where beam spread is suppressed by the guiding action, the width can be reduced to the order of a wavelength and in certain cases the guide can be oriented at will (as for fibers, for example). Such acoustic waveguides are being used increasingly in sensor and NDE applications, and they have good potential for multifunction devices and acoustic nonlinear applications due to the inherent possibility of having a high power density. There are some technological difficulties to be overcome such as loss reduction, increase of excitation efficiency, and fabrication problems for fibers, but these are soluble in principle. General overview of acoustic waveguides have been given in [32, 70, and 71].

There are two general considerations that enter into the design of acoustic waveguides. The first is the degree of field confinement or, viewed otherwise, the rate of decay of the acoustic field in the substrate. This must be controlled to suit the application. For example, a strong decay is desirable to reduce crosstalk but in other applications some degree of coupling between guides may be desirable. A second consideration is dispersion. Since such guides are made using thin films or thin topographical structures there is always an intrinsic length scale involved, which introduces dispersion. It is generally desirable to design for a dispersionless or low dispersion region within the bandwidth, especially for long delay lines. Three different approaches to acoustic waveguides will be discussed briefly.

10.8.1 Overlay Waveguides

The basic principle here is to deposit a film or films on the substrate to lower the sound velocity in the guide region. As seen previously, this can lead to trapping in the guide and evanescent decay of this mode in the substrate. The most direct way to do this is with the strip guide as shown in Figure 10.14. The material of the overlay is chosen so that it corresponds to acoustic loading of the substrate beneath it.

The strip guide is dispersive and its behavior with frequency follows the behavior expected from Section 10.5.2. As $\beta b \to 0$, the phase velocity approaches the Rayleigh wave velocity in the substrate. Hence, there is almost no guiding action and the wave is spread out over the substrate. As the frequency increases, the velocity decreases toward the Rayleigh wave velocity in the strip and at sufficiently high frequencies the wave is confined to the strip.

10.8.1.1 Slot Waveguide

The slot waveguide is the complementary configuration to the strip. The wave is guided along the bare substrate with strips on either side as shown in Figure 10.14. The material of the strips is chosen so that it stiffens the

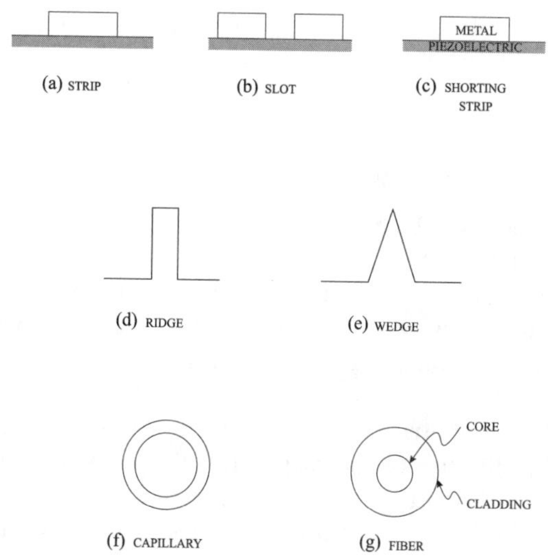

FIGURE 10.14
Acoustic waveguide configurations: (a), (b), and (c) are flat overlay waveguides; (d) and (e) are topographic waveguide configurations; (f) and (g) are two types of circular fiber waveguide.

Rayleigh velocity of the substrate. As a result, the acoustic wave is trapped in the slot that forms an acoustic waveguide.

The basis of the analytical treatment for the slot is similar to that for the strip [70], but the dispersion curve is clearly different now as the phase velocity increases with frequency. At low frequency, as before, the wave is spread out and the guiding action is weak. The velocity increases with frequency but at sufficiently high frequency it must again equal the Rayleigh velocity in the substrate, so it passes by a broad maximum. At high frequencies, the acoustic wave is confined to the slot.

10.8.1.2 Shorting Strip Waveguide

In the spirit of the strip and slot waveguides, an acoustic waveguide can be formed by any means that produces a local change in substrate velocity. The strip and slot accomplish this by the use of loading and stiffening layers, respectively. Another way to do this is with the metallic shorting strips on a piezoelectric substrate. The metal shorts the piezoelectric field, which results in a lowering of the Rayleigh wave velocity, the decrease depending on the magnitude of the piezoelectric constant. The actual behavior of the device is then very similar to that of the strip waveguide. Since the change in velocity is typically only 1 or 2% the wave is only weakly guided in this configuration.

Finally, a conceptually equivalent approach to the shorting strip would be to use diffusion, in implantation or local depoling, to produce the requisite local velocity changes. There are indications [70] that waveguides produced by diffusion can produce significant velocity changes of the order of several percent but with no accompanying attenuation increase, thus overcoming one of the main drawbacks of the overlay technique.

10.8.2 Topographic Waveguides

These are produced by a local deformation of the substrate, in the form of a protuberance such as a ridge, wedge, etc. Unlike overlay waveguides, where the binding is loose and governed by horizontal reflections, topographical waveguiding is vertical and strong. Hence, the two cases are qualitatively different. The main types are as follows:

1. Ridge, antisymmetrical, or flexural modes. The ultrasonic field is strongest at the top of the ridge and dies away exponentially into the substrate. This type of guide is strongly dispersive. It follows the A_0 mode down to a minimum, then rises to a cutoff at a frequency determined by the height-width aspect ratio of the ridge.
2. Ridge, symmetric, or pseudo-Rayleigh mode. In this case, there is almost no dispersion. The displacement is a combination of that due to S_0 and SH modes. It is close to acting like an acoustic co-ax line, with tight confinement, almost zero dispersion and propagating down to zero frequency.
3. Wedge waveguide. An ideal wedge (i.e., one with no substrate effects) has no length scale and so should be dispersionless. The waveguiding properties should then be controlled uniquely by the apex angle. This idealized condition can be approached in practice. Mainly flexural waves are excited in the wedge and they are tightly bound to the apex. In practice, the structure has very low dispersion.

10.8.3 Circular Fiber Waveguides

The aim of this development originally was to obtain low loss, low dispersion, and very long delay lines. Two types generally developed historically:

1. Capillary waveguides [69], where the acoustic wave propagates as a Rayleigh wave along the inside surface of the capillary. Relatively constant group velocity can be obtained over a limited frequency range. One big advantage is that the structure can be made by drawing standard fused silica tubing.
2. A second approach is that of cladded acoustic fibers [72], based on the principle of clad optical fibers. If the velocity of the cladding is greater than that of the case, then the acoustic mode can be trapped in the case and propagated over large distances. This principle has been adapted to the development of cladded delay lines as described in Chapter 15.

Summary

Partial waves are the components of the plane wave solutions appropriate to a particular guided wave problem. They are oriented so that they have a common wave vector β along the waveguide. Possible modes are determined by transverse resonance in the guide.

The waveguide equation encompasses the concepts of transverse resonance and cutoff in an acoustic waveguide. Its geometrical formulation involves the slowness curve of the waveguide material, which can be used to determine whether a given mode is trapped, propagating, or evanescent.

Love waves are SH modes propagating in a layer on a substrate. They can only occur if the transverse wave velocity in the layer is less than that in the substrate.

Stoneley waves are interface waves between two solids, which propagate without leakage. Their existence conditions are quite stringent, depending on the ratios of densities and elastic constants.

Questions

1. Describe qualitatively the differences between the acoustic modes possible in a plate and a tube, for all ranges of the ratio of thickness to wavelength and of thickness to tube diameter.
2. Compare the full range of acoustic modes to be found in a fluid-loaded rod to those for a tube with fluid loading inside and outside.
3. Explain the physical connection between Rayleigh and Lamb waves by considering a plate at a given value of fb and varying the thickness from zero to infinity.
4. Which acoustic waveguide configuration would be the most appropriate for detecting the difference between ice and water on the surface of a material structure? Explain.
5. "Theoretically, problems involving SH modes are much simpler to solve than those for saggital modes, but experimentally it is much easier to excite and study saggital modes than SH modes." Discuss.
6. Describe qualitatively the changes in the dispersion curve for the SH waveguide when the slowness curve is changed from a circle to an ellipse with major axis oriented along the guide.
7. Explain why Stoneley waves are nondispersive and have no higher-order modes.

8. Give a qualitative description of guided waves in a fluid contained between two solids. Sketch the expected dispersion curve.
9. Sketch the form of the displacements in the fundamental mode for compressional, torsional, and flexural waves in a cylinder of radius a.
10. Work out the quarter wavelength matching layer problem between a liquid and a solid for the case where two layers are used. Suggest suitable materials and thicknesses when the liquid is liquid helium and the solid is sapphire.
11. Under what conditions will a Love wave be most sensitive to conditions on the surface? This will determine the suitability of this mode as an acoustic sensor.

11

Crystal Acoustics

11.1 Introduction

Hooke's law for a three-dimensional solid gave the previous result

$$T_{ij} = c_{ijkl}S_{kl} \tag{4.47}$$

Using the definition of the strain tensor S_{kl}, this becomes

$$T_{ij} = \frac{c_{ijkl}}{2}\frac{\partial u_k}{\partial x_l} + \frac{c_{ijkl}}{2}\frac{\partial u_l}{\partial x_k} \tag{11.1}$$

Since $c_{ijkl} = c_{ijlk}$, the two terms on the right-hand side are equal, so that

$$T_{ij} = c_{ijkl}\frac{\partial u_l}{\partial x_k} \tag{11.2}$$

The equation of motion was also shown to be

$$\frac{\partial T_{ij}}{\partial x_j} = \rho\frac{\partial^2 u_i}{\partial t^2} \tag{5.11}$$

which now becomes

$$\rho\frac{\partial^2 u_i}{\partial t^2} = c_{ijkl}\frac{\partial^2 u_l}{\partial x_j \partial x_k} \tag{11.3}$$

For a bulk medium in three dimensions, we look for plane wave solutions in the form

$$u_l = u_{0l}\exp j(\omega t - \vec{k} \bullet \vec{r}) \tag{11.4}$$

where the propagation vector $\vec{k}$ is normal to planes of constant phase. Writing $\vec{n}(n_1, n_2, n_3)$ as a unit vector perpendicular to the wave front, we have

$$\vec{k} = \left(\frac{\omega}{V}\right)\vec{n} \tag{11.5}$$

where V is the phase velocity. For simplicity the subscript P is dropped in this section.

We also have $\vec{u}(u_1, u_2, u_3)$ is the particle displacement vector. To summarize, for a plane wave propagating in any direction, the direction of propagation is given by the components of $\vec{n}$ and the displacement (hence, the polarization) is determined by $\vec{u}$.

For bulk waves in isotropic media it was seen that there is one longitudinal mode and two transverse modes. It turns out that for crystalline media the corresponding general treatment that one can make is that for a given direction of propagation three independent waves may be propagated, each at a particular phase velocity and whose displacements are mutually orthogonal.

In general, these waves will be neither longitudinal nor transverse, and their displacements will have no specific orientation with respect to the wavefront. As will be shown for a given crystal structure, there are, however, certain directions in which "pure" modes (i.e., pure longitudinal or pure transverse) can be propagated.

Returning to the equation of motion, we now outline the standard procedure for determining phase velocities and displacements for a given propagation direction. Substituting the solution Equation 11.4 in the wave equation we obtain directly

$$\rho V^2 u_{oi} = c_{ijkl} n_k n_j u_{ol} \tag{11.6}$$

which is called Christoffel's equation. It is the very basis for subsequent determinations of the phase velocity. It is put in standard form by defining

$$\Gamma_{il} \equiv c_{ijkl} n_k n_j \tag{11.7}$$

so that

$$\Gamma_{il} u_{ol} = \rho V^2 u_{oi} \tag{11.8}$$

This means that u_{0i} is an eigenfunction of Γ_{il} and ρV^2 are its eigenvalues, determined by

$$\left|\Gamma_{il} - \rho V^2 \delta_{il}\right| = 0 \tag{11.9}$$

Since Γ_{il} is symmetric by its definition, it follows that the eigenvalues are real and the eigenfunctions orthogonal, which proves the statement made earlier on the three modes of propagation for a given direction. Application of the Γ tensor is a straightforward and a powerful way of determining the phase velocities for any direction in any crystal structure [26]. We will content ourselves here with two simple examples for the cubic system based on direct application of Equation 11.9.

11.1.1 Cubic System

The elastic constants for different crystal lattices are determined by the corresponding symmetry properties of those systems. Equation 11.6 for the cubic system yields the following three equations:

$$\begin{aligned}(c_{11}-c_{44})u_{01}n_1^2+(c_{12}+c_{44})u_{02}n_1n_2+(c_{12}+c_{44})u_{03}n_1n_3 &= (\rho V^2-c_{44})u_{01}\\ (c_{12}+c_{44})u_{01}n_2n_1+(c_{11}-c_{44})u_{02}n_2^2+(c_{12}+c_{44})u_{03}n_2n_3 &= (\rho V^2-c_{44})u_{02}\\ (c_{12}+c_{44})u_{01}n_3n_1+(c_{12}+c_{44})u_{02}n_3n_2+(c_{11}-c_{44})u_{03}n_3^2 &= (\rho V^2-c_{44})u_{03}\end{aligned} \quad (11.10)$$

As mentioned previously, this system can be solved formally for any particular direction $\vec{u}$ to give the three orthogonal polarizations and their corresponding phase velocities. Here we will rather look for the conditions for the existence of pure modes and then solve the simplified equations for three special directions.

For longitudinal waves, $\vec{u}$ is, by definition, parallel to $\vec{n}$. A necessary condition for this is $\vec{u}\times\vec{n}=0$. It follows that

$$\begin{aligned}u_{02}n_3-u_{03}n_2 &= 0\\ u_{03}n_1-u_{01}n_3 &= 0\\ u_{01}n_2-u_{02}n_1 &= 0\end{aligned} \quad (11.11)$$

which leads to

$$n_1:n_2:n_3=u_{01}:u_{02}:u_{03} \quad (11.12)$$

For the principal directions involving 0 or 1, this relation can be satisfied by the following combinations:

1. $n_1=n_2=n_3=1$: Direction [111]
2. one index zero and the other two unity: Equivalent directions [110], [101], [011]
3. two indices zero and the other unity: Equivalent directions [100], [010], [001]

(11.13)

This result tells us that pure longitudinal waves propagate in the [100], [110], and [111] directions or their equivalent. To determine the phase velocity in the [100] direction, for example, we substitute these values of $\vec{n}$ in Equation 11.10. Rearranging the terms gives

$$\begin{aligned} u_{01}(c_{11}-\rho V^2)+u_{02}(0)+u_{03}(0) &= 0 \\ u_{01}(0)+u_{02}(c_{44}-\rho V^2)+u_{03}(0) &= 0 \\ u_{01}(0)+u_{02}(0)+u_{03}(c_{44}-\rho V^2) &= 0 \end{aligned} \tag{11.14}$$

In this case, the calculation of the determinant is trivial leading to

$$\rho V_{[100]}^2 = c_{11} \tag{11.15}$$

for longitudinal waves. This direction also supports transverse waves, which by inspection have the phase velocity

$$\rho V_{[100]}^2 = c_{44} \tag{11.16}$$

Transverse wave phase velocities can be calculated in a similar way, although now the appropriate relation between $\vec{n}$ and $\vec{u}$ is

$$\vec{u} \bullet \vec{n} = 0 \tag{11.17}$$

or

$$u_{01}n_1 + u_{02}n_2 + u_{03}n_3 = 0$$

In a fashion similar to that for longitudinal waves it can be shown that the same directions also support transverse waves. It is a more complicated task to show that these are the only pure mode directions for cubic systems. This is carried out in more advanced treatments [26, 73, 74]; our goal here is to introduce the concept of propagation in anisotropic media and not to give a complete treatment.

11.2 Group Velocity and Characteristic Surfaces

The crystal structure does more than impose severe restrictions on the allowed directions for the propagation of pure modes. It also has profound implications on the direction of propagation of energy, which may be quite different from the direction of the official wave propagation unit vector $\vec{n}$. In order to uncover these implications of crystallinity, we will establish the link between the energy propagation velocity and phase velocity vectors. This can be done by rewriting the equation for the acoustic Poynting vector. It was previously shown that the Poynting vector can be written as

$$\mathcal{P}_i = -T_{ij}\frac{\partial u_j}{\partial t} = -C_{ijkl}\frac{\partial u_l}{\partial x_k}\frac{\partial u_j}{\partial t} \tag{11.18}$$

For plane waves,

$$u_i = u_{0i}f\left(t - \frac{n_j x_j}{V}\right) \tag{11.19}$$

where the equation for the wave front is $n_j x_j$ = constant. We then have directly

$$\frac{\partial u_i}{\partial t} = u_{0i}f' \tag{11.20}$$

$$\frac{\partial u_i}{\partial x_j} = -\frac{n_j u_{0i} f'}{V} \tag{11.21}$$

$\mathcal{P}_i$ can then be written as

$$\mathcal{P}_i = c_{ijkl}u_{0j}u_{0l}\frac{n_k}{V}f'^2 \tag{11.22}$$

The general form for the Poynting vector for plane waves is $P_i = u_a V_{ei}$, where V_e is the energy propagation velocity and the energy density $u_a = u_K + u_P$.

It is a well-known result that $\bar{u}_K = \bar{u}_P$ so $\bar{u}_a = 2u_K$. Hence,

$$\bar{u}_a = 2\cdot\frac{1}{2}\rho\left(\frac{\partial u_i}{\partial t}\right)^2 = \rho_{u_{0i}}f'^2 \tag{11.23}$$

and finally

$$V_{ei} = \frac{c_{ijkl}u_{0j}u_{0l}n_k}{\rho V} \tag{11.24}$$

where we have put $u_{0i}^2 = 1$.

We want to simplify the above relation between V_{ei} and V. This can be done by multiplying both sides of Christoffel's equation by u_{0i} to obtain $c_{ijkl}n_j n_k u_{0i} u_{0l} = \rho V u_{0i}^2$. Finally, we form

$$\vec{V}_{ei}\bullet\vec{n} = V_{ei}n_i = \frac{c_{ijkl}u_{0j}u_{0l}n_i n_k}{\rho V u_{0i}^2} = V \tag{11.25}$$

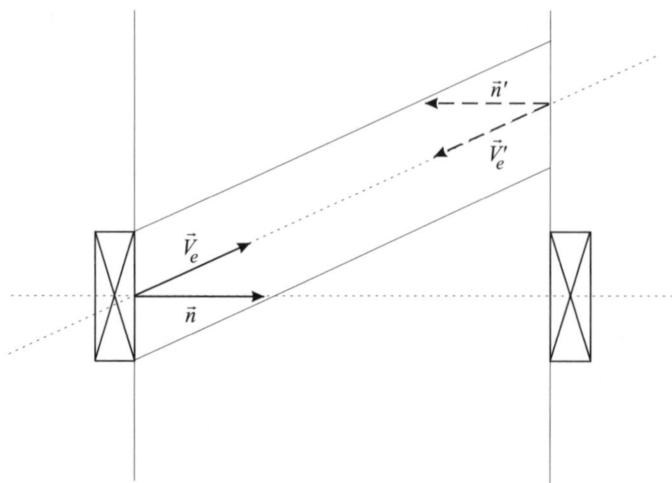

FIGURE 11.1
Transducer on a misoriented anisotropic buffer rod. The ultrasonic pulse will propagate off-axis in the direction of the group velocity as shown, thus missing a receiving transducer placed opposite the emitter. The reflected signal returns to the latter.

which means that the projection of the energy propagation velocity on the propagation direction gives the phase velocity.

This result has immediate practical consequences. In Figure 11.1, we show a crystal with plane parallel faces fitted with an emitting transducer for longitudinal waves on the left face and a receiving transducer on the right (the latter might also be a spherical cavity of an acoustic microscope used to focus the ultrasonic beam). The propagation direction is chosen to be a pure mode direction so that the energy propagation direction should correspond exactly with $\vec{n}$ if everything has been designed correctly. But if a mistake has been made in the choice of crystal orientation, then while $\vec{n}$ is still perpendicular to the transducer face the energy of the ultrasonic beam will propagate crabwise as shown in the figure. In a worst-case scenario, it may miss the receiver completely! Perversely, the reflected beam from the right-hand face will have $\vec{n}'$ antiparallel to $\vec{n}$, and the acoustic energy will retrace its path crabwise to the emitting transducer.

To gain further physical insight into this relation between V_e and V_P, we use the well-known result [26] that in linear acoustics the energy propagation velocity is equal to the group velocity V_G where

$$V_{Gj} \equiv \frac{\partial \omega}{\partial k_j} = \frac{\partial V}{\partial n_j} \tag{11.27}$$

or in vectorial form

$$\vec{V}_G = \vec{\nabla}_k \omega. \tag{11.28}$$

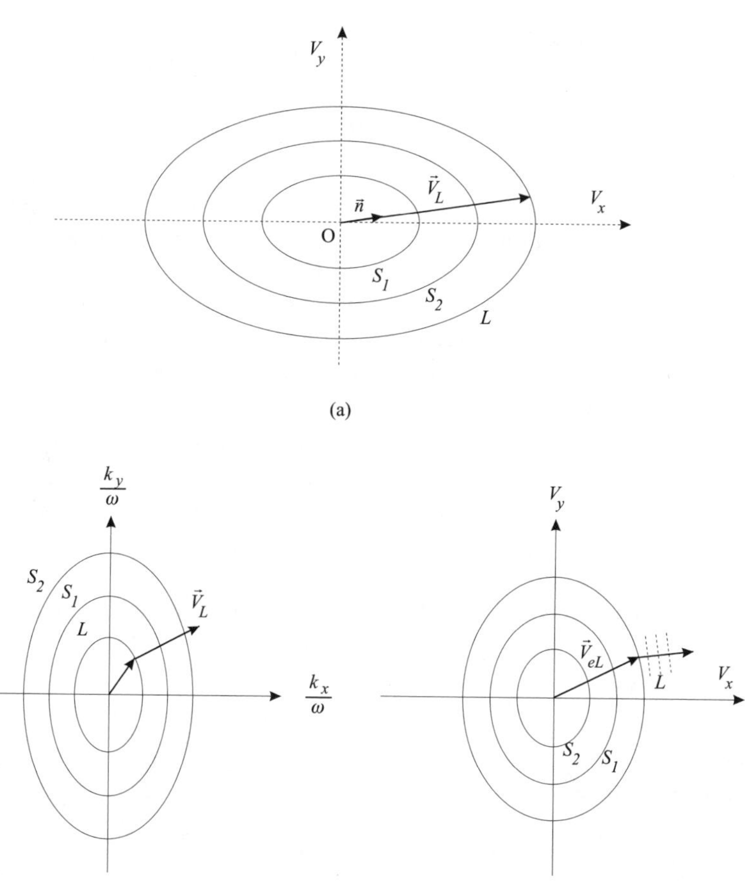

FIGURE 11.2
Schematic view of the characteristic surfaces for acoustic wave propagation in anisotropic solids. In all cases there are three shells, one quasi-longitudinal and two quasi-shear. (a) The velocity surface, which gives the phase velocity as a function of direction. (b) The slowness surface, which gives the variation of $1/V_P$ in $\vec{k}/\omega$ space. (c) The wave surface, which is the locus of points traced out by V_e as a function of propagation direction.

By vector analysis, the second form for $\vec{V}_G$ shows explicitly that $\vec{V}_G$ is perpendicular to a constant energy surface in $\vec{k}$ space.

In analogy with optics and the propagation of electromagnetic waves, several different surfaces can be constructed to describe the wave propagation. These have been described in detail in [26].

1. Velocity surface

   As shown in Figure 11.2(a), the velocity surface for a crystal is formed by tracing out the phase velocity variation $\vec{V}_P = V_P\vec{n}$ as a function of direction from a fixed origin O. There are three sheets corresponding to one quasi-longitudinal mode and two quasi-shear modes.

2. Slowness surface

   As shown in Figure 11.2(b), this surface has already been constructed for isotropic systems; in $\vec{k}/\omega$ space, it gives the variation of $1/V_P$ with direction for the three branches. A slowness surface is a surface of constant ω. Hence, for a point P on the surface the radius vector OP gives $1/V_P$ for that direction, and the group velocity for that direction is normal to the slowness surface at point P. Since this is a "reciprocal" space, the *L* surface is now inside the two *S* surfaces.

3. Wave surface

   As shown in Figure 11.2(c), this is the locus of the group velocity vector $\vec{V}_e$ as a function of direction starting from a fixed origin. Therefore, it gives the distance traveled by a wave emitted from O for different directions during a fixed time *t*. Since the wave arrives at all points on the surface at the same time, it is also an equiphase surface. For a given point P on the surface the propagation vector $\vec{n}$ for a plane wave with that value of $\vec{V}_e$ is perpendicular to the surface.

11.3 Piezoelectricity

11.3.1 Introduction

There are several different methods for exciting ultrasonic waves, including piezoelectricity, electrostriction, magnetostriction, electromagnetic (EMAT), laser generation, etc. Of these, the piezoelectric effect is by far the most widely used. The subject is covered at an advanced level, for materials and transducers design perspective in many sources [20, 31]. In the following, we give a general overview of the subject and introduce the parameters that come into play when piezoelectric materials are used to make ultrasonic transducers.

Piezoelectricity means that when we apply a stress to a crystal, not only a strain is produced but also a difference of potential between opposing faces of the crystal. This is called the direct piezoelectric effect. Conversely, the indirect effect corresponds to applying a difference of potential, which induces a strain in the crystal. Since the process is known to work at extremely high frequency (piezoelectric generation of sound has been reported up to 10^{12} Hz), piezoelectric crystals can be used to generate (inverse effect) and detect (direct effect) ultrasonic waves. The key to the phenomenon lies in the absence of a center of symmetry in piezoelectric crystals. This is, in fact, a necessary but not a sufficient condition for piezoelectricity; of the 21 crystal systems lacking a center of symmetry, 20 are piezoelectric.

The physics of the piezoelectric effect can be understood by referring to the case of quartz [75]. The piezoelectric crystal is placed between two metallic plates, which can support a stress and also serve as electrodes. If no stress is applied, the system of positive and negative charges share a common center of gravity. This means that there is no molecular dipole moment so the polarization is zero. If the crystal is subjected to compressive or tensile stress, the unsymmetrical distribution of positive and negative charge means that the centers of gravity of positive and negative charges no longer coincide. This creates a molecular dipole moment, hence a net polarization, the sign depending on whether compression or expansion took place. This leads to a corresponding accumulation of charge on the electrodes and hence to a potential difference between them. If an AC stress is applied, an AC potential difference is created at the same frequency with magnitude proportional to that of the applied stress.

The previous example can be made more concrete using a simple one-dimensional model, which will be retained, for simplicity, in this and the following section. Suppose that $\pm q$ are the charges of the positive and negative ions, and a is the charge in dimension of the unit cell. Again, for simplicity, we suppose one atom of piezoelectrical material per unit cell. Then the induced polarization can be expressed as qa/unit cell volume = eS, where e is the piezoelectric stress constant and S is the strain. Then the usual relation for dielectric media can be written

$$D = \varepsilon_0 E + P \tag{11.29}$$

$$= \varepsilon^S E + eS \tag{11.30}$$

where D and E are the electric displacement and electric field, respectively. The superscript S is standard in the literature for such relations and corresponds to permittivity at constant or zero strain. In a similar way, it can be shown that

$$T = c^E S - eE \tag{11.31}$$

These two relations are known as the piezoelectric constitutive relations; they will be examined in more detail in the next section.

11.3.2 Piezoelectric Constitutive Relations

Since there are two electrical variables (D, E) and two mechanical variables (T, S), there are several different possible ways of writing the constitutive relations introduced in the last section. In fact, choosing one electrical and one mechanical quantity as independent variables, we easily find that there are four different sets of constitutive relations that can be written. If, for example, we choose T and E as independent variables we can write $S = S(T, E)$ and $D = D(T, E)$. For small variations one can make a Taylor expansion of S and D about the equilibrium values and retain only the linear terms, resulting in

$$S = \left(\frac{\partial S}{\partial T}\right)T + \left(\frac{\partial S}{\partial E}\right)E \tag{11.32}$$

$$D = \left(\frac{\partial D}{\partial T}\right)T + \left(\frac{\partial D}{\partial E}\right)E \tag{11.33}$$

The proportionality constants are defined by

$$s^E = \left(\frac{\partial S}{\partial T}\right)_E, \quad d = \left(\frac{\partial S}{\partial E}\right)_T = \left(\frac{\partial D}{\partial T}\right)_E \quad \text{and} \quad \varepsilon^T = \left(\frac{\partial D}{\partial E}\right)_T \tag{11.34}$$

where the equality for d (and similar conditions for the other constitutive relations) can be obtained by thermodynamic considerations. Thus we have

$$S = s^E T + dE \tag{11.35}$$

$$D = dT + \varepsilon^T E \tag{11.36}$$

In a similar way for the other constitutive relations, we have

$$S = s^D T + gD \tag{11.37}$$

$$E = -gT + \beta^T D \tag{11.38}$$

$$T = e^E S - eE \tag{11.39}$$

$$D = eS + \varepsilon^S E \tag{11.40}$$

$$T = c^D S - hD \tag{11.41}$$

$$E = -hS + \beta^S D \tag{11.42}$$

Two of these constants merit particular attention for transducer applications [20]:

1. Receiver constant g, which determines the potential drop across the transducer for a given applied stress. We use Equations 11.37 and 11.38

$$S = s^D T + gD \tag{11.43}$$

$$E = -gT + \beta^T D \tag{11.44}$$

For a high-impedance receiver, the input current is small so the displacement current i_D in the electroded piezoelectric transducer goes to zero. With $i_D = \partial D/\partial t$, this gives D = constant or zero. Hence,

$$E = gT, \quad S = s^D T \tag{11.45}$$

For a given input stress T to the receiving transducer, the potential difference across the transducer is proportional to g. In this connection, a useful relation obtained from [20] gives

$$g = \frac{d}{\varepsilon^T} \tag{11.46}$$

2. Transmitting constant h. Another set of constitutive relations gives

$$T = c^D S - hD \tag{11.47}$$

$$E = -hS + \beta^S D \tag{11.48}$$

and we see that h gives the electric field (hence, potential difference) required to produce a given strain S. It can be shown that $h = e/\varepsilon^S$. All of the above has been done for a simple one-dimensional model. Of course, real crystals are three-dimensional so instead of constants linking $\vec{E}$, $\vec{D}$ (first-order tensors) to T_{ij}, S_{kl} (second-order tensors) the piezoelectric constants now become third-order tensors, e.g., $e \rightarrow e_{ijk}$. In reduced notation, this becomes e_{iJ}, $i = (x, y, z$ or $1, 2, 3)$ and $J = 1, 2, \ldots, 6$ as for the elastic constants. Thus we can write one of the constitutive relations as

$$T_I = c_{IJ}^E S_J - e_{Ij} E_j \tag{11.49}$$

$$D_i = \varepsilon_{ij}^S E_j + e_{iJ} S_J \tag{11.50}$$

For a given crystal, the nonzero values of the c_{IJ}, e_{Ij} and ε_{ij} are determined by symmetry as shown in detail in advanced treatises [e.g., 26].

An important result is that of the PZT, which is transversely isotropic about the z axis. We consider longitudinal propagation along the z axis, normal to the surface of a wide plate of PZT. If the width of the plate is much greater than the wavelength, the edges can be considered clamped so that $S_1 = S_2 = 0$, $T_1 \neq 0$, and $T_2 \neq 0$. We wish to determine T_3 for an applied electric field E_z. The two parameters are related by

$$T_3 = c_{33}^E S_3 - e_{3z} E_z \tag{11.51}$$

$$D_z = \varepsilon_{zz}^S E_z - e_{z3} S_3 \tag{11.52}$$

so the important constants are c_{33} and e_{3z}. These are among the nonzero constants for the case of transverse isotropy (hexagonal), which are

$$c_{11} = c_{22}, \quad c_{11} - c_{12} = 2c_{66}, \quad c_{44} = c_{55}, \ c_{33} \tag{11.53}$$

$$e_{z3}, e_{z1}, e_{z2}, e_{z4} = e_{z5} \tag{11.54}$$

$$\varepsilon_{xx} = \varepsilon_{yy} \tag{11.55}$$

In the next section, we consider the specific case of a transducer and define a simple coupling constant, which is the one simple parameter retained for practical characterization of piezoelectric transducers.

11.3.3 Piezoelectric Coupling Factor

The concept of coupling factor is used to determine the efficiency of coupling of electrical to mechanical energy. The coupling factor is also useful to compare the efficiency of different piezoelectric materials. The subject is fully treated in the IEEE standard of piezoelectricity [76] and we give only an overview for the one-dimensional case with propagation along the z axis.

For an infinite piezoelectric dielectric medium with no free charges and $B = 0$

$$\begin{aligned} \vec{\nabla} \bullet \vec{D} &= 0 \\ \vec{\nabla} \times \vec{E} &= -\dot{\vec{B}} = 0 \\ \vec{J} &= \dot{\vec{D}} = 0 \end{aligned} \tag{11.56}$$

which in one dimension leads to

$$E = -\frac{\partial \phi}{\partial z} \tag{11.57}$$

and

$$\frac{\partial D_z}{\partial z} = 0 \tag{11.58}$$

For longitudinal waves, we use the constitutive relations for T and D and use $S = \partial u_z/\partial z$

$$T_{zz} = c^E \frac{\partial u_z}{\partial z} - eE_z \tag{11.59}$$

$$D_z = e\frac{\partial u_z}{\partial z} + \varepsilon^S E_z \tag{11.60}$$

and the equation of motion

$$\frac{\partial T_{zz}}{\partial z} = \rho\frac{\partial^2 u_z}{\partial t^2} \tag{11.61}$$

Putting the two relations together immediately gives a new equation of motion for u_z in the piezoelectric medium

$$\rho\frac{\partial^2 u_z}{\partial t^2} = c^E\left(1 + \frac{e^2}{c^E \varepsilon^S}\right)\frac{\partial^2 u_z}{\partial z^2} \tag{11.62}$$

This shows that in the piezoelectric medium the sound velocity is stiffened compared to the nonpiezoelectric case

$$V_L = \sqrt{\frac{c^E}{\rho}} \quad \text{(unstiffened)} \tag{11.63}$$

$$V_L^D = V_L\sqrt{1+K^2} \quad \text{(stiffened)} \tag{11.64}$$

corresponding to

$$c^D = c^E(1+K^2) \tag{11.65}$$

with

$$K^2 = \frac{e^2}{c^E \varepsilon^S}, \quad \text{the piezoelectric coupling constant} \tag{11.66}$$

Note that this result is only valid for $D = 0$. Values of K^2 typically range from 10^{-2} to 0.5 so that the correction can be important for strongly piezoelectric materials. The formulation of K^2 is only valid for transversely clamped transducers (width much greater than the wavelength), which is generally true in ultrasonics. In practical transducer analysis, the impedance is determined by a related and oft-quoted parameter, k_T^2, the effective coupling constant

$$k_T^2 = \frac{K^2}{1+K^2} \tag{11.67}$$

For $K^2 \ll 1$, $k_T^2 \approx K^2$ but otherwise the difference may be significant.

On a more fundamental level, it can be shown that the piezoelectric coupling factor is given by

$$K^2 = \frac{U_{elec}}{U_{elas}} \tag{11.68}$$

where

$$U_{elec} = \text{stored electrical energy} \tag{11.69}$$

$$U_{elas} = \text{stored elastic energy} \tag{11.70}$$

This relation clearly reveals K^2 as being a parameter reflecting the coupling from electrical to mechanical energy. This relation may also be expressed as

$$\frac{U_{elec}}{U} = \frac{K^2}{2(1+K^2)} \tag{11.71}$$

where U is the total stored energy (the sum of kinetic, elastic, and electrical components).

12

Piezoelectric Transducers, Delay Lines, and Analog Signal Processing

12.1 Bulk Acoustic Wave Transducers

There are almost an infinite number of ways that ultrasonic transducers can be used in widely diverse applications. For purposes of this chapter, in specifying the medium that the transducer is coupled to, we group the use of BAW transducers into four main application areas:

1. Excitation for delay lines
2. General coupling into solids for NDE
3. Emission into aqueous media
4. Resonators in air for sensors and timing applications

The section starts with a summary of the representation of a BAW transducer by an equivalent circuit, followed by a discussion of transducer behavior in the neighborhood of the resonance frequency. Various cases of impedance matching and backing of the transducer are discussed and it is shown how these parameters influence the transducer operation in the time and frequency domain. Piezoelectric tranducers are discussed in most books on ultrasonics. One of the most comprehensive and useful contributions is given by Kino [20] and much of this section summarizes his treatment.

The transducer element is a cut from an oriented piezoelectric crystal, chosen so longitudinal or transverse waves are emitted perpendicular to the flat faces. Electrodes are applied to the opposite faces as shown in Figure 12.1(a), so an applied difference of potential gives rise to a uniform electric field in the z direction with $E_x = E_y = 0$. It is standard procedure to ground the lower electrode and attach the active lead to the top electrode, either by bonding, silver paste, or spring contact for laboratory R&D applications. Field applications transducers are packaged with an integral BNC connector so that these connections are made automatically. For the unpacked case, the ground connection is established easily by a (metallic) sample holder for

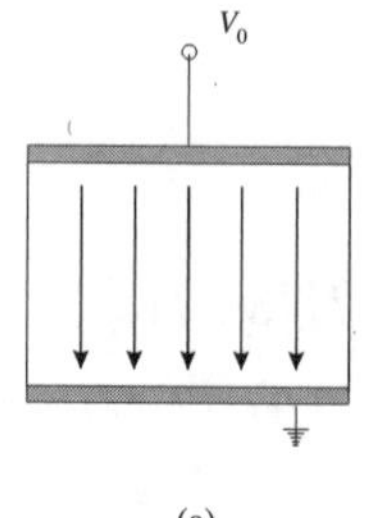

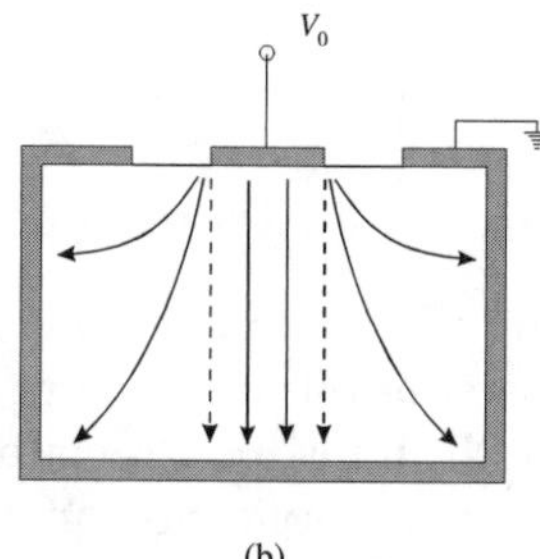

FIGURE 12.1
(a) Electric field in an ideal thickness mode piezoelectric transducer.
(b) Electric field in a thickness mode piezoelectric transducer with coaxial electrode configuration.

metallic samples, but the ground connection presents a problem for insulating samples. This is readily resolved by use of the coaxial configuration shown in Figure 12.1(b) in which the ground electrode is wrapped around to the front surface, so the ground and center connections can be made from above, at the price of a weak fringing field. The electrodes are generally vacuum-deposited gold, approximately 0.5 μm thick, on a thin chromium film for adhesion. In the equivalent circuit representation, the presence of the electrodes is generally neglected below 100 MHz as their thickness is much less than the wavelength. This is not the case at much higher frequencies where they must be included.

It is reasonably clear from Figure 12.1 that the transducer must be described, in general, by a three-port network. There are two acoustic ports corresponding to the media on either side of the transducer and one electrical port furnished by the two electrodes. The acoustic parameters conventionally chosen for such a description agree with the boundary conditions described earlier, namely the force F (stress) and particle velocity v, which correspond to the potential difference V and current I in electrical parlance. The sign conventions are chosen so that F is positive in the $+z$ direction and the particle velocity is positive when pointing toward the piezoelectric material. Appropriate values of F and v at the acoustic ports are represented in Figure 12.2(a). For the interior of the transducer we have

$$v = v_F \exp(-j\beta z) + v_B \exp(j\beta z) \tag{12.1}$$

$$T = T_F \exp(-j\beta z) + T_B \exp(j\beta z) - hD \tag{12.2}$$

These can be re-expressed in terms of equivalent circuit parameters as follows:

$$\bar{\beta} = \omega\sqrt{\frac{\rho_0}{c^D}} \tag{12.3}$$

$$\bar{Z}_0 = \sqrt{\rho_0 c^D} \tag{12.4}$$

$$T_F = -\bar{Z}_0 v_F \tag{12.5}$$

$$T_B = \bar{Z}_0 v_B \tag{12.6}$$

The electrical parameters are given directly by V_3 and I_3 from Figure 12.2.

Finally, the transducer can be described by a 3×3 impedance matrix coupling the current and voltage parameters as follows:

$$\begin{bmatrix} F_1 \\ F_2 \\ V_3 \end{bmatrix} = \begin{bmatrix} \bar{Z}_C \cot \bar{\beta}_a l & \bar{Z}_C \operatorname{cosec} \bar{\beta}_a l & \frac{h}{\omega} \\ \bar{Z}_C \operatorname{cosec} \bar{\beta}_a l & \bar{Z}_C \cot \bar{\beta}_a l & \frac{h}{\omega} \\ \frac{h}{\omega} & \frac{h}{\omega} & \frac{1}{\omega C_0} \end{bmatrix} \begin{bmatrix} v_1 \\ v_2 \\ I_3 \end{bmatrix} \tag{12.7}$$

where $Z_C = \bar{Z}_0 A$ is the impedance for a transducer of area A.

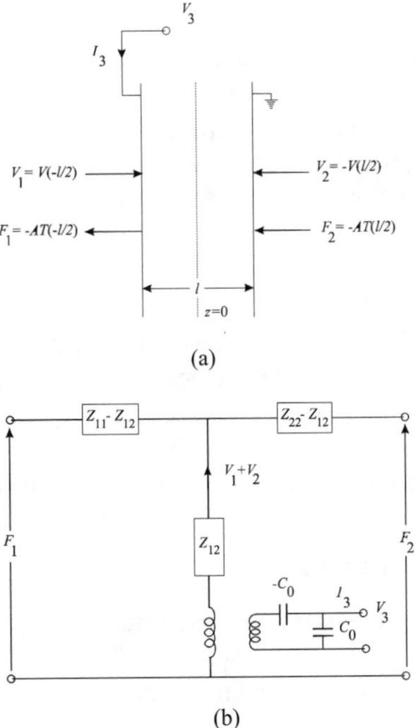

FIGURE 12.2
(a) Physical parameters of a transducer as a three-port device. (b) Mason equivalent circuit for (a). The transformer turns ratio is hC.

$C_0 = \varepsilon^S A/l$ is the clamped capacitance. C_0 comes up everywhere in the equations and is a dominant term in the transducer's behavior. This is entirely normal, as to zero order the transducer looks like a capacitor. The overall behavior of the transducer for practical applications can be characterized by the equivalent circuit, which can be derived from Equation 12.7 above, and the input impedance. The latter can be given by

$$Z_3 = \frac{V_3}{I_3} = \frac{1}{j\omega C_0}\left[1 + k_T^2 \frac{j(Z_1 + Z_2)Z_C \sin\beta_a l - 2Z_C^2(1 - \cos\beta_a l)}{[(Z_C^2 + Z_1 Z_2)\sin\beta_a l - j(Z_1 + Z_2)Z_C \cos\beta_a l]\beta_a l}\right] \tag{12.8}$$

where

$$Z_1 = -\frac{F_1}{v_1}$$

$$Z_2 = -\frac{F_2}{v_2}$$

$$k_T^2 = \frac{K^2}{1 + K^2}$$

and it is assumed that the transducer is in clamped conditions, that is, that the transducer width is much greater than the wavelength. The formula highlights the importance of the coupling constant as the main piezoelectric parameter to characterize the transducer. It is equally clear that $1/\omega C_0$ modulates the overall input impedance response as a function of frequency.

Finally, the full Mason equivalent circuit is given for reference in Figure 12.2(b). The two acoustic ports and their associated impedances are clearly identified. The electrical connection is made via a transducer with turns ratio $N = hC_0$ involving a piezoelectric constant as it should. The Mason equivalent circuit is the first and best-known equivalent circuit for the piezoelectric transducer. It has been criticized, for example, for the use of an unphysical negative capacitor, and modified, leading to the Redwood, KLM, and other models, which are described in detail by Kino [20].

12.1.1 Unloaded Transducer

The basic transducer operation near resonance is best understood for the case of the unloaded transducer ($Z_1 = Z_2 = 0$), which also provides the basis for its use as an acoustic resonator. From Equation 12.8 with $Z_1 = Z_2 = 0$ we have

$$Z_3 = \frac{V_3}{I_3} = \frac{1}{j\omega C_0}\left(1 - \frac{k_T^2 \tan\frac{\beta_a l}{2}}{\frac{\beta_a l}{2}}\right) \tag{12.9}$$

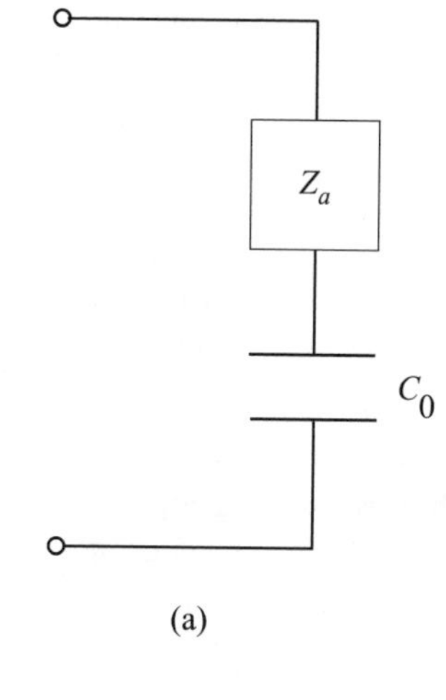

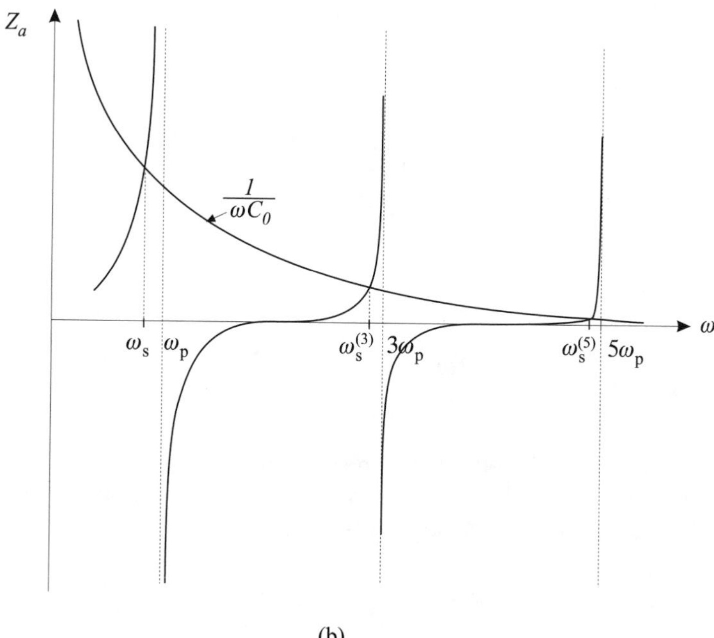

FIGURE 12.3
(a) Equivalent circuit and (b) impedance of a piezoelectric transducer as a function of frequency showing series (ω_S) and parallel (ω_P) resonant frequencies.

which can be represented as in Figure 12.3 as a capacitance C_0 in series with a motional impedance Z_a

$$Z_a = -\frac{k_T^2}{\omega C_0} \frac{\tan \frac{\beta_a l}{2}}{\frac{\beta_a l}{2}} \tag{12.10}$$

The circuit exhibits both series and parallel resonances which can be described as follows:

1. Parallel resonance

   This corresponds to $Z_a \to \infty$, so that $\beta_a l = (2n + 1)\pi$ or resonances at $n(\lambda/2)$ where n is odd. Even resonances do not exist, as they would for a nonpiezoelectric plate, due to the odd symmetry of the associated RF field. The fundamental resonance is labeled ω_p.

2. Series resonance

   There is another nearby series resonance taking into account both Z_a and C_0, so that the transducer looks like a series LCR circuit. In this case, the total impedance Z_3 is equal to zero at $\omega = \omega_s$

$$\frac{\tan\frac{\bar{\beta}_a l}{2}}{\frac{\bar{\beta}_a l}{2}} = \frac{1}{k_T^2}$$

   which yields

$$\frac{\tan\frac{\pi\omega_s}{2\omega_p}}{\frac{\pi\omega_s}{2\omega_p}} = \frac{1}{k_T^2}$$

   or

$$\frac{\omega_s}{\omega_p} = \left(1 + \frac{8K^2}{\pi}\right)^{-\frac{1}{2}} \tag{12.11}$$

so that k_T^2 and K_2 can be obtained directly from measurements of ω_s and ω_p. It is shown by Kino [20] that the transducer can be excited in higher harmonic modes (n) with an effective coupling constant

$$k_{\text{eff},n}^2 = \frac{8}{[(2n+1)\pi]^2} k_T^2 \tag{12.12}$$

12.1.2 Loaded Transducer

The opposite limit of the loaded transducer is a complex subject and the frequency response of the transducer depends on the ratio of the acoustic impedances of the transducer and those of the sample and backing medium. In addition, the choice of the best transducer configuration depends on the particular application and is inevitably a compromise between bandwidth, pulse response sensitivity, and insertion loss considerations. We provide a brief description of the problem and some general guidelines. The most important configurations are shown in Figure 12.4.

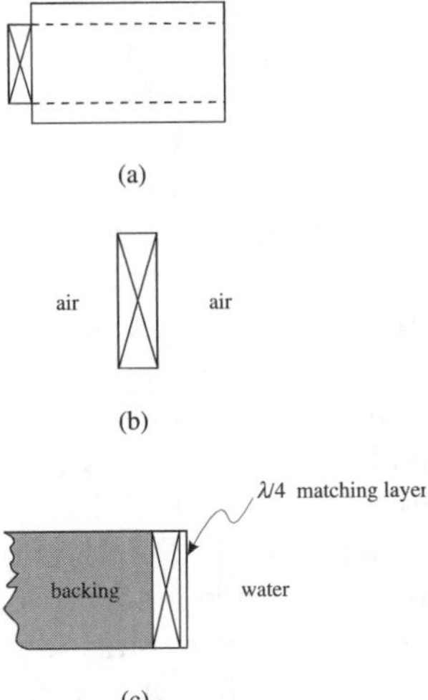

FIGURE 12.4
Schematic view of thickness mode piezoelectric transducers in different configurations. (a) Generation of acoustic waves in a solid sample or buffer rod. (b) Resonator. (c) Emission into a liquid.

Let us regard the frequency response of the transducer first. We have seen that for a free resonator (Z_1, $Z_2 \ll Z_C$), the thickness resonance is at $d = \lambda/2$ This is consistent with stress-free boundary conditions and an antinode for the displacement at each surface. If either Z_1 or Z_2 is greater than Z_C, then there is a $\lambda/4$ resonance; again, if we consider an air-backed transducer roughly matched to a substrate, this is consistent with the boundary conditions of a displacement node at the interface and an antinode at the free surface. For all other cases, the bandwidth is quite broad and the resonance is smeared out. An air-backed transducer on a reasonably well-matched buffer rod will normally give adequate bandwidth and good pulse response for most applications.

The time domain (pulse) response is important for many applications. A sharp, narrow acoustic pulse is required for doing NDE in the pulse echo mode where one wishes to detect echoes from small scattering objects. In very general terms, the time and frequency response are connected by Fourier analysis, i.e., a wide bandwidth provides a sharp temporal response while a sharp resonance will provide an extended response in the time domain. For example, a resonant transducer uncoupled in air will exhibit ringing over a long time scale. If such a transducer is poorly bonded to a buffer rod, it will likewise exhibit ringing; in fact, for an experimentalist, this is a good indicator for a transducer bond of poor quality.

From the preceding, it follows that if a transducer is matched well on at least one face then it will give good pulse response. For practical reasons, a specially designed acoustic load is put on the back face ("backing") to accomplish this at the small cost of an extra 3 dB in insertion loss. Evidently, the worst case would be to have good matching on the back surface with a badly mismatched front surface. In this case, all of the energy would go into the backing and be dissipated there as heat.

The ideal transducer would be optimized in the way shown by Figure 12.4(c). An epoxy layer can be loaded with Ti particles to provide the desired impulse response. We assume the transducer is designed to be used for a particular working medium, for example, doing NDE of steel or concrete blocks. Then appropriate $\lambda/4$ matching layers can be chosen to maximize energy transfer into the working medium, which will come at the price, of course, of reduced bandwidth.

The question of maximum power transfer from the electrical source to the sample via the transducer will clearly involve matching the electrical and acoustic impedance. Even if the acoustic impedance of the medium is matched to the transducer well, at resonance the transducer will have no material reactive impedance but it will have a relatively high capacitance reactance, $1/\omega C_0$, especially for low k_T^2 materials. This means the transducer will be badly electrically mismatched to the RF source. One solution to this would be to tune out the clamped capacitance with a series inductance.

12.2 Bulk Acoustic Wave Delay Lines

12.2.1 Pulse Echo Mode

Although some specialized ultrasonic measurements are made in continuous wave (CW) mode the majority are made in pulse echo (PE) mode where an ultrasonic pulse is emitted from the transducer and echoes from various obstacles are received by the same transducer. A variant is the so-called pitch catch mode or transmission configuration, where the pulse is launched by one transducer and received by another. In either case, sound velocity and attenuation can be obtained by measuring the amplitude variation and travel time. The problems encountered in ultrasonic PE measurements are conceptually and practically similar to those of radar, in part due to the similar frequency ranges used and because they are simple and similar ranging operations. In the following, we consider the standard problem of a transducer emitting into a delay line or buffer rod. Most of the problems encountered in a more general propagation problem can best be described and studied in this "test bench" configuration.

The configuration is shown in Figure 12.4(a). The buffer is typically 5 to 15 mm in length and perhaps 5 to 10 mm in diameter. The transducer will have a slightly smaller diameter to avoid edge effects. In physical acoustics and in cases where quantitative data on the materials used is required, a so-called tone-burst is used. A tone-burst is formed by gating the output of a CW oscillator to the desired pulse width and amplifying it as necessary.

We need at least five, preferably ten or more, cycles in the pulse envelope; otherwise, the finite pulse width (and shape) will overly affect the frequency content. The tone-burst offers an interesting tradeoff: it allows use of high-sensitivity superheterodyne detection, accurate knowledge of the frequency used, and simultaneously the temporal resolution provided by pulse techniques. Moreover, the tone-burst method is not limited to use with only the fundamental resonance of the transducer. It was seen earlier that there is an effective coupling constant to the odd harmonic resonances, and these can be excited by turning the RF source at their frequencies, provided that the transducer faces have received the special "overtone polish" to give the required transduction efficiency. Due to the fragility of transducer materials, it is not practical to mass produce transducer resonators with a fundamental resonance greater than 30 MHz. In typical applications, the harmonics can be excited at room temperature up to 200 or 300 MHz. At liquid helium temperatures (~4 K), where the intrinsic attenuation of the buffer rod becomes vanishingly small, at least one case is known where a 10-MHz quartz transducer glued on the end of a quartz buffer rod has produced echoes up to 10 GHz.

An alternative approach to the oft-used tone-burst in more qualitative NDE work, such as thickness measurement of plates, is to use a DC or video pulse which can be made very sharp. Here, assuming that the medium is nondispersive, the frequency content is of no interest, and one is solely concerned with accurate measurements of time of flight. In fact, one could loosely say that CW, tone-burst, and DC pulse form part of a tradeoff continuum, depending on whether the frequency or propagation time information is of most importance in the application at hand. Alternatively, the approaches can be combined by using a sharp DC pulse at the source and doing a spectral analysis of the received signal. This approach was used in much of the earlier work [13].

Returning to the tone-burst signal in the buffer rod, the ultrasonic pulse will travel down the rod, and multiple echoes will occur between the end faces. Each time the pulse hits the transducer it will be detected by the receiver, leading to the echo pattern shown in Figure 12.5. If the sample is perfect (homogeneous, isotropic, flat, and parallel end faces) and diffraction effects can be ignored, we will get a pure exponential decay reflecting the losses in the system. This echo pattern can be used to determine the velocity and attenuation of the ultrasonic wave.

1. Velocity

   We assume for the moment that $V_P = V_G = V$. Then the velocity can be obtained by simple time of flight measurement between selected echoes. Hence, $V = 2l/t$ where l is the appropriate propagation path = nL, n is the difference in echo number, and L the length of the buffer rod. For measurement of very small velocity changes, which is a typical problem, the fine structure of the RF wave inside the pulse can be used as a fine time scale as used in the pulse echo overlap method.

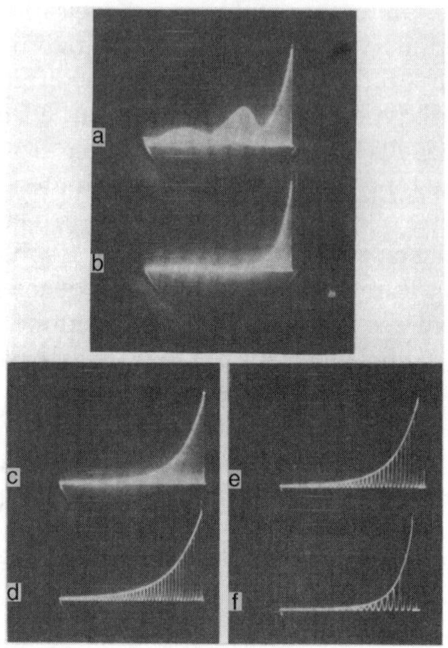

FIGURE 12.5
Series of echo patterns as a function of increasing frequency showing sidewall reflection effects at lower frequencies and attenuation increase at higher frequencies. Germanium single crystal sample. Compressional waves propagating in the <111> direction. (a) 10 MHz. (b) 30 MHz. (c) 50 MHz. (d) 90 MHz. (e) 130 MHz. (f) 170 MHz. (From Truell, R., Elbaum, C., and Chick, B.B., *Ultrasonic Methods in Solid State Physics*, Academic Press, New York, 1969. With permission.)

In fact, there are many embellishments of high-sensitivity velocity measurements but the latter is now the accepted approach.

In this connection, two very different types of velocity measurement are generally required. *Absolute* measurements are needed, mainly to determine elastic constants when combined with the density. Typical experimental accuracy is of the order 1 to 2%, and if exceptional precautions are taken (either with buffer rods or acoustic microscopy) accuracy of the order of 10^{-3}% can be obtained. *Relative* velocity measurements are used to monitor relatively small changes in velocity with variation of an external parameter such as pressure or temperature, and the measurements are the effects of principal interest in physical acoustics and much of NDE. Great care must be taken for velocity measurements in dispersive media. In this case, time of flight measurements always give the group velocity while special phase comparison techniques are needed to measure the phase velocity.

2. Attenuation

Attenuation is much more difficult to determine than velocity, and the absolute attenuation of a sample is a tenuous concept of little interest since it is so sample dependent and sensitive to the presence of small and usually poorly characterized defects. Hence, when attenuation is of interest it is usually its relative variation for problems in physical acoustics as a function of temperature, pressure, magnetic field, etc. Care must be taken to extract the intrinsic attenuation, which is of interest from apparent losses due to the transducer, bond phase effects, diffraction, nonparallelism, inhomogeneity, etc. Some of these points will be covered in the next section.

Historically, actual attenuation measurements were made using an exponential comparator superimposed on the multi-echo decay pattern. More recently, relative attenuation measurements have been made using a two-gate boxcar integrator, keeping the height of one echo fixed and monitoring the variation of the amplitude of a later echo with that of an external parameter. Attenuation is, of course, important in actual acoustic devices. In this case, it is not a question of measuring its absolute value but rather of minimizing it and keeping it constant to reduce the insertion loss of the device.

12.2.2 Buffer Rod Materials

In order to obtain long and reproducible delays, buffer rods should be made of low loss materials of reproducible characteristics. A simple crystal oriented along a pure mode direction is a good choice, especially if the Debye temperature is high so that attenuation due to phonon-phonon interactions is reduced. Polycrystals are generally to be avoided as grain boundary scattering can be severe and in any case is most reproducible between different samples. Some glasses make very good delay rods as they can have very low attenuations as well as being isotropic and quite cheap. Based on the above considerations duraluminum is very good as a makeshift delay line at a few MHz, fused quartz is excellent up to at least 100 MHz, and *c* axis sapphire is very good for higher frequencies.

Lengths are usually chosen in the range 5 to 15 mm so that individual echoes can be clearly distinguished. If a crystal is used, then the pure mode axis must be carefully aligned; otherwise, crablike propagation will occur. Of prime importance are the flatness of the end faces and their parallelism. The requirements are, of course, highly frequency dependent, as any roughness or deviation from parallelism must be much smaller than an acoustic wavelength. For example, at 10 GHz the best optical polish is required (at least $\lambda_{Na}/5$ where $\lambda_{Na} \sim 600$ nm) with a parallelism of a few seconds of arc. At 10 MHz, these requirements can be relaxed by a factor of 1000 to obtain the same signal quality.

Transducer bonding is always a preoccupation in any ultrasonic applications with solid samples, critically so for physical acoustic measurements as a function of temperature. The general requirements are that the bond be as thin as possible to avoid parasitic phase and attenuation errors, have high transmission, and be perfectly stable and reproducible. For permanently mounted buffer rods, epoxy resin is a good choice if the probability of successful bonding is high, otherwise the transducer is invariably lost as the bond is essentially permanent. More temporary and demountable solutions usually include the use of vacuum grease, silicon oil, and variants thereof. In this case, the transducer should be "wrung" onto the buffer rod if the latter is of hard material, for example, by pressing the transducer with the eraser end of an old-fashioned typewriter brush. Use of bonded transducers for low-temperature work is particularly exacting due to differential contraction between the transducer, bonding agent, and sample. This can be very high, leading even to breakage of one or more of the above. One solution, elaborate in its execution, has been to condense volatile organic components at low temperatures (~100 K) where much of the differential contraction has already occurred in cooling down. For low and room temperatures above 100 MHz, the ideal solution is to use transducers made of ZnO or AlN sputter deposited directly onto the sample, thus eliminating the bond altogether.

12.2.3 Acoustic Losses in Buffer Rods

From a physical acoustics standpoint, losses are important in buffer rods as they must be understood, controlled, and quantified if one is to make accurate attenuation measurements. From a device standpoint, they must be controlled in order to reduce the insertion loss. So far, we have traced the ultrasonic chain from the RF source to the transducer across the bond and into the buffer rod to maximize power transfer and minimize loss. Now we must consider the buffer geometry. Assuming that the buffer rod has been chosen to have the lowest intrinsic attenuation possible, there remain two additional components of loss related to geometrical considerations that superficially resemble each other in their consequences: diffraction and loss due to lack of parallelism.

12.2.3.1 Diffraction

An ultrasonic wave in a buffer rod is not like a laser beam in that there is no intrinsic collimation in the generation process. Since the wavelength is of the same order as the transducer and rod dimensions at low frequencies, significant diffraction effects occur. In the near field, up to distances of the order of $z_F = a^2/\lambda$, the beam is approximately collimated. Further out, it spreads and eventually bounces off the sidewall and is reflected back into the main beam where interference effects occur. For a low frequency buffer rod $a \sim 5$ mm, $f = 5$ MHz and $\lambda \leq 1$ mm so $z_F \sim 25$ mm.

Since tens or hundreds of reflections can occur, diffraction will be an issue in this case. The interference effects give rise to a modulation of the echo pattern, which is most pronounced at low frequencies. The effect will clearly be most visible for samples of low attenuation and the first maximum will occur around the Fresnel distance. Figure 12.5 shows the effect for a germanium single crystal as the frequency is raised, as well as the effect of increased attenuation at the higher frequencies. A detailed analysis [13] shows that the earliest peaks occur at $z/z_0 = 0.73; 1.05; 2.04$. As a rule of thumb, diffraction effects give rise to an attenuation of 1 dB for each Fresnel distance traveled.

12.2.3.2 Parallelism

An important fact that is not always appreciated is that the piezoelectric transducer output is sensitive to the variation of the phase of the ultrasonic wave across the wavefront. Special steps can be taken to randomize the phase but these are rarely used in practice. This means that interference effects are possible and to reduce them the wavefront must be made as parallel as possible to the transducer faces. One way that dephasing can occur is if the end faces of the buffer rod are not parallel, leading to a tilt of the wavefront of the returning wave. If the axial displacement of the wavefront across the beam is l, then obviously we want $kl \ll 1$, or $l \ll \lambda$ to reduce the associated phase change. For a given buffer rod, hence fixed l, the effect becomes more important at high frequencies. At a given frequency, the modulation can be shown to be proportional to $2J_1(2kan\theta)/2kan\theta$ which is the same as the diffraction modulation; this is observed in Figure 12.6. Although the two effects give the same manifestation at a given frequency, they are easily distinguished; if the effect decreases when the frequency is raised it is due to diffraction, and if it decreases at low frequency it is due to parallelism (of course, the two can always be present at the same time to complicate matters). Detailed calculations show that the associated attenuation $\alpha \sim 9.10^{-5} fa\theta$. For f = 10 MHz, a = 5 mm, and $\theta \sim 4.10^{-4}$ rad this gives α = 1.7 dB/echo, which is quite significant. Finally, other effects such as dislocation networks, temperature gradients, and other inhomogeneities can give rise to attenuations of the same order, so care must be taken to reduce them as much as possible.

12.2.4 BAW Buffer Rod Applications

BAW buffer rods have rather specialized uses in niche applications. They are ubiquitous in the research laboratory where studies are carried out on large crystals for purposes of echo separation. They are also useful for certain

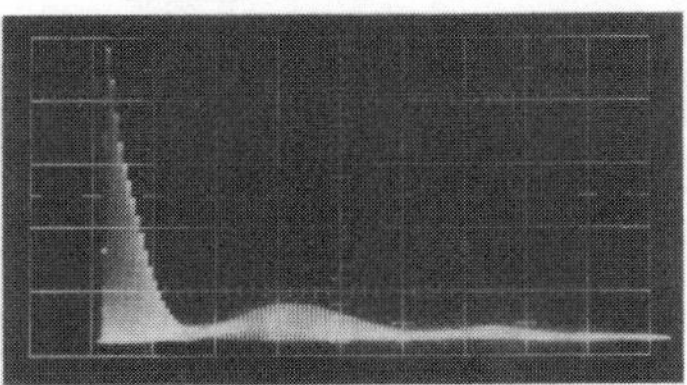

FIGURE 12.6
Echo pattern for a silicon sample at 30 MHz with a nonparallelism angle $\theta = 2 \times 10^{-4}$ rad. The envelope clearly follows the curve jinc($2kan\theta$) where a is the transducer radius and n is the echo number. (From Truell, R., Elbaum, C., and Chick, B.B., *Ultrasonic Methods in Solid State Physics*, Academic Press, New York, 1969. With permission.)

applications in NDE especially where access to sample in hostile environments is required. Some of these applications are described in Chapter 15. BAW buffer rods also have continued application for use with Quate-type and other acoustic lenses for imaging purposes, as described in detail in Chapter 14.

Historically, one of the chief uses of BAW buffer rods was for dynamic delay lines for storage and signal processing. The technology used up to 1965 has been reviewed in [77]. Polygonal delay lines made of low loss fused silica with 30 or 40 faces were used to produce multiple echoes around the polygon. With careful design, such delay lines could produce delays up to 10,000 μsec; they could also be tapped to allow signal processing functions. Since the advent of SAW planar technology, however, these BAW delay lines are now only of historical interest and will not be considered further.

12.3 Surface Acoustic Wave Transducers

12.3.1 Introduction

Historically, BAW components were the first acoustic applications in electronics mainly as delay lines, filters, and oscillators during and after World War II. Since the introduction of the integrated circuit their place has largely been taken over by SAW. SAW has many advantages, including the following:

1. The SAW geometry provides a convenient and accessible length scale. Since the velocity of ultrasonic waves is about 10^{-5} that of light the wavelength at a given frequency is 10^{-5} that of electromagnetic waves. This makes it easy to sample and perform operations on the signal in the time and spatial domain. It also allows significant miniaturization compared to bulky electromagnetic devices in the microwave range.
2. The surface of a piezoelectric substrate provides a nondispersive, guided, and accessible medium for the propagation of an acoustic wave that is within approximately 100 μm of the surface.
3. Modern microelectronic fabrication technology is ideally suited to SAW devices, including fabrication and characterization of thin films and the application of high-resolution photolithography to produce very fine and precise electrode configurations.
4. The SAW delay line forms an almost perfect approximation to a transversal filter, which is at the basis of modern signal processing.

5. The long delay available means that large values of $d\phi/df$ are available, where ϕ is the phase, making the technology adaptable to forming oscillators.
6. The tapped delay line configuration is amenable to adding signals and the substrate nonlinearity permits operations involving multiplications.

For these reasons, in the last 30 years SAW has progressively replaced BAW devices in microelectronic signal processing, with the notable exception of the ubiquitous 5-MHz quartz resonator. In the overall scheme of things, the victory may be short-lived, however, as the next emerging technology is seen to be MEMS-based filters and oscillators which, at least initially, will be BAW based.

12.3.2 Interdigital Transducers (IDT)

Surface acoustic waves can be generated by many ingenious ways [32, 35] but the IDT has proved to be ideally adapted to SAW device and signal processing applications. The principle of two neighboring electrodes (finger pair) of an IDT is derived simply from the BAW thickness mode resonance configuration. For the SAW device, the electrodes are now two metallic strips positioned on the surface of a piezoelectric substrate separated by a distance l creating an electric field in the surface region. A surface acoustic wave is then generated by the piezoelectric effect in the usual way. For a single pair the Q is small, and the response is broadband. The resonance can be sharpened by adding many finger pairs in interdigital fashion with alternating polarity. The system is resonant with the wavelength equal to the distance between finger pairs, so that the contributions from all of these add up in phase. If the frequency is off resonance then the different contributions are no longer in phase and the response is small. Thus, with many finger pairs the resonance is sharp and the Q is high.

12.3.3 Simple Model of SAW Transducer

We describe the simplest available model, the delta function model, for the IDT transducer. Comprehensive summaries of IDT are available in many sources, in particular [20, 70, 78, 79]. The model described here follows [20].

The basic assumption is that the acoustic signal generated by a finger is proportional to the charge Q on it. Then for N fingers of width w and propagation in the z direction, the amplitude at z due to a source element dz' is

$$dA(z, z', w) = \alpha\sigma(w, z')e^{-jk(z-z')}dz' \tag{12.13}$$

where

α = coupling factor
σ = charge per unit length

and the exponential is a phase factor.

If $\sigma(z) \equiv 0$ outside the region of the transducer, which is normally the case, then this expression can be integrated over all space to give the IDT response

$$A(z, w) = \alpha \int_{-\infty}^{\infty} \sigma(\omega, z') e^{-jk(z-z')} dz' \quad (12.14)$$

This fundamental result shows that the frequency response of the IDT is the Fourier transform of the charge density on the fingers.

Applying this to a uniform transducer with N fingers, pair spacing l individual width l_1, and charge Q per finger ($Q = \sigma l_1$), one obtains the total transducer response as

$$A(z) = j\alpha Q e^{-jkz} \frac{\sin\left(\frac{kNl}{2}\right)}{\cos\left(\frac{kl}{4}\right)} \operatorname{sinc} \frac{l_1}{\lambda} e^{jk(N-1)\frac{l}{2}} \quad (12.15)$$

For $N \to \infty$, this gives

1. Values of kl for zero response

$$kl = 2\pi\left(1 \pm \frac{1}{N}\right)$$

2. Bandwidth

$$\frac{\Delta\omega(\text{between zeros})}{\omega_0} = \frac{2}{N} \quad (12.16)$$

$$\frac{\Delta\omega(3\,\text{dB})}{\omega_0} = \frac{0.89}{N} \quad (12.17)$$

An IDT pair and the transducer response are shown in Figure 12.7. Like the thickness mode BAW transducer, the IDT can be described by several different models and equivalent circuits. These are discussed in detail in the specialized literature and will not be covered here. In what follows, we give an overview of several important signal processing applications.

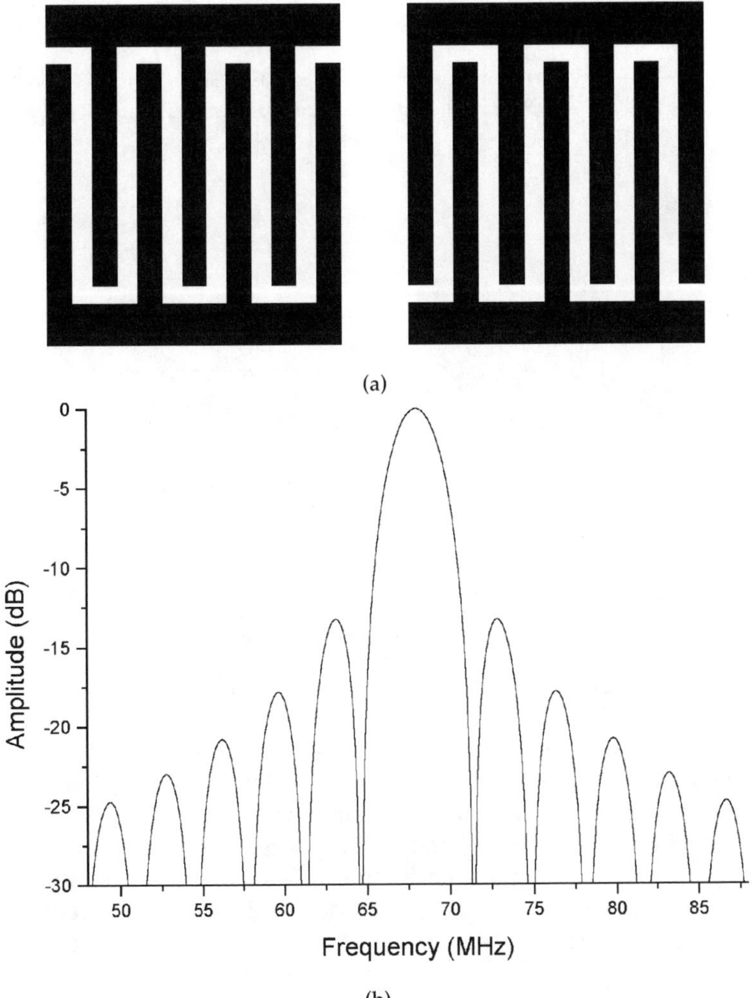

FIGURE 12.7
(a) Schematic of a pair of SAW IDT electrodes. (b) Response function for one of the transducers shown in (a).

12.4 Signal Processing

12.4.1 SAW Filters

SAW devices owe much of their widespread use in signal processing to the concept of the transverse filter. A transverse filter is basically a tapped delay line where each tap is connected to a common input or output. Such filters are particularly useful in radar and communications where they give a

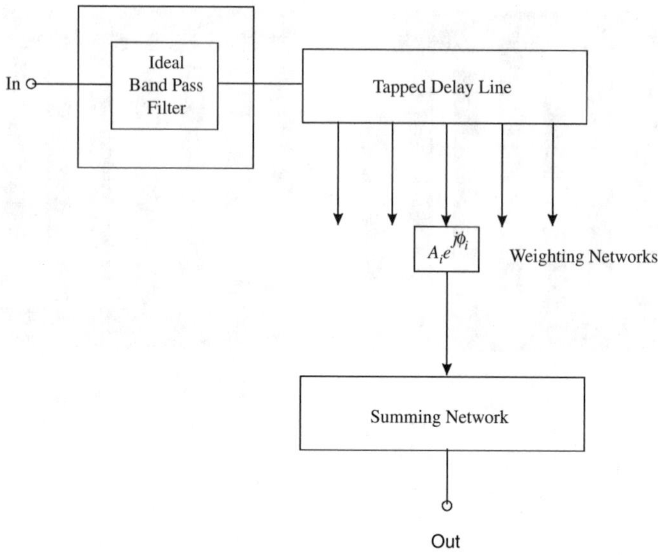

FIGURE 12.8
Schematic representation of a transversal filter using a tapped delay line.

coherent response to a known form of input signal to which they are matched and reject the noise that is unmatched, thereby improving the *S/N* ratio. They are also highly adapted to equalization techniques to reduce distortion; equalization uses an inverse filter to cancel out known, unwanted distortion.

The basic form of a transversal filter is shown in Figure 12.8. It is built around an ideal band pass filter with bandwidth $\Delta f = BW$. This feeds into a uniform tapped delay line with N taps. Each tap can be connected to an independent weighing element where either amplitude or phase can be modified. The outputs of all of these weighing elements are summed to provide the output of the transversal filter. The transfer function of the filter can be written as

$$\begin{aligned} S(f) &= \sum_{n=1}^{N} A_n \exp(i\phi_n)\exp(-j2\pi f n\tau), \quad |f - f_0| \le \frac{BW}{2} \\ &= 0, \quad \text{elsewhere} \end{aligned} \tag{12.18}$$

Thus the tap weights turn out to be the coefficients of an N-term Fourier series. The IDT transducer has all of the major elements of the transversal filter. The pair finger spacing determines the bandwidth, which can be made close to that of an ideal band pass filter. The finger pairs act as taps and their contact pads act automatically as a summing network. Amplitude weighing can be accomplished by apodizing (Greek, meaning *shape*) the electrodes, i.e., varying their overlap length. Of course, in practice, the IDT departs appreciably from the ideal of a transverse filter, for example, in such areas as cross-talk between electrodes, diffraction loss, beam steering (off-axis propagation), velocity change due to metallization (shorting effect on piezoelectric substrate and mass loading), and velocity change due to temperature variations. In fact, all of these effects can be solved in large measure by the very sophisticated SAW filter design procedures now available. These have been applied to band pass filters, which have found widespread use in communications and television circuitry. Such filters have good shape factors and at least 50 dB out-of-band rejection.

12.4.2 Delay Lines

A delay line is a two-port system in which the output signal can be time delayed with respect to the input. In the classic BAW or SAW delay line configuration, the delay can be controlled by "time of flight" by simply adjusting the path length between generating and receiving transducers.

The delay line is one of the oldest and simplest of the signal processing functions. Conceptually and technically, it provides the basic building blocks for almost all of the other functions. It is also important in its own right, particularly for communications and radar applications. We touch briefly on the principal parameters including delay, bandwidth stability, and loss.

Regarding loss, there is an intrinsic loss of 6 dB for a SAW transducer even if it is perfectly matched as it generates waves in forward and backward directions and there is a 3 dB loss for perfectly matched unidirectional generation. Loss in itself is not a problem for short delay lines of less than 10 or 20 μsec, and it can usually be reduced or compensated in such cases. Of more importance is the presence of spurious signals, which can degrade the dynamic range. For SAW delay lines these include "triple transit echoes" and spurious bulk waves. The former can be suppressed by a variety of design techniques, including use of unidirectional transducers and multistrip couplers. The latter can be avoided by careful choice of crystal propagation direction and the judicious use of absorbing materials.

Maximum attainable bandwidth is basically determined by the piezoelectric material used (K^2) and is a tradeoff with the permitted insertion loss. For a given insertion loss, the maximum attainable bandwidth is larger for high K^2 materials. In other terms, low K^2 materials can be electrically matched to the source but at the price of reduced bandwidth. For some low K^2 materials (e.g., ST quartz), improved temperature stability may be an acceptable compromise. In fact, intrinsic temperature stability gives rise to a similar compromise with K^2 as the bandwidth, in that high K^2 materials invariably have a high-frequency temperature coefficient (10^{-4} ppm/°C for $LiNbO_3$). There are other solutions to reduce the temperature coefficient such as the use of compensating multilayer structures or even controlled temperature, but this is at the cost of the simplicity of the device.

Long delay lines (up to several milliseconds) are needed for specialized applications, such as storing TV frames. There is a practical size limit (≤10 cm) for inline structures that puts an upper limit of 100 μsec even for slow materials such as bismuth germanium oxide (BGO). Various ingenious geometrical paths have been devised [80] and in one case this led to a delay of over 900 μsec for a BGO structure at 83 MHz. An alternative solution is provided by acoustic waveguides using capillaries and fibers as discussed in Chapter 10.

12.4.3 SAW Resonators

It is well known that cheap, high Q BAW resonators can be made from selected cuts of quartz at low frequencies (approximately 5 MHz). Various attempts have been made to achieve a similar result with SAW but there is no natural equivalent to the thickness resonator, and SAW-BAW conversion has defeated many attempts in this direction. However, experience has shown that the principle of the grating reflector can be used to produce moderately high Q resonators (≤2000) for SAW. In this case, for example, metallic lines are placed in an array such that their spacing is $\lambda/2$. Even though the reflection coefficient of each line is small, the reflected waves are in phase leading to a large cumulative effect of the whole array. Quantitatively, if Z_0 is the characteristic impedance of the substrate and Z_1 is that of one grating line, the transmission and reflection coefficients are

$$T = \frac{2\beta}{1+\beta^2} \tag{12.19}$$

$$R = (1-T^2)^{\frac{1}{4}} \tag{12.20}$$

where $\beta = (Z_1/Z_0)^{N_R}$ and N_R is the number of grating periods. For $\beta > 1.01$, the value of R approaches unity for N_R of the order of several hundreds, and R approaches unity even faster for larger values of β.

Such reflection gratings have been used to form a resonant cavity around an IDT source-receiver pair. Since the grating is sharply resonant compared to the IDT, it gives rise to a sharp resonant spike superposed on the broad maximum IDT insertion loss curve. This configuration can be used to form a high Q oscillator, as described in the next section.

12.4.4 Oscillators

The resonator configuration consisting of an IDT generator-receiver pair centered inside a reflective array can be transformed into an oscillator by employing a positive feedback loop. Two conditions (one for the amplitude and one for the phase) must be satisfied:

1. Total round trip gain >1. The amplifier in the feedback loop must have sufficient gain to overcome the accumulated losses around the loop.
2. Total phase shift around the loop = $(\omega L/V_R) + \phi_e = 2\pi n$

where ϕ_e is the phase change associated with the transducers and the electronics and L is the acoustic path length.

Single-mode operation is generally desired. This will occur at the frequency that has the lowest insertion loss (IL) and for which the phase condition is satisfied. Evidently, the gain must be adjusted so that the round trip gain is greater than unity for this mode but less than unity for all other modes. The situation can be optimized by choosing the dimensions such that the frequencies for the other modes satisfying the phase condition fall on the zeros of the insertion loss curve.

Oscillator stability is one of the main criteria for most applications. Short-term stability (<1 s) is determined by noise, principally the phase noise (due to phase fluctuations) of the amplifier. From the oscillator phase condition, we have immediately

$$\frac{\partial f}{f} = -\frac{\partial \phi_e}{2\pi N} \tag{12.21}$$

where N is the number of acoustic wavelengths in the acoustic transmission line. In principle, for a given phase fluctuation $\partial\phi_e$, the required frequency stability can be attained by increasing N sufficiently. Beyond the practical limit for this, Johnson noise in the amplifier and losses in the circuit, which require increased gain, come into play.

Over the medium term (minutes) and long term (hours), temperature variations are the main cause of instability and drift. The standard solution is to choose ST-cut quartz, which can have a temperature variation close to zero over a reasonable range. The residual frequency drift is given by $\Delta f(\text{ppm}) = 0.03(\Delta T)^2$ with ΔT in °C. If temperature variations are then minimized stability of 1 ppm or better can be easily attained.

12.4.5 Coded Time Domain Structures

The frequency domain filters and other devices discussed up to now were essentially nondispersive in nature. We now look at selected time domain structures that depend on dispersion for their functionality.

1. Chirp

   The chirp IDT is a geometrically broad band transducer which was developed to obtain pulse compression, principally for radar applications. A basic radar system has two main characteristics: the detection sensitivity is proportional to the excited power and the resolution sensitivity is determined by the pulse width. In its simplest form, the system is then optimized by the use of very narrow high-peak power pulses, which can become very expensive and tricky to operate. An alternative and cheaper solution lies in using the strategem of pulse compression. A DC pulse is applied to the sending IDT, which has a staggered electrode spacing corresponding to the bandwidth of the pulse. The lower frequencies travel the furthest distance and hence are at the tail end of the emitted pulse. The now long pulse is re-emitted with the low frequencies now at the leading edge and the pulse is detected by an IDT structure complementary to the source. The final result is that a very narrow pulse is reconstituted and it is the width of this pulse that determines the time resolution. The effective temporal resolution is the reciprocal of the bandwidth. More formally, an important parameter is the time-bandwidth product $T(BW)$, where T is the width of the input pulse. Then the effective pulse width is the actual pulse width divided by the time-bandwidth product.

2. Reflective array compression (RAC)

   The principle comes from the multireflected polygonal BAW delay line as a way to increase the delay significantly compared to an inline configuration. In this case, the chirp device is half as long as the equivalent inline device. There are additional advantages, including a reduced spurious BAW component due to the complex trajectory followed, no electrical connection between the reflecting elements, a possibility to use phase compensation using additional thin films, and an adjustment of amplitude weighing by varying the depth of the grooves of the reflecting elements.

   The chirp filter has obvious potential for Fourier transform applications, as discussed in detail in [80]. It is also a useful RF component for instances where a wide band tone-burst capability is desired. For example, it has been used to measure the dispersion curves for guided waves in a tube using a single transducer simply by linearly scanning the frequency of the source generator.

12.4.6 Convolvers

Applications up to now have involved linear summing operations. There is also a need for a multiplication capability, for example. This capability can be provided in acoustic devices by exploiting the nonlinearity of the medium. The product of two input functions, 1 and 2, can be provided by the wave-wave interaction where the power density of the resultant wave at frequency ω_3 is given by

$$P(\omega_3) = KP(\omega_1)P(\omega_2) \tag{12.22}$$

where K is the acoustic nonlinearity constant. The basic physics of the interaction is best seen from a phonon point of view where the interaction can be described by conservation of energy and momentum as

$$\omega_3 = \omega_1 + \omega_2 \tag{12.23}$$

$$k_3 = k_1 + k_2 \tag{12.24}$$

In general, the waves involved must be nondispersive as is the case for SAW. In addition, the SAW configuration provides the high-power densities characteristic of Rayleigh waves, which facilitate the obtaining of nonlinear effects.

The product or convolution integral is usually obtained in the counter-propagating configuration. In the case where $\omega_1 = \omega_2$, ω_3 is a spatially uniform RF voltage that can be detected by a simple setup of metallization of length L in the central region. If the input waves are of the form $S_i(t)e^{j(\omega_i t - \beta_i z)}$ then the output signal

$$v_3(t) = V_3(t)e^{2j\omega_1 t}\int_0^L S_1\left(t - \frac{z}{v}\right)S_2\left(t - \frac{L-z}{v}\right)dz \qquad (12.25)$$

which can finally be simplified to [80]

$$V_3(t) = \int_{-\infty}^{\infty} S_1(\tau)S_2(2t - \tau)d\tau \qquad (12.26)$$

which is the convolution of signals 1 and 2 with time compressed by a factor of two. The device can be directly transformed into a correlator by using an inverter in one of the sources.

12.4.7 Multistrip Couplers (MSC)

From the preceding, it is easy to imagine geometries using slanting arrays or grooves that could be used to split up an acoustic beam and could be used to make acoustic power splitters or multiplexers. However, there is a simpler and more effective way to do this by electrical connections and this is the multistrip coupler. The basic idea is shown in Figure 12.9 where an IDT is used to launch a SAW wave; this is picked up by a receiver T_1, which is electrically coupled to an identical IDT T_2. The acoustic energy arriving at R_1 will be partially converted to electrical energy that will then generate an identical acoustic wave in T_2. Thus we have found a way to split the acoustic beam into two separate channels. In fact, the whole process can be done with a uniform grating as shown in Figure 12.9(b). The resulting device is known as a multistrip coupler (MSC).

In practice the two halves of the grating in Figure 12.9(b) act as coupled resonators, so like a pair of coupled pendula, acoustic energy is transferred from one resonator to the other as a function of time. In this case, the degree of coupling depends on the length of the grating. A simple analysis shows that complete transfer of acoustic energy from the first channel to the second occurs after a length

$$L_T \approx \frac{2\lambda}{K^2}$$

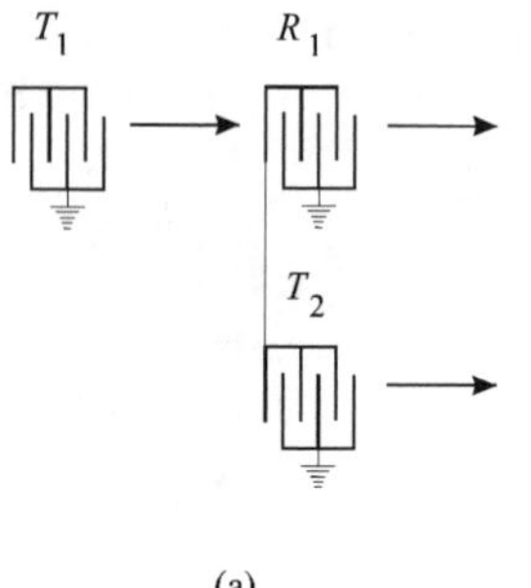

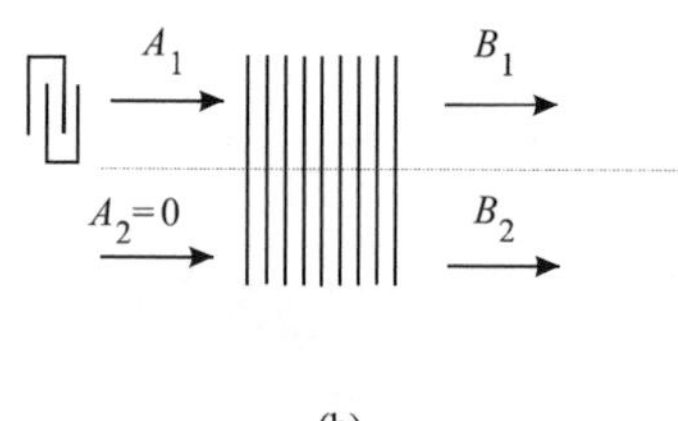

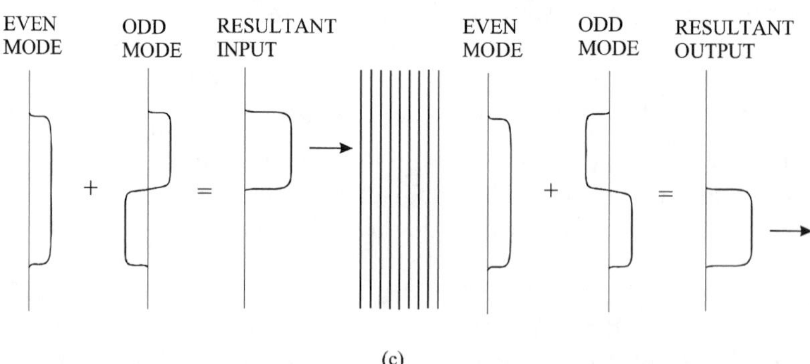

FIGURE 12.9
(a) Acoustic beam splitter. (b) Multistrip coupler. (c) Analysis of multistrip coupler in terms of symmetric and antisymmetric modes. (From Ash, E.A., Fundamentals of signal processing, in *Acoustic Surface Waves*, Oliner, A.A., Ed., Springer-Verlag, Berlin, 1978, chap. 4. With permission.)

For $K^2 \sim 0.05$, as for a strong piezoelectric like $LiNbO_3$, this gives $L_T \sim 40\lambda$. For a weaker piezoelectric, the length would be much longer, so for practical reasons the application is limited to strong piezoelectric materials. It should also be noted that the MSC structure is identical to that of the reflection grating used for acoustic resonators. The difference is that in the latter application the spacing is adjusted to a frequency f_0 corresponding to a spacing $d = \lambda/2$. The multistrip coupler, however, is typically used in the range $0.3f_0 < f_0 < 0.9f_0$.

An alternative way to look at the cyclic energy transfer between the two channels is to recognize that the rectangular SAW pulse input of Figure 12.9(c) can be seen as a superposition of symmetric and antisymmetric modes as shown, in the spirit of Lamb waves. These modes have slightly different velocities, and if for the material chosen after propagation a distance L_T the phase difference between them is π, then the emerging phases for the two modes are as shown in Figure 12.9(c). In this case the acoustic pulse is switched from the first to the second channel by the MSC.

The MSC is a versatile device that is widely used in SAW applications, including:

1. Use in band pass filters to allow full use of apodized IDTs in both elements of the filter
2. Reflecting grating
3. Unidirectional IDT with low IL and few finger pairs
4. Beam compression (10:1 or greater)
5. SAW multipliers (5 or 10 channels at several hundred megahertz)

13

Acoustic Sensors

A sensor is a device for detecting the presence of a physical, chemical, or biological property and, by appropriate transduction, transforming the detected quantity into an electrical signal. Sensors, in general, are of many types, based on sensing mechanisms that may be electrical, optical, acoustic, magnetic, etc. in nature. We limit the discussion here to acoustic sensors, which are devices where the environmental property perturbs the acoustic wave. Traditionally, sensors have varied in size from very small to very large instruments. We make a further distinction here in that we will concentrate almost exclusively on acoustic microsensors, that is, those fabricated by microelectronic techniques and integrated into silicon or hybrid circuits. Some of the more common microsensors that will be discussed in this chapter are based on acoustic waveguide geometries discussed in Chapter 10.

Acoustic microsensors may be configured as one- or two-port devices. A one-port device, which is active, contains a feedback loop that converts the device into an oscillator. The external perturbation is then manifested as a frequency shift, as shown for a gas sensor in Figure 13.1. The one-port device has the advantage of simplicity but some information is lost. This situation is rectified in the two-port passive device, which has an input and output. In this case, amplitude and phase can be measured, but the disadvantage is that extra, bulky instrumentation is needed to convert this into a practical sensor. For this reason, most practical sensors are configured in the oscillator mode. Figure 13.1 shows the response curve of a typical gas sensor. Some of the more important parameters characterizing the sensor can be appreciated from this figure. It is desirable to operate within the linear range for simplicity although a nonlinear response could be handled by adding a look-up calibration table. The resolution of the sensor, the smallest signal that can be measured, is specified by the minimum detectable mass (MDM); it is often determined by the electrical noise in the measuring system. It is essential to distinguish the resolution from the sensitivity; the latter is proportional to the slope of Figure 13.1. The sensitivity of acoustic microsensors will be treated in detail later in this chapter. Finally, parameters such as reversibility and cyclability will be important practical considerations.

This chapter deals primarily with the different types of acoustic configurations used for sensing. All of these configurations involve the propagation

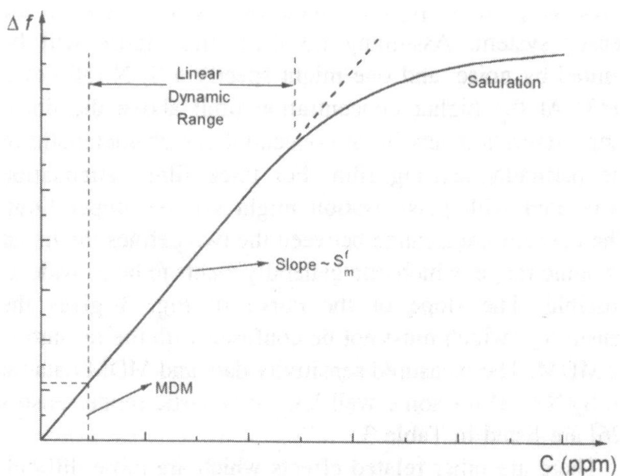

FIGURE 13.1
Sensitivity, resolution, and range parameters for a typical acoustic sensor.

of guided waves. Representative applications will be discussed, mainly in the realm of physical sensors. Chemical and biological acoustic microsensors use similar sensing platforms, but their practical applications involve problems in chemistry and biology, which are not central to the discussion here.

13.1 Thickness-Shear Mode (TSM) Resonators

These have been traditionally called the quartz crystal microbalance (QCM), but the present terminology is applied in line with Ballantine et al. [81], where emphasis is placed on the particular acoustic mode employed. The TSM was originally developed by Sauerbrey [82] for measuring the thickness of metal films deposited on substrates in vacuum. In its simplest form, the TSM resonator is simply a shear wave cut piezoelectric thickness resonator with free surfaces. The transducer is active over a region defined by the electric field set up between upper and lower electrodes. The resonant frequency can be determined from the condition that the total phase change for reflection across the bare substrate is 2π, corresponding to constructive interference between incident and reflected waves. A superposition of these waves leads to the displacement

$$u_x(z,t) = (Ae^{jkz} + Be^{-jkz})e^{j\omega t} \tag{13.1}$$

where x is the coordinate in the plane of the resonator and z is along the thickness direction. Constructive interference corresponds to the condition

$$2kh_q = 2\pi n \tag{13.2}$$

which leads to

$$h_q = n\left(\frac{\lambda}{2}\right) \tag{13.3}$$

and hence

$$f_n = \frac{nv_s}{2h_s} \tag{13.4}$$

where

$$v_s = \sqrt{\frac{\mu_q}{\rho_q}} \tag{13.5}$$

where μ_q and ρ_q are the shear modulus and density of the quartz, respectively.

Hence, acoustic resonances occur for odd multiples of half the acoustic wavelength. In fact, it is well known that for piezoelectric resonators only the odd harmonics ($n = 1, 3, 5,\ldots$) can be excited.

Determination of the resonant frequency allows us to calculate the $\vec{k}$ and the displacement profile in the crystal. At resonance, this will consist of standing waves, due to total reflection at the free surfaces. Since the stress and displacement are in quadrature and the stress has a node at the stress-free surface, the displacement has an antinode at the surface. This determines the form of the standing waves as

$$u_x(z, t) = u_{xo}\cos(k_N z)e^{j\omega t} \tag{13.6}$$

Examples of the displacement for two low-order modes are shown in Figure 13.2. Since the displacement is a maximum at the surface, the resonator will be very sensitive to surface conditions (e.g., adsorbed atoms or thin layers) and this is the basis for its use as a sensor. Some general properties follow from the form of the solution. The motion is entirely shear, so that there is no change in thickness for the bare crystal. The fact that the mode is shear means that when the resonator is immersed in a liquid, the coupling of the transverse displacements takes place by the viscosity, which is very weak compared to coupling of absent normal or compressional vibrations. So, the sensor can be used in liquid phase if care is taken. Some shear modes in quartz are very stable with respect to temperature variations, which is a very desirable property for a mass loading detecting device.

The mass loading sensitivity of the device is one of its most important characteristics as a sensor. This will be considered on three different levels, all of which give insight into the physics of the problem. The first, and simplest treatment, assumes that the adsorption of added mass is equivalent to increasing the thickness of the resonator, which increases the wavelength

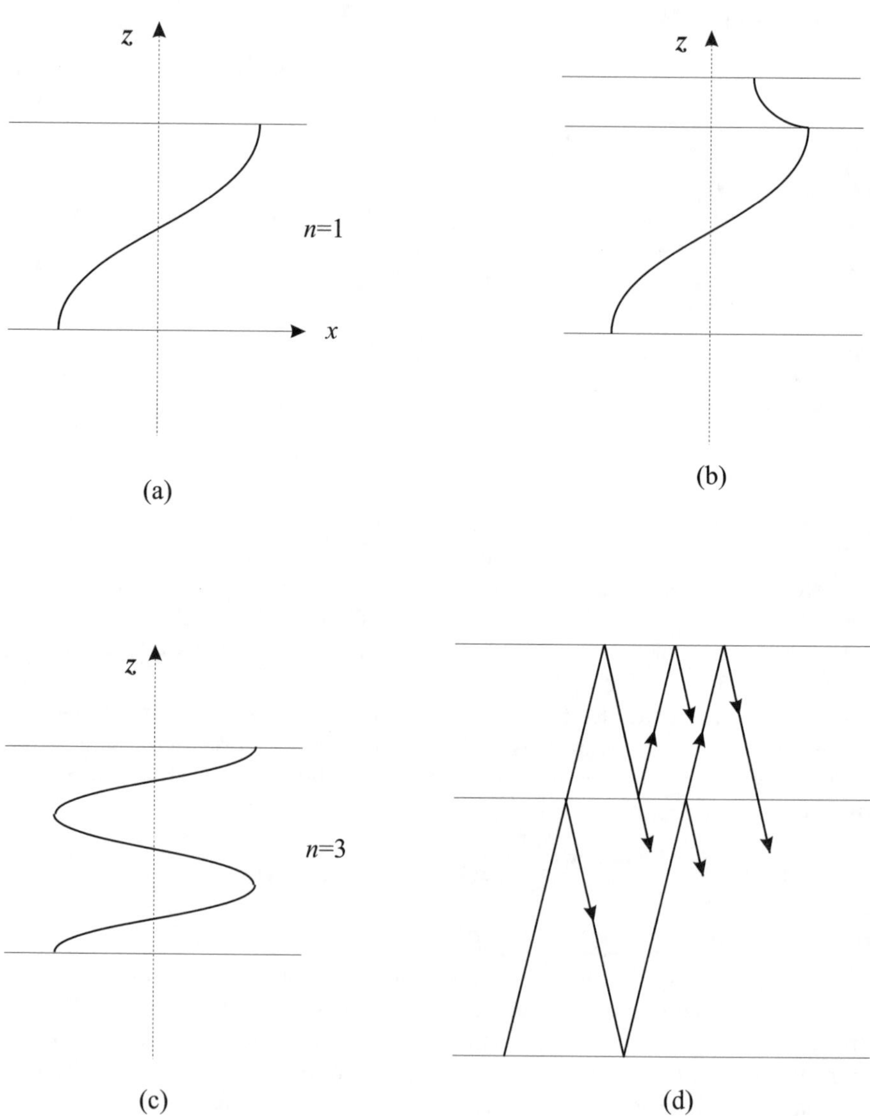

FIGURE 13.2
(a) Displacement profile in the fundamental mode for a TSM sensor. (b) Modification of the displacement profile by an adsorbed thin film. (c) Third harmonic displacement profile for the TSM sensor. (d) Multiple reflection analysis for acoustic waves in a BAW resonator with an adsorbed layer.

at resonance and hence lowers the frequency. These considerations lead to a frequency change

$$\frac{\Delta f}{f} = \frac{\Delta h_q}{h_q} \tag{13.7}$$

For a quartz resonator in their fundamental mode this can be rewritten as

$$\frac{\Delta f}{f_0} = -\frac{2f_0\Delta m}{A\sqrt{\rho_q\mu_q}} \tag{13.8}$$

where

Δf = measured frequency shift
f_0 = unloaded resonance frequency
Δm = added mass
A = piezoelectrically active area

This result was first inferred by Sauerbrey [82]. This result is only strictly valid if the deposited mass layer is thin enough so that the mass is effectively deposited at the antinode and also if the added material resembles quartz in its mechanical properties. Of course, this second requirement is seldom met, and the treatment gives no physical insight into the mechanism involved, nor how sensitive the result is to the properties of the added mass.

These deficiencies have been remedied by a detailed acoustic treatment by Miller and Bolef [83]. Without going into the detailed mathematical analysis, we can profitably examine the acoustics in this approach. Miller and Bolef treat the composite resonator consisting of the quartz crystal and the adsorbed thin film as shown in Figure 13.2(d). The density, acoustic phase velocity, and length are described by ρ_q, V_q, and l_q for the crystal and by ρ_f, V_f, and l_f for the thin film, respectively. CW acoustic waves are incident from the left and set up multiple reflections in the system, governed by partial velocity transmission (T) and reflection (R) at the interface where, as in Chapter 7,

$$R_{q\to f} = \frac{Z_q - Z_f}{Z_q + Z_f} \quad \text{and} \quad T_{q\to f} = \frac{2Z_q}{Z_q + Z_f}$$

and $Z_q = \rho_q V_q$ and $Z_f = \rho_f V_f$ are the characteristic impedances per unit area. Total reflection at the two extremities is assumed. The effect of the electrodes is ignored in this treatment, although they could easily be included at the expense of some additional complexity. The detailed analysis, which can be found in [83], then proceeds to determine the resonant frequencies (fundamental and harmonics) of the composite resonator and the frequency shifts by a Taylor's series expansion. The result reproduces Sauerbrey's equation as given previously in Equation 13.8. However, there is much more physics in this treatment, in particular the demonstration that the wave must propagate in the film for there to be a frequency shift, i.e., it is found explicitly that $\Delta f = 0$ when $R = 1$ at the interface. The authors also point out that including terms greater than second order cannot be justified due to corrections to their one-dimensional model due to crystal mounting, temperature, gas pressure, and three-dimensional propagation effects.

Another physically based derivation is based on the Rayleigh principle. This is a general approach and can be applied to acoustic sensors of quite different geometry, and thus is in some sense a universal theory for sensor sensitivity. According to this hypothesis, a mechanical resonant system oscillates at a frequency at which the peak kinetic energy U_K is equal to the peak potential energy U_P in the same volume. Analogous to a pendulum, mass on a spring, or other oscillating system, there is an interchange of potential to kinetic energy every quarter cycle. The two components for the TSM resonator can be calculated as follows:

By definition,

$$u_K = \frac{1}{2}\rho\sum_{i=1}^{3}\dot{u}_i^2 \tag{13.9}$$

so that

$$\begin{aligned} u_K &= \frac{\omega^2}{2}\left(\rho_s u_{xo}^2 + \rho_q\int_0^{h_q}|u_x(y)|^2 dy\right) \\ &= \frac{\omega^2 u_{xo}^2}{2}\left(\rho_s + \frac{\rho_q h_q}{2}\right) \end{aligned} \tag{13.10}$$

where ρ_s is the areal mass density (mass/area) of the surface mass layer.

The peak potential energy U_P occurs as usual at points in the cycle where displacement is maximum and the velocity is zero. Again, by definition of the instantaneous strain energy u_S

$$u_S = \sum_{IJ=1}^{6}\frac{1}{2}S_I c'_{IJ} S_J = \sum_{i,j,k,l=1}^{3}\frac{1}{2}c'_{ijkl}\frac{\partial u_i}{\partial x_j}\frac{\partial u_k}{\partial x_l} \tag{13.11}$$

where c' is the piezoelectrically stiffened elastic constant. It follows that

$$u_P = \frac{1}{2}\mu_q k^2 u_{xo}^2\int_0^{h_s}\sin^2(kz)dz = \frac{\mu_q k^2 u_{xo}^2 h_s}{4} \tag{13.12}$$

Applying Rayleigh's hypothesis, one directly obtains

$$\left(\frac{\omega_o}{\omega}\right)^2 = 1 + \frac{2\rho_s}{h_s\rho_q} \tag{13.13}$$

where

$$\omega_o = \frac{n\pi}{h_s}\sqrt{\frac{\mu_q}{\rho_q}}$$

For weak loading $\rho_s \ll h_q\rho_q$ and the above becomes

$$\frac{\Delta f}{f_o} = -\frac{\rho_s}{h_q\rho_q}$$

eliminating h_q

$$\Delta f = -\frac{2f_1^2 \rho_s}{\sqrt{\mu_q \rho_q}} \tag{13.14}$$

13.1.1 TSM Resonator in Liquid

There are two cases of interest:

1. Biosensing, where a principal objective is to detect the added mass of species adsorbed from the liquid phase
2. The study of the properties of a homogeneous bulk liquid

Consider a Newtonian liquid, characterized by a constant shear viscosity η

$$T_{xz} = -\eta \frac{\partial v_x}{\partial z} \tag{13.15}$$

The velocity and displacement field in the liquid can be obtained by solving the Navier-Stokes equations for one-dimensional planar flow using the previous coordinate system:

$$\eta \frac{\partial^2 v_x}{\partial z^2} = \rho \dot{v}_x \tag{13.16}$$

Assuming an oscillatory driving force at the interface, which gives a velocity field at $z = 0$ of $v_x = v_{xo} \cos \omega t$, the following solutions are obtained

$$v_x(z,t) = v_{xo} e^{-z/\delta} \cos\left(\frac{z}{\delta} - \omega t\right) \tag{13.17}$$

where a viscous penetration depth δ is defined.

δ can be interpreted in the following way. A liquid with zero viscosity does not support propagation of a shear wave at all. However, a shear wave will be propagated in a liquid of finite viscosity, but it will be very highly damped, being attenuated in a characteristic distance δ obtained from Equation 13.13 as

$$\delta = \sqrt{\frac{2\eta}{\omega \rho}} \tag{13.18}$$

The frequency dependence of δ is of interest, bearing in mind that biological macromolecules are of the order of 1 μm in dimension. For sufficiently high frequencies, δ can become of this order, so that such a sensor should be very sensitive to molecules of this size.

The presence of the liquid leads to a decrease in the resonant frequency, analogous to the effect of added mass. The results for the frequency shift for a quartz TSM in contact with a liquid have been given by Kanezawa and Gordon [84]. They use for the instantaneous particle velocity in the liquid

$$v_x(z,t) = Ae^{-kz}\cos(kz - \omega t) \tag{13.19}$$

where A is the amplitude of the wave at the interface at $z = 0$. The characteristic distance for the decay profile is $1/k$, the viscous penetration depth δ where, as in Equation 13.18, $\delta = \sqrt{2\eta/\omega\rho}$.

Using rigid boundary conditions (continuity of transverse stress and velocity), the authors obtain for the frequency shift

$$\Delta f = f_0^{3/2}\sqrt{\frac{\eta\rho}{\pi\mu_Q\rho_Q}} \tag{13.20}$$

which was found to be in very good agreement with experiment for sucrose-water solutions.

13.1.2 TSM Resonator with a Viscoelastic Film

The response of resonators with a viscoelastic film deposited on one face is important because viscoelastic polymer films are often used as chemically selective layers on TSMs and other acoustic sensors for gas sensing. The film is attached to the substrate sufficiently strongly that it follows the surface motion of the latter. However, typically a phase lag occurs in the movement of the upper surface of the film. Different regimes can be defined depending on the phase shift of the acoustic wave across the film.

For very thin films, the motion of the film is synchronous with that of the substrate, corresponding to very small phase shifts $\phi \ll \pi/2$. The film moves as a unit with the upper surface of the resonator, as in the model for mass loading in the Sauerbrey microbalance treatment. For thicker films there is a phase shift, as the free upper surface tends to lag the lower driven surface. For $\phi < \pi/2$, the movement is still in phase but overshoot occurs. Film resonance occurs for $\phi = \pi/2$. For $\phi > \pi/2$, the upper surface is π out of phase, and the damping is very high near resonance. For a given sensor, these effects must be understood, and the correct regime identified, in order to make a valid analysis of the sensor operation.

13.2 SAW Sensors

From what has been seen so far on Rayleigh waves and the TSM sensor, SAW is potentially an interesting sensing configuration. It has been seen that the displacements are confined within one or two wavelengths of the surface, so that SAW should be very sensitive to the surface environment. One important difference with the TSM resonator is immediately apparent; as the operating

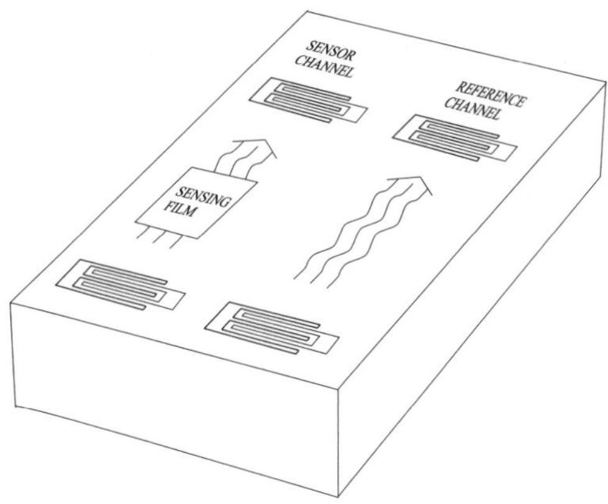

FIGURE 13.3
Dual-channel SAW sensor; the sensor channel contains the chemically selective film while the reference channel is used to compensate for the effect of temperature variations.

frequency of the SAW device is raised, then acoustic energy flux is trapped closer and closer to the surface. Hence, it is expected that the fractional frequency shift will increase with frequency for a SAW sensor; this is indeed the case as will be shown in more detail later in the chapter. A second difference with the TSM is that with SAW there are displacement components both perpendicular and parallel to the surface, so that high loss compressional acoustic radiation into the fluid will occur with SAW for most liquids. Except for very rare exceptions, SAW devices cannot be operated in liquids.

SAW can be generated in a piezoelectric substrate in a number of different ways, but the interdigital transducer is the method of choice for acoustic microsensors. The main reason is that the planar electrodes can be fabricated by standard microelectronic techniques, which allows great flexibility in design and high volume production. IDT techniques have also become the basis for generating and detecting other guided modes to be discussed in this chapter. The basics of IDT theory and technology have been described in the previous chapter and this will be sufficient for present purposes. A typical dual-channel SAW sensor is shown in Figure 13.3. A difference measurement between sensing and reference channels eliminates the temperature dependence to first order.

13.2.1 SAW Interactions

As will be seen in what follows, guided modes are much richer than resonators in their behavior as sensors in that they have more ways of interacting with the adjacent medium. For SAW and the other modes that follow, mass loading, acoustoelectric, and viscoelastic interactions are the most common.

For propagating modes the problem can be posed generally as follows, following Ballantine et al. [81]. In general, for a wave propagating in a homogeneous medium, $I = u_a V_e$, where I is the power flux density, u_a the energy density, and V_e the energy propagation velocity in the medium. Implicit differentiation gives

$$\frac{\Delta V}{V_0} = -\frac{\Delta u_a}{u_{a0}} \tag{13.21}$$

where V_0 and u_{a0} are the unperturbed sound velocity and energy density, respectively.

Thus an increase in the kinetic energy density will cause a decrease in wave velocity. For SAW, a small adsorbed mass moving synchronously with the surface causes an increase in kinetic energy with no dissipation. Then, directly from Equation 13.9, the change in average kinetic energy per unit surface area is

$$\Delta u_K = \frac{\rho_s}{4}(v_{xo}^2 + v_{yo}^2 + v_{zo}^2) \tag{13.22}$$

where v_{xo}, v_{yo}, and v_{zo} are SAW particle velocities.

Combining this with Equation 13.21 gives

$$\frac{\Delta V}{V_0} = -\frac{\omega v_o \rho_s}{4I}\left(\frac{v_{xo}^2}{\omega I} + \frac{v_{yo}^2}{\omega I} + \frac{v_{zo}^2}{\omega I}\right) \tag{13.23}$$

Since the components $v_{io}^2 \propto \omega I$, the term in brackets is a constant for the material.

Hence, Equation 13.23 can be rewritten as

$$\frac{\Delta V}{V_0} = -c_m f_o \rho_s \tag{13.24}$$

where

$$c_m = \frac{\pi v_o}{4I}\left(\frac{v_{xo}^2}{\omega I} + \frac{v_{yo}^2}{\omega I} + \frac{v_{zo}^2}{\omega I}\right) \tag{13.25}$$

This result clearly shows that the fractional velocity change increases linearly with the operating frequency. This form of the result will also be useful in comparing with other sensing modes.

13.2.2 Acoustoelectric Interaction

The calculation of the perturbation to the velocity and attenuation of the SAW can be carried out by elementary electrical theory [85, 86]. We first calculate the power transferred from the SAW to mobile carriers in the film and subsequently the effect of this power loss on the propagation of the SAW itself. The coordinate system is the same as that for Rayleigh wave propagation (Figure 8.1) for the case of a film of thickness b much less than the acoustic wavelength.

A SAW propagating on the surface of a piezoelectric substrate has an associated electric field $\vec{E}$ and hence an electric potential

$$\phi(x,t) = \phi_0 \exp j(\omega t - \beta x) \tag{13.26}$$

The electric field sets up a current density J_x, which is related to the induced sheet charge density ρ_s (charge per unit area) by the continuity equation

$$\frac{\partial J_x}{\partial x} = -\frac{\partial}{\partial t}\left(\frac{\rho_s}{b}\right) \tag{13.27}$$

or

$$\rho_s = \frac{kbJ_x}{\omega}$$

This induced charge density in turn modifies the potential to a net self-consistent surface potential $\phi + \psi$, so that

$$J_x = \sigma E_x = -\sigma\frac{\partial}{\partial x}(\phi + \psi) = jk\sigma(\phi + \psi) \tag{13.28}$$

The extra surface potential can be found by the relation $Q = C\psi$ as

$$\psi = \frac{\rho_s}{C_s k} = \frac{\rho_s}{k(\varepsilon_0 + \varepsilon_1)} \tag{13.29}$$

where $C_s k$ is the capacitance per unit area at the interface for wave number k

$$C_s = \varepsilon_0 + \varepsilon_1 \qquad (13.30$$

ε_0 = permittivity for the region above the substrate
ε_1 = permittivity of the substrate

From Equations 13.27, 13.28, and 13.29

$$\rho_s = \frac{jk\sigma_s C_s \phi}{V_0 C_s - j\sigma_s} \tag{13.31}$$

where $\sigma_s = \sigma b$ is the sheet conductivity of the film.

Finally, the power flow to the mobile carriers is

$$P_{az} = -\frac{\partial \rho_s}{\partial t}\cdot\frac{\phi_0}{2} = -\frac{j\omega\rho_s\phi_0}{2} \tag{13.32}$$

where ρ_s is given by Equation 13.31.

The effect of this power loss is to change the velocity and attenuation of the SAW, i.e., it changes k. This change can be calculated by putting the changed potential ϕ_0 as

$$\phi_0(x) = \phi_0(0)\exp(-\gamma x) \tag{13.33}$$

with $\gamma = \alpha + j\beta$. Thus, β is the new wave number and α is the attenuation of the wave. They can be related to the complex power by Equations 13.31 and 13.32. To do this, we define an acoustoelectric impedance Z_0

$$Z_0 = \frac{\phi_0^2}{2P_x} \tag{13.34}$$

The power flow for a beam of width w can be written in standard form

$$P_{ax} = u_a w V_0 \tag{13.35}$$

where u_a is the total energy density (electrical plus mechanical) per unit surface area:

$$u_a = \frac{\phi_0^2}{2Z_0 w V_0} \tag{13.36}$$

By definition the power supplied to the carrier is

$$P_{az} = -V_0\frac{\partial u_a}{\partial x} = -\frac{1}{\omega}\frac{\partial P_x}{\partial x} \tag{13.37}$$

so that by Equations 13.35 and 13.36

$$P_{az} = -\frac{\phi_0}{Z_0 w}\cdot\frac{\partial \phi_0}{\partial x} = \frac{\gamma\phi_0^2}{Z_0 w} \tag{13.38}$$

Equations 13.34 and 13.38 are the key results to relate γ to I_z, and it remains to determine Z_0. This is done from the definition of K^2

$$K^2 \equiv \frac{u_{aE}}{u_a} = \frac{\frac{1}{2}kC_s\phi_0^2}{\frac{P_{ax}}{wV_0}} \tag{13.39}$$

Using Equation 13.34, we conclude that

$$Z_0 = \frac{K^2}{kC_s w V_0} \tag{13.40}$$

Writing $\gamma = \alpha + j\beta$ and equating real and imaginary parts in Equations 13.34 and 13.38 with $V_0 = \omega/k$ we obtain

$$\frac{\alpha}{k} = \frac{K^2}{2}\left(\frac{V_0 C_s \sigma_s}{\sigma_s^2 + V_0^2 C_s^2}\right) \tag{13.41}$$

$$\frac{\Delta V}{V_0} = -\frac{\beta}{k} = -\frac{K^2}{2}\left(\frac{\sigma_{s_s}^2}{\sigma_s^2 + V_0^2 C_s^2}\right) \tag{13.42}$$

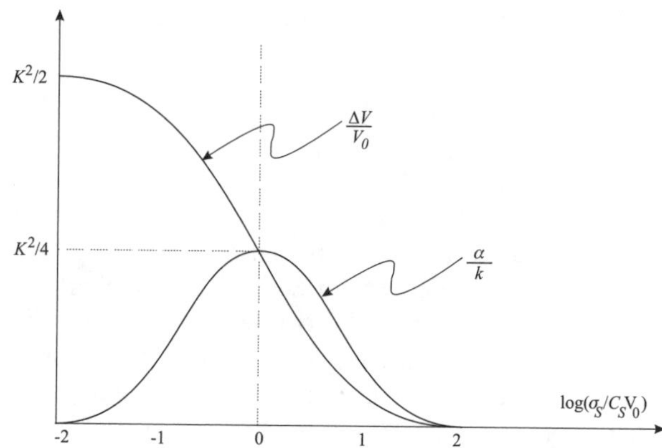

FIGURE 13.4
Schematic representation of the calculated variation of the velocity and attenuation changes as a function of sheet conductivity of the film due to the acoustoelectric interaction.

These changes are plotted in Figure 13.4. They are seen to give rise to a relaxation peak for the attenuation and a step-like increase in the sound velocity when the conductivity is sufficiently small. This is referred to as "piezoelectric stiffening."

The value of K^2 used above pertaining to SAW can be obtained by measuring the change in $\Delta V/V$ when the acoustoelectric interaction is shorted out by a conducting film on the surface. However, such a film only shorts out longitudinal fields along the surface but not those in the interior; therefore, the value of K^2 so obtained is less than that pertaining to BAW in bulk materials. Finally, it should also be noted that the theory can be modified to describe the interaction with conducting solutions [87].

13.2.3 Elastic and Viscoelastic Films on SAW Substrates

The description of the mass loading regime assumed the presence of an extremely thin film, so that the mass moved in synchronism with the SAW leading to a velocity shift with no attenuation. Films of finite thickness are considered here; they can be deformed and absorb energy. The film can be described by bulk modulus $K = K' + K''$ and shear modulus $G = G' + G''$ where the in-phase components K' and G' are known as storage moduli and the quadrature components K'' and G'' are loss moduli. Two distinct regimes can be identified. Acoustically thin films are those where the displacement is constant across the film in the saggital plane but displacement gradients in the plane of the film give rise to compression, tension, and bending of the film.

In the simplest case, the film is perfectly elastic, so that the moduli are real and there is no loss. Tiersten and Sinha [88] have shown that the SAW velocity perturbation due to the film is then given by

$$\frac{\Delta V}{V_0} = -\omega h\left[c_1\left(\rho - \frac{\mu}{v_o^2}\right) + c_2\rho + c_3\left(\rho - \frac{4\mu}{v_o^2}\frac{\lambda+\mu}{\lambda+2\mu}\right)\right] \tag{13.43}$$

where the c_i are the SAW-film coupling parameters, ρh is the mass loading term, and λ and μ are the film Lamé constants.

Acoustically thick films are characterized by the displacement of the free surface of the film lagging that of the driven film-substrate interface. This leads to a shear displacement in the film, which can be described by the film's viscoelastic response. This case has been studied in great detail by Martin et al. [89]. It is relevant to the characterization of polymer films on SAW sensors in different regimes (glassy or rubbery). The measured attenuation and velocity shifts follow closely those calculated.

13.3 Shear Horizontal (SH) Type Sensors

The TSM occupies a unique place in acoustic sensors; apart from its historical importance it is also intrinsically a resonator and is, in its classical configuration, uniquely sensitive to mass loading. However, there do exist a number of other transverse mode sensors, SH mode, that are based on guided waves and that allow other interaction mechanisms with ambient media. All of these sensors are very well adapted to use in liquid media (for example, biosensing) in that the transverse wave motion is weakly coupled to the liquid. There is a fundamental difference between the various SH mode sensors and Rayleigh mode devices. Rayleigh modes are intrinsically bound to the surface. However, this is not true for SH modes as evidenced by the fact that the fundamental SH mode for a plate has a displacement that is uniform throughout the plate. This has important implications from a device point of view. Whereas both types of waves are usually generated using IDTs, SH mode devices require some sort of geometrical confining mechanism that guides the waves and defines their mode structure. The way in which this is done is the principal fundamental difference between the four types of devices that are described in this section.

13.3.1 Acoustic Plate Mode (APM) Sensors

The device is configured so that it provides a textbook example of an SH guided wave system shown below. The wave is launched by an IDT, diffracts into the plate, and standing waves are set up. Thus the opposing face of the crystal is the confining mechanism for APM modes. While the device superficially resembles a SAW sensor, it has in fact several strong points of similarity

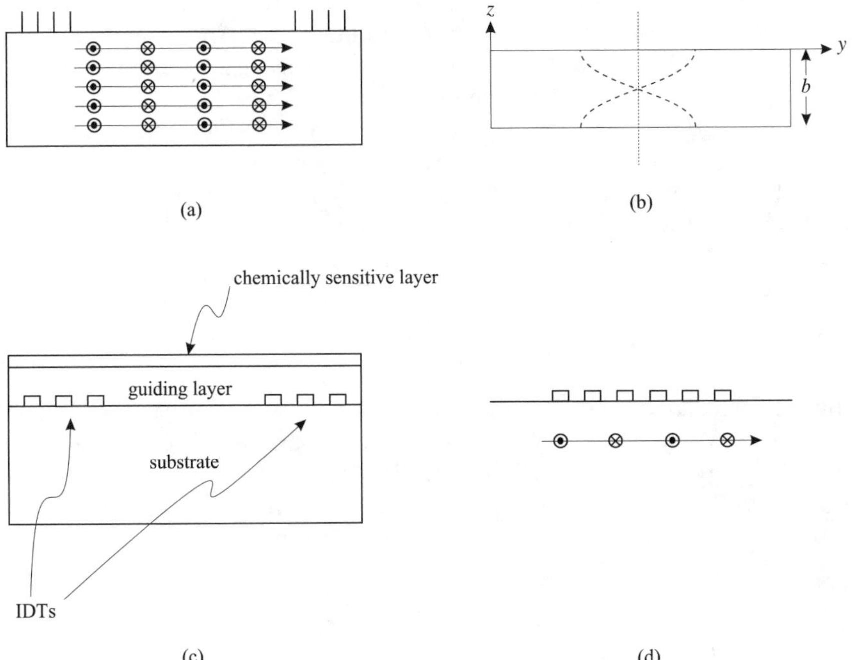

FIGURE 13.5
Acoustic plate mode (APM) sensor. (a) Side view, showing IDTs for generation and detection. (b) End view showing acoustic displacement profile with thickness. (c) Love mode. (d) STW device.

with the TSM sensor, as will be shown in the following, and as can be seen from Figure 13.5.

Assuming for the moment that an SH wave has been set up in the plate, the displacement (perpendicular to the sagittal plane) can be described by

$$u_y(z) = u_{yo}\cos\left(\frac{n\pi z}{b}\right)e^{(j\omega t-\gamma_n x)} \tag{13.44}$$

Clearly this solution describes standing waves in the transversal direction (by the cosine term) and propagation down the waveguide with wave number γ_n (by the exponential term)

$$\gamma_n = j\sqrt{\left(\frac{\omega}{v_o}\right)^2 - \left(\frac{n\pi}{b}\right)^2} \tag{13.45}$$

The situation thus corresponds to the SH mode waveguide of Chapter 10, which was described by the slowness diagram and the waveguide equation. In Equation 13.45, n is the mode number giving the number of nodes across the plate section.

In practice, the acoustic waves are excited by an IDT deposited on a piezoelectric substrate of appropriate crystallographic cut. An excited mode will have a wavelength equal to the IDT periodicity d, i.e., $f = V/d$. In addition, Hou and van de Vaart [90] showed that the coupling efficiency is a function of b/λ. In addition to the frequency condition, single-mode operation requires that the bandwidth be inferior to the spacing between modes. Although the piezoelectric plate is anisotropic, it can be modeled as isotropic for simplicity, in which case the excited frequency is

$$f_n = \frac{v_o}{h_s}\sqrt{1+\left(\frac{nd}{2b}\right)^2} \tag{13.46}$$

The mass sensitivity is determined by considerations similar to those for the TSM and SAW devices. As before, surface perturbations will affect the guided wave in proportion to the term v_{xo}^2/I. The modes with $n > 1$ have higher sensitivity than the lowest mode $n = 0$. The mass sensitivity can be calculated as usual by the Rayleigh hypothesis, and Martin et al. [91] obtain

$$\frac{\Delta V}{V_0} = -\frac{J_n \rho_s}{b\rho_q} \equiv -c_f \rho_s \tag{13.47}$$

where $J_0 = 1/2$ and $J_n = 1$ for $n > 1$.

It is instructive to compare the $\Delta V/V_0$ expressions with those for the TSM and SAW. Except for the $n = 0$ mode, which does not exist for the TSM, the results for the APM and TSM are identical with $\Delta V/V_0 \sim 1/b$. This makes sense physically, as in both cases the acoustic energy is distributed throughout the interior of the plate, which acts as a dead volume for the sensor, while the sensing efficiency is determined by what happens at the surface. Although APM and TSM have a reduced sensitivity due to the $1/b$ factor, Martin et al. [91] show that the APM has a lower minimum detectable mass due to the higher operating frequency. In contrast to this, for the SAW sensor, the sensitivity varies as $1/\lambda$. The thickness of the plate does not intervene directly; rather, the acoustic energy is concentrated within a wavelength of the surface. This distance becomes shorter at higher frequencies, leading to an increase in mass sensitivity.

Due to its transverse polarization, the APM sensor is particularly interesting for applications in the liquid phase. As with the TSM and the other guided wave sensors, the oscillating surface of the APM entrains a thin layer of liquid at the interface. Typical liquids can be treated as Maxwellian with a single relaxation time τ. At low frequencies, $\omega\tau \ll 1$, and the liquid is in the Newtonian regime characterized by a shear viscosity η. In the opposing limit, $\omega\tau \gg 1$, the liquid molecules are unable to follow the motion at sufficiently high frequencies, so that the dissipation decreases to zero and the energy is stored elastically. In this limit, the liquid acts like an amorphous solid with shear modulus μ and $\tau = \eta/\mu$.

Martin et al. [91] have calculated the plate-to-fluid coupling by a perturbation analysis. They find

$$\frac{\Delta\alpha}{k} = \frac{c_f\eta}{2v_o}\mathrm{Re}\left(\frac{\xi}{1+j\omega t}\right) \tag{13.48}$$

and

$$\frac{\Delta V}{V_0} = -\frac{c_f\eta}{2\omega}\mathrm{Im}\left(\frac{\xi}{1+j\omega t}\right) \tag{13.49}$$

where

$$\xi^2 = \left(k^2 - \frac{\omega^2\rho}{\mu}\right) + j\frac{\omega\rho}{\eta} \tag{13.50}$$

For $\omega\tau < 1$, these results give the well-known transverse decay length in the fluid, i.e., the thickness of fluid entrained by the surface. Attenuation and velocity change vary as $\sqrt{\eta}$ at low viscosities leading to possible application as a microviscometer. The complete viscoelastic theory was shown to give a good description of the data over a wide range of viscosities.

The APM device also exhibits an acoustoelectric interaction, similar to SAW. In this case, however, the interaction is between the acoustic SH mode and a conducting liquid. The result is

$$\frac{\Delta V}{V_0} = -\frac{K^2}{2}\left(\frac{\varepsilon_S + \varepsilon_0}{\varepsilon_S + \varepsilon_l}\right)\frac{\sigma^2}{\sigma^2 + \omega^2(\varepsilon_S + \varepsilon_l)^2} \tag{13.51}$$

$$\frac{\Delta\alpha}{k_0} = \frac{K^2}{2}\left(\frac{\varepsilon_S + \varepsilon_0}{\varepsilon_S + \varepsilon_l}\right)\frac{\omega\sigma(\varepsilon_S + \varepsilon_l)}{\sigma^2 + \omega^2(\varepsilon_S + \varepsilon_l)^2} \tag{13.52}$$

where ε_S, ε_l, and ε_0 are the dielectric permittivities of the substrate, liquid, and free space, respectively, and σ is the bulk conductivity of the liquid. The above results are in good agreement with experiment.

13.3.2 SH-SAW Sensor

This type of sensor is very similar to the APM device, while sharing some important characteristics with SAW. SAW calculations for anisotropic substrates show that surface waves, or those confined closely to the surface, in general have displacement components in all three directions [92]. For certain crystal cuts, close to those for pure Rayleigh wave propagation, there exist solutions for pseudo-surface waves with an SH component larger than the other two. While this wave leaks into the substrate, certain favorable directions of propagation keep the energy flux very close to the surface. Such modes have been shown to be very useful to form sensing devices.

Clearly the SH-SAW modes share some characteristics with APM and SAW. Like APM, they are essentially SH modes and so are very well adapted to liquid phase applications, where their interaction characteristics with the liquid are virtually identical to APM. At the same time, like SAW, they can potentially be operated at much higher frequencies than APM. Moreover, their mass sensitivity is identical to that of a SAW device and is independent of the plate thickness. The acoustoelectric interaction is similar to those of the two other devices. Applications of SH-SAW devices have been summarized in detail in [87].

13.3.3 Love Mode Sensors

Love waves were discussed in detail in Chapter 10. Basically, they are SH-type waves trapped in a layer on a substrate with the property that the shear velocity in the layer is less than that in the substrate, so that the superposed layer provides the confining mechanism. Again, such modes are particularly useful for liquid phase applications. The general configuration for a Love wave oscillator is shown in Figure 13.5(c). X-cut quartz is typically chosen as the substrate ($V_S \sim 5000\ \mathrm{m \cdot s^{-1}}$) while SiO_2 provides a suitable guiding layer ($V_S \sim 3800\ \mathrm{m \cdot s^{-1}}$). A thin chemically sensitive layer (PMMA with $V_S \sim 1100\ \mathrm{m \cdot s^{-1}}$) is deposited on the guiding layer for chemical sensing. With adjustment of the parameters, a compromise can be attained whereby a significant fraction of the acoustic energy flux can be made to propagate in this sensing layer. It will turn out that Love mode sensors have a great deal of design flexibility, and they constitute potentially the acoustic sensors with the highest mass loading sensitivities.

Jacoby and Vellekoop [93] have made a quantitative study of the configuration of Figure 13.5(a). Using a perturbation approach [32], they calculate the power flux and the attenuation for modes in the substrate, guiding layer, and chemical overlayer. For simplification, an isotropic model and pure shear loss are considered. Thus viscous losses are introduced by using a complex shear modulus $\mu = \mu_0 + j\omega\eta_\mu$, so that they can be described by a loss tangent $\tan\delta = \omega\eta_\mu/\mu$.

A model system was considered consisting of an ST quartz substrate, assumed isotropic for simplicity, with density $2650\ \mathrm{kg \cdot m^{-3}}$ and $V_s \sim 5060\ \mathrm{m \cdot s^{-1}}$. The guiding layer was a 5-μm layer of fused quartz with a 1-μm overlayer of PMMA ($\mu = 1.43 \times 10^9\ \mathrm{N \cdot m^{-2}}$, $\eta_\mu = 0.01\ \mathrm{Ns \cdot m^{-2}}$). A perturbation method was used and found to be accurate up to loss angles of $\tan\delta = 0.1$ at $\omega = 1.5 \times 10^9\ \mathrm{s^{-1}}$, which was eventually found to be near the optimal frequency. The distribution of power flux in the layered system and the system attenuation were calculated as a function of frequency. As the frequency was increased, the wave was much more concentrated in the layer but the attenuation rose toward a maximum attainable value for device operation. A compromise optimal operating frequency of $\omega = 1.4 \times 10^9$ s^{-1} was determined. This also allowed a significant fraction of the power flux to be in the PMMA layer where the actual sensing function was carried out.

13.3.4 Slow Transverse Wave (STW) Sensors

For STW sensors, the confining mechanism for SH waves is provided by a metallic grating deposited on the surface. Physically, trapping occurs due to planar resonance in the grating. *X*-cut quartz plates are generally used as substrates, as in the other SH devices. The STW device resembles a SAW device with a grating between the IDTs, Figure 13.5(d), and its general behavior is that of a SAW sensor operating in the SH mode.

Baer et al. [94] have described a prototype STW device, the "Attila" operating at 230 MHz. They characterize the trapping characteristics of this and related sensors in terms of the same trapping parameter used previously, namely v^2/I. They show that trapping efficiency depends on material (gold is much superior to Al and SiO_2 due to its higher density) and increases with grating thickness for a given material. A specific comparison for trapping efficiency between gratings and plates for the same mass of material was carried out.

Since the plate corresponds to Love waves, the results showed that in this case the STW mode has better trapping and hence, *a priori*, should have higher mass sensitivity. In fact, a more detailed treatment by Wang et al. [95] shows that the optimal sensitivity for Love mode sensors is significantly better than for the STW mode. Nevertheless, the STW mode sensor displays good trapping, minimal reflections from a well-designed grating, and improved coupling for the IDT. The sensor was shown to be particularly well suited to biosensing.

13.4 Flexural Plate Wave (FPW) Sensors

The FPW sensor [96] is a direct application of Lamb wave propagation in a thin membrane. It is potentially the simplest, most sensitive, and with the lowest cost of all of the family of acoustic sensors. It has the interesting characteristic that it is especially well adapted to liquid phase sensing, yet it is not based on the SH mode. The sensing principle can be understood directly from the form of the Lamb wave dispersion curves of a thin plate, which will be assumed to be immersed in water.

The A_0 mode is of particular interest. If fd (or d/λ) is small enough, then the velocity of this flexural mode is less than the velocity of sound in water. It then follows from Snell's law that the acoustic wave is trapped in the plate and does not leak into the fluid, much as light waves are totally reflected inside an optical fiber. This is the basis for the interest of the device for liquid phase application since loss due to the liquid is small. For general applications

$$V_P = \sqrt{\frac{B}{M}} \tag{13.53}$$

$$B = \left(\frac{\lambda}{2\pi}\right)^2 \frac{E'd^3}{12}, \quad \text{the bending modulus} \tag{13.54}$$

$$E' = \frac{E}{1 - v^2}, \quad \text{the effective Young's modulus} \tag{13.55}$$

In practical circumstances, for membranes (5 to 10 μm thick), the previous considerations lead to operating frequencies of 1 to 10 MHz. This is a definite advantage over SAW-based devices, as at these low frequencies the associated electronic circuitry is simple and low cost. One of the advantages of the FPW sensor is that it can be fabricated by standard microelectronic techniques. A silicon nitride membrane is made self-supporting by etching away the silicon substrate behind it. A piezoelectric ZnO film is deposited on the silicon nitride, and IDTs are used to generate Lamb waves in the structure. The geometry of the device has an additional advantage in that etching forms a natural cavity for containing liquids to be studied, at the same time allowing transducers, electronics, etc. on the other face to be protected from the liquid. The sensing configuration, gas handling system, and a typical result are shown in Figures 13.6 through 13.8, respectively.

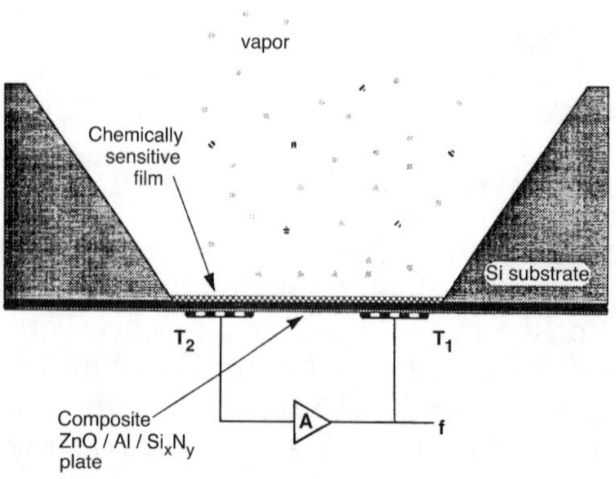

FIGURE 13.6
Lamb wave (FPW) chemical vapor sensor. (From Wenzel, S.W., Applications of Ultrasonic Lamb Waves, Ph.D. thesis, University of California, Berkeley, 1992. With permission.)

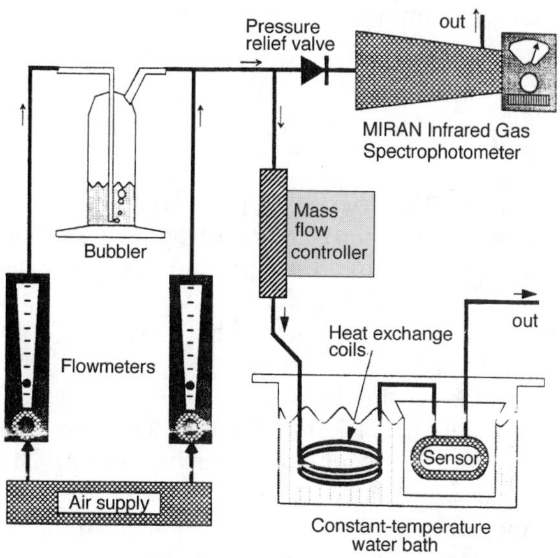

FIGURE 13.7
Typical vapor flow system used for acoustic sensor measurements. (From Wenzel, S.W., Applications of Ultrasonic Lamb Waves, Ph.D. thesis, University of California, Berkeley, 1992. With permission.)

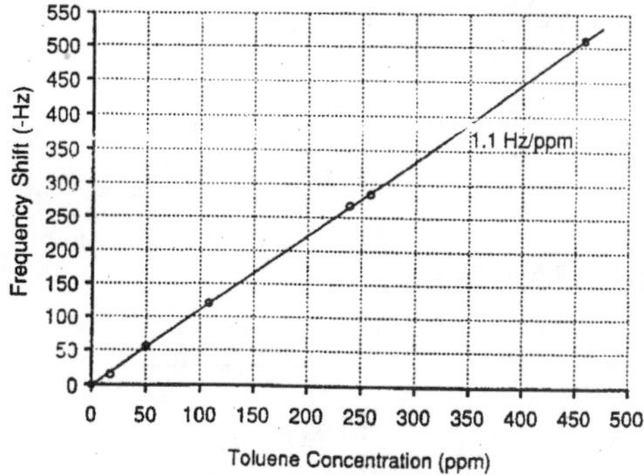

FIGURE 13.8
Frequency shift vs. concentration for toluene vapor. (From Wenzel, S.W., Applications of Ultrasonic Lamb Waves, Ph.D. thesis, University of California, Berkeley, 1992. With permission.)

The mass loading sensitivity follows from Equation 13.53. Inevitably there will be residual stress induced in the membrane during the fabrication process. Including this in the velocity equation we have:

$$V_P = \sqrt{\frac{T_x + B}{M + m'}} \tag{13.56}$$

A Taylor expansion gives immediately for the perturbed mass loaded system

$$\frac{\Delta V_P}{V_P} = \frac{(V_P)_{final} - (V_P)_{original}}{(V_P)_{original}} \cong -\frac{m'}{2M} \tag{13.57}$$

Thus the sensitivity $\Delta V_P/V_P$ varies as $1/d$ ($M = rd$) and becomes very high as d becomes very small; quantitative comparisons will be made with other sensors in the next section. Interestingly enough, the sensitivity can be increased by reducing the thickness d and hence the frequency f. This counterintuitive behavior is due to the form of the A_0 dispersion curve and is of course the opposite to that of SAW and related devices where the frequency is raised to increase the sensitivity.

As already mentioned, operation in fluids is the main application of the FPW sensor. The question of coupled modes between a plate immersed in a fluid and the latter was treated in detail in Chapter 9. The lowest-order interface mode, the Scholte wave, becomes coupled to the unperturbed A_0 mode, leading to dispersion curve repulsion and mode character interchange as discussed previously. In the subsonic regime ($V_P < V_0$), the lossless Scholte mode travels mainly in the plate, and it is this mode that is involved in the sensing mechanism. A very lossy wave, the upper branch, exists in the liquid. It is unobserved, as the attenuation is extremely high [97].

It was shown by Wenzel and White [97] that for a plate loaded by a fluid on one side the phase velocity becomes

$$V_P = \sqrt{\frac{T_x + B}{M + \rho_F \delta_E}} \tag{13.58}$$

where the evanescent decay length

$$\delta_E = \left(\frac{\lambda}{2\pi}\right)\sqrt{1 - \left(\frac{V_P}{V_F}\right)^{-\frac{1}{2}}} \tag{13.59}$$

δ_E represents the thickness of the fluid that effectively loads the plate.

For the case of viscous fluid loading, the description for viscous fluids given earlier applies. To first order, the effect of the fluid is that of a layer of fluid of thickness equal to the viscous decay length, effectively clamped to the plate, giving

$$M_\eta = \frac{\rho_F \delta_v}{2} \tag{13.60}$$

where δ_v is the viscous decay length, Equation 13.18.

This gives a viscosity-dependent phase velocity and an attenuation varying as $\sqrt{\eta\rho}$ so that in principle the device can be used as a microviscometer by accurate attenuation measurements [81].

13.5 Thin Rod Acoustic Sensors

The thin rod or fiber acoustic wave sensor [98] is a one-dimensional analog to the FPW sensor. It shows one of the latter's main features, namely that the mass sensitivity varies as the reciprocal of the thickness dimension, so that very thin rods or fibers can have very high gravimetric sensitivity.

The mass sensitivity, $S_m^V = (1/V_p)(\partial V_p/\partial m_s)$ was previously found to be $-2/\rho a$, $-1/\rho a$, or $-1/2\rho a$ for the first torsional, radial-axial (extensional), or flexural thin rod modes, respectively [99]. However, these results were obtained with the assumption that the radius and the elasticity of the fiber are not affected by the added material. In many cases, this assumption is valid, but in other cases, the theoretical model which includes the effect of the elasticity and the inertia of the loading material has to be used.

Analytic relations including contributions of elasticity and inertia of added material on mass sensitivity of a thin rod sensor can be derived from the dispersion relation related to the acoustic mode of interest. Since it is desirable to operate such a sensor at low frequency, the long wavelength limit approximations can be used. By assuming that the thickness of the loading material is kept small, as compared to the fiber radius, we can simplify the formulae for the dispersion relations for thin rods [99] where $R = 1 + h/a$ with h = thickness of loading material. We neglect terms of order $(h/a)^2$ and higher, which are small compared to one. For the lowest torsional wave mode T_{00} this leads to:

$$V \approx V_{S1}\sqrt{\frac{1+4\left(\frac{\rho_2 h}{\rho_1 a}\right)C_S}{1+4\left(\frac{\rho_2 h}{\rho_1 a}\right)}} \tag{13.61}$$

for the lowest extensional wave mode, R_{01}

$$V \approx V_{E1}\sqrt{\frac{1+2\left(\frac{\rho_2 h}{\rho_1 a}\right)C_E}{1+2\left(\frac{\rho_2 h}{\rho_1 a}\right)}} \tag{13.62}$$

for the lowest flexural wave mode, F_{11}

$$V \approx \sqrt{\frac{\omega a}{2}}\sqrt{V_{E1}}\left[\frac{1+4\left(\frac{\rho_2 h}{\rho_1 a}\right)C_E}{1+2\left(\frac{\rho_2 h}{\rho_1 a}\right)}\right] \tag{13.63}$$

where

$$V_{S1} = \sqrt{\frac{\mu_1}{\rho_1}} \tag{13.64}$$

is the shear wave velocity

$$V_{E1} = \sqrt{\frac{E_1}{\rho_1}} \tag{13.65}$$

is the extensional wave velocity in the low frequency limit:

$$C_S = \frac{\mu_2\rho_1}{\mu_1\rho_2}; \quad \text{and} \quad C_E = \frac{E_2\rho_1}{E_1\rho_2} = \left(\frac{V_{E2}}{V_{E1}}\right)^2 \tag{13.66}$$

By substituting Equations 13.61 through 13.63 into the sensitivity formula Equation 13.68 in which $m_s = \rho_2 h$, the relation describing the mass sensitivities of a thin rod sensor in the three modes can be derived. These results are summarized in Table 13.1. These formulae indicate that the effect of inertia of the loading material (related to $\rho_2 h$) gives rise to a smaller magnitude in mass sensitivity of a thin rod sensor although its effect is usually small as compared with that of elasticity. The effect of elasticity of the loading material (related to C_S and C_E) can give rise to either a smaller or larger magnitude in mass sensitivity of a thin rod sensor. This depends on whether the absolute value of $(1 - C_S)$, $(1 - C_E)$, or $(1 - 2C_E)$ is smaller or larger than one.

These results can be compared to those published previously for plate wave sensors [100]. In fact, we can observe similarities between T_{00} and SH_0 modes, between R_{01} and S_0 modes, and between F_{11} and A_0 modes. Nevertheless, the plate thickness has to be four or eight times thinner than the fiber diameter in order for a plate wave sensor to reach the same magnitude of mass sensitivity of a thin rod acoustic wave sensor.

In order to verify experimentally the theoretical derivations of mass sensitivities listed in Table 13.1, a known amount of mass was deposited on a fiber, and the influence of that deposition on the phase of the thin rod

TABLE 13.1

Theoretical Mass Sensitivity of a Thin Rod Sensor

Acoustical Modes	S_m^V (Mass Loading Only)	S_m^V (Mass, Elasticity, and Inertia Loading)
T_{00} (torsional)	$-\frac{2}{\rho_1 a}$	$-\frac{2}{\rho_1 a}\frac{(1-C_S)}{\left(1+\frac{4\rho_2 h}{\rho_1 a}\right)\cdot\left(1+\frac{4\rho_2 h}{\rho_1 a}\cdot C_S\right)}$
R_{01} (extensional)	$-\frac{1}{\rho_1 a}$	$-\frac{1}{\rho_1 a}\frac{(1-C_E)}{\left(1+\frac{2\rho_2 h}{\rho_1 a}\right)\cdot\left(1+\frac{2\rho_2 h}{\rho_1 a}\cdot C_E\right)}$
F_{11} (flexural)	$-\frac{1}{2\rho_1 a}$	$-\frac{1}{2\rho_1 a}\frac{(1-2C_E)}{\left(1+\frac{2\rho_2 h}{\rho_1 a}\right)\cdot\left(1+\frac{4\rho_2 h}{\rho_1 a}\cdot C_E\right)}$

Note: Subscripts 1 and 2 denote the fiber and loading material, respectively. The other parameters are defined in the text.

Source: From Viens, M. et al., *IEEE Trans. UFFC*, 43, 852, 1996. ©IEEE. With permission.

acoustical delay line was monitored. Acoustic waves were excited in the fiber by using a hollow glass horn designed to concentrate the energy of a 2-MHz PZT compressional transducer into the gold fiber. By choosing the orientation of the hollow horns compared to the axis of the fiber, either extensional or flexural wave modes could be excited and received, as shown in Figure 13.9(a). The deposited copper formed a solid and uniform layer on the surface of the gold fiber. Such a layer is desirable since it does not significantly attenuate the amplitude of the acoustical signal propagated in the fiber. Figure 13.9(b) shows the variation in phase of extensional wave in a gold fiber, 25 μm in diameter, with mass of copper deposited per unit surface area during deposition and dissolution. The phase angle decreases during deposition and increases during dissolution. The average slope of the curves in Figure 13.9(b) is directly related to the mass sensitivity of the thin rod delay line in the extensional mode. In fact, the relation can be expressed as

$$S_m^V = -\frac{1}{360°}\cdot\frac{V}{lf}\cdot\frac{\partial\phi}{\partial m_s} \tag{13.67}$$

where l is the fiber length immersed in the electrolyte (8 cm), V is the thin rod wave phase velocity (2099 m·s^{-1}), and f is the wave frequency (1.95 MHz). The mass sensitivity of gold fiber, 25 μm in diameter, is evaluated to be 95 cm^2/g.

Considering the related equation in Table 13.1 in which h is assumed to be close to zero, our theory predicts a mass sensitivity of 95.2 cm^2/g. This theoretical value is in good agreement with the experimental one (95 cm^2/g) predicted by pure mass loading effect. The flexural wave mode has also been

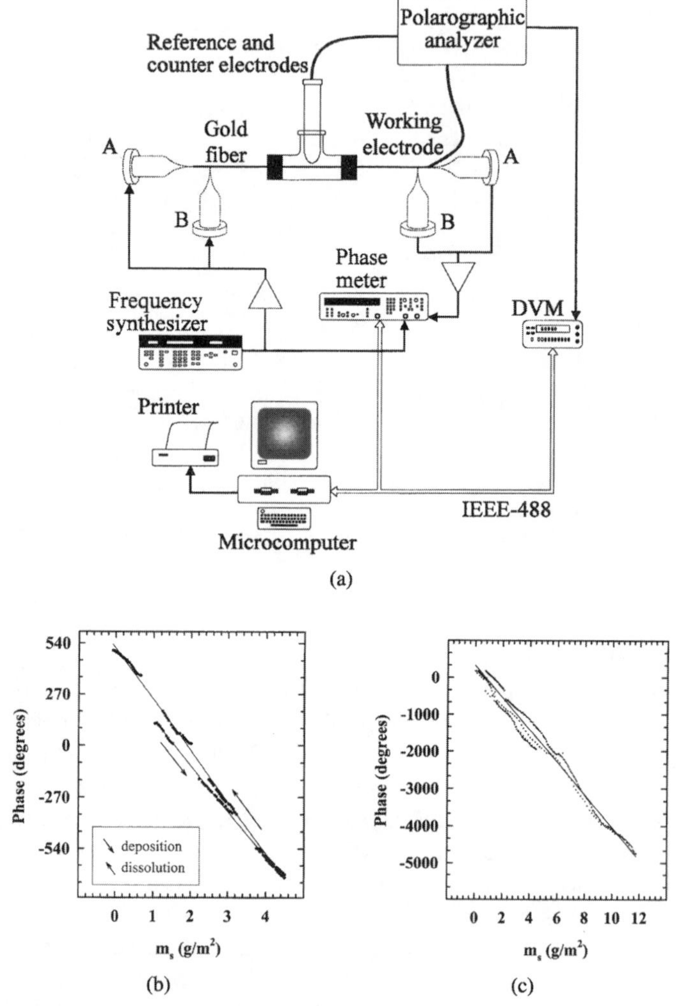

FIGURE 13.9
(a) Instrumentation and sample configuration for thin rod sensor mass sensitivity determination. (b) Variation in phase of extensional wave in a 25-μm diameter gold fiber with mass per unit area of deposited copper, during deposition and dissolution. (c) Variation in phase of flexural wave in a 50-μm diameter gold fiber with mass per unit area of deposited copper, during deposition and dissolution. (From Viens, M. et al., *IEEE Trans. UFFC*, 43, 852, 1996. ©IEEE. With permission.)

excited in a gold fiber of 50 μm in diameter using a similar system. The variation in phase as a function of mass of copper deposited per unit surface area is shown in Figure 13.9(c). In fact, the experimental mass sensitivity was evaluated to be 44 cm^2/g. This agrees reasonably well with the value predicted by a model that neglects the effect of inertia of the copper layer (57.9 cm^2/g).

13.6 Gravimetric Sensitivity Analysis and Comparison

The standard structure for this analysis consists of a layer of thickness h deposited on a plate of thickness b. For BAW sensors, this corresponds directly to the resonator and coating layer. For the other configurations, the substrate corresponds to the substrate for guided waves along it (FPW, APM, and thin rod) or to a semi-infinite substrate (SAW, Love, etc.). In the present theoretical analysis, apart from BAW and SAW, we describe the behavior by the acoustic modes directly rather than by the type of device. The modes considered are the lowest-order flexural Lamb mode (A_0), the lowest-order extensional Lamb mode (S_0), the lowest-order shear horizontal plate mode (SH_0), the surface transverse mode (STW), and the lowest-order Love mode. The definition of the mass sensitivity of a sensor configured in resonant frequency measurements is given by

$$S_m^f = \frac{1}{f_0} \lim_{\Delta m_s \to 0} \left(\frac{\Delta f}{\Delta m_s} \right) \tag{13.68}$$

where Δf is the change in resonant frequency for adsorbed mass per unit area Δm_s and $m_s = \rho_2 h$. The sensitivities for all of the acoustic modes have been calculated and the results are collected in Table 13.2 [101].

The parameters are defined as follows.

$$m_r = \frac{\rho_2 h}{\rho_1 b} \tag{13.69}$$

is the mass ratio of the coating layer to the substrate.

$$C_E = \frac{E_2'/\rho_2}{E_1'/\rho_1} = \frac{V_{02}^2}{V_{01}^2} \tag{13.70}$$

is the ratio of the extensional plate wave velocities of the coating layer and of the substrate. Subscripts 1 and 2 are for the substrate and coating, respectively, and E' is the flexural modulus given by

$$\frac{E_i'}{\rho_i} = \frac{E_i}{(1-\sigma_i^2)\rho_i} = 4\mu_i \frac{(\lambda_i + \mu_i)}{\rho_i(\lambda_i + 2\mu_i)} \tag{13.71}$$

and

$$C_S = \frac{\mu_2/\rho_2}{\mu_1/\rho_1} = \frac{V_{S2}^2}{V_{S1}^2} \tag{13.72}$$

TABLE 13.2

Sensitivity Formulae and Numerical Examples for Different Acoustic Modes

Mode	Simple Theory Result, S_m^f	S_m^f from Two-Layer Composite Resonator or Waveguide Theory Results (Mass and Elasticity Effects Considered)	Numerical Examples for Simple Theory Formulae ($cm^2/\mu g$)	Example of Mass Effect of the Sensing Layer	Example of Elasticity Effect of the Sensing Layer
BAW	$\frac{-1}{\rho_1 b}$ or $\frac{-2}{\rho_1 \lambda}$	$\frac{-1}{\rho_1 b} \cdot \frac{1}{(1+m_r)}$	15.1 (10 MHz)	–0.4%	No effect
$D_2(S_0)$	$\frac{-1}{2\rho_1 b}$	$\frac{-1}{2\rho_1 b} \cdot \frac{(1-C_E)}{(1+m_r)(1+m_r C_E)}$	626.6 (4.72 MHz)	–31.3%	–3.94%
$D_1(A_0)$	$\frac{-1}{2\rho_1 b}$	$\frac{-1}{2\rho_1 b} \cdot \frac{(1-3C_E)}{(1+m_r)(1+3m_r C_E)}$	626.6 (4.72 MHz)	–31.3%	–11.8%
SH_0	$\frac{-1}{2\rho_1 b}$	$\frac{-1}{2\rho_1 b} \cdot \frac{1-C_S}{(1+m_r)(1+m_r C_S)}$	22.77 (<18.8 MHz)	–1.1%	–10.2%
SH_m	$\frac{-1}{\rho_1 b}$	$\frac{-1}{2\rho_1 b} \cdot \frac{1-C_S(1-Q_m)}{(1+2m_r)(1+2m_r C_S)}$	45.4 (>18.8 MHz)	–1.1%	–10.2%
SAW	$\frac{-K(\sigma)}{\rho_1 \lambda}$	$\frac{-K(\sigma)}{\rho_1 \lambda} \cdot \frac{1-C_R}{1+(\rho_2 h)/(\rho_1 \lambda)}$	370.0 (200 MHz)	A sensing layer on free surface will increase mass sensitivity usually	
STW	$\frac{-K_{STW}}{\rho_1 \lambda}$		180.0 (250 MHz)		
Love Wave	$\frac{-K(\rho_{1,2},\mu_{1,2},h)}{\rho_1 \lambda}$		3944.8 (250 MHz)		

Note: The last two columns give the effect of taking into account the properties of the coating layer on the sensitivity of the acoustic sensor. The various parameters are defined in the text.

Source: From Cheeke, J.D.N. and Wang, Z., *Sensors Actuators*, 59, 146, 1999. With permission.

is the shear wave velocity ratio of the coating layer to substrate

$$Q_m = \frac{m^2\pi^2 V_{S1}^2}{\omega^2 b^2} \tag{13.73}$$

$$C_R = \frac{V_{02}^2}{V_R^2}\frac{1}{(1+\eta)}, \qquad \eta = \sqrt{\frac{1 - V_R^2/V_{L1}^2}{1 - V_R^2/V_{S1}^2}} \tag{13.74}$$

where V_R is the Rayleigh wave velocity. V_{01} and V_{02} are the extensional plate mode velocity at the low-frequency limit for the two materials in the thin plate case, respectively. E_i, σ_i, ρ_i, λ_i, and μ_i are the Young's moduli, Poisson ratios, densities of the materials, and the Lamé coefficients, respectively.

In the sensitivity formula of the Rayleigh SAW mode sensors, λ is the SAW wavelength, and the coefficient $K(\sigma)$ is a coefficient between 1 and 2. In the formula of the STW sensor sensitivity, λ is the STW wavelength and the factor K_{STW} is of the order of unity. In the lowest Love mode sensor sensitivity formula, the coefficient value, $K(\rho_{1,2}, \mu_{1,2}, h)$, not only depends on the material and the thickness of the over layer but also on the optimal design condition.

By using the simple theory, i.e., Rayleigh hypothesis or simple perturbation method, where the mass loading is assumed infinitesimal and the inertia and elasticity of the sensing layer are ignored, the sensitivity formulae of the acoustic sensor can be obtained as shown in the first column of Table 13.2. In the following numerical analysis, only the magnitude will be used to show the sensitivity. It is clear that the magnitude of the sensitivity is inversely proportional to the density of the substrate for any kind of sensor. The equivalent thickness of the acoustic coating layer depends on the acoustic mode. For the two lowest Lamb wave modes and the lowest SH mode, the equivalent thickness is the same as the physical thickness of the layer; for the BAW and high-order SH modes, it is half of their physical thickness. For the SAW, STW, and Love wave sensors, it is quite complicated and we will discuss this problem later by numerical examples.

A typical example for each type of sensor is given in the following. The substrates of the different kinds of sensors, the parameters of the samples and the sensitivity values of the sensors evaluated by using the formulae shown in the second column are listed in the fourth column of Table 13.2.

It is seen that the lowest Lamb mode sensors have a high sensitivity. The S_0 mode has not been used, perhaps due to excitation difficulties. It is well known that the lowest flexural mode is a promising sensor [96]. SAW sensors have a high sensitivity at high frequency, in this example, $K(\sigma) = 1.7$ is used. The STW sensor is not really better than its SH counterpart; however, when a layer of film is overlaid on the surface of the substrate, the STW becomes a Love wave and its sensitivity will greatly increase.

When the inertia and elasticity of the coating layer are taken into account, the sensitivity formulae are modified as shown in the third column of Table 13.2. As mentioned above for the STW case, the acoustic mode will

TABLE 13.3
Parameters Used to Calculate the Numerical Values of Table 13.2

Mode	Substrate	Frequency (MHz)	ρ_1 (10^3 kg/m^3)	V_{L1} (km/s)	V_{S1} (km/s)	Thickness (μm)
BAW	X-quartz	10	2.65	5.00		1000
FPW (S_0)	SiN film	4.72	3.99	11.0	6.04	2.0
FPW (A_0)	SiN film	4.72	3.99	11.0	6.04	2.0
SH_0	Fused quartz	<18.8	2.20		3.76	1000
SH_m	Fused quartz	>18.8	2.20		3.76	1000
SAW	Fused quartz	200	2.20	5.96	3.76	Semi-infinite
STW	ST-quartz	250	2.65		5.06	Semi-infinite
Love Wave	2-μm Polymer on ST-quartz	250	2.65		5.06	Semi-infinite

Source: From Cheeke, J.D.N. and Wang, Z., *Sensors Actuators*, 59, 146, 1999. With permission.

become the Love wave when a layer of soft material is overlaid on the surface of the substrate. It is shown that the mass loading of the coating layer will decrease the sensitivity of the sensors by a factor of $1/(1 + m_r)$ or $1/(1 + 2m_r)$. The elasticity of the coating layer decreases the sensitivity of plate mode sensors but has no effect for the BAW. For the SAW sensors, the elasticity effects on the sensitivity are not simple. Usually, the energy will be trapped in a thinner layer near the surface and the sensitivity will increase. The quantitative relation for the sensitivity of the Rayleigh-type wave sensors is quite complicated when the coating layer is involved.

We use the same coating layer for all sensors, a 2-μm-thick polymer layer with following parameters:

$$V_{L2} = 2170\ \text{m}\cdot\text{s}^{-1} \qquad V_{S2} = 1200\ \text{m}\cdot\text{s}^{-1} \qquad \rho_2 = 1250\ \text{kg}\cdot\text{m}^{-3}$$
$$V_{02} = 2000\ \text{m}\cdot\text{s}^{-1} \qquad E_2' = 5\times 10^{9}\ \text{N}\cdot\text{m}^{-2} \qquad h = 2.0\ \mu\text{m}$$

The substrates for the sensors are specified in Table 13.3. The evaluated results for the inertia terms and elasticity term are listed in the sixth and seventh columns, respectively, of Table 13.2. The minus sign in the numerical examples means that the sensitivities are decreased. It is shown that for the BAW and for the four kinds of plate modes, the sensitivity decreases due to the effects of inertia and elasticity. Two kinds of sensors will be emphasized:

1. SAW sensors. For the SAW sensor case, the sensitivity could increase or decrease depending on the specific case. If the fields have no change when the coating layer is overlaid, the sensitivity decreases as shown in the formula. However, it is believed that the field distribution will usually be changed, the energy will be trapped in a thinner layer near the surface and this effect will increase the sensitivity. These two effects are in competition. In addition, higher-order SAW modes may have to be considered when the sensing layer is not very thin.

2. STW as Love waves. STW modes cannot be supported by a free surface. The STW can be trapped along the surface by a metal grating, for example, on ST-cut quartz coated with an aluminum grating to trap the energy near the surface [95]. In this sample, $\lambda_0 = 20\ \mu m$ is the period of the IDT, which is the wavelength of the SAW operating in the center frequency 250 MHz as reported. The measured sensitivity was 180 ($cm^2 \cdot g^{-1}$) [95]. Based on a more general formula referred to in [101], the sensitivity value would be: $|S^f_{ml}| = 1/\rho_1\lambda_0 = 1/(2/65 \times 0.002) = 188.7$ ($cm^2 \cdot g^{-1}$). It was believed that the liquid loading decreases the sensitivity a little (5%, for example) [95]. This means that the energy trapping of the STW by a grating is weaker than in the Rayleigh wave case. However, when a layer of polymer is overlaid on the surface, the energy trapping may significantly improve and the sensitivity may greatly increase. This is the Love wave case.

The sensitivity of the Love mode sensor was calculated for the same substrate and polymer coating as the STW example. Substituting the given parameters (taking the same frequency 250 MHz) into the dispersion equation of the Love waves [94], the phase velocity is $V = 1.489\ km \cdot s^{-1} \cong 1.24V_{S2}$. It is interesting that it corresponds to an optimal design, i.e., $\beta_2 h = 1.550 = 0.4934\pi \cong \pi/2$. β_2 = propagation constant along the thickness direction. The sensitivity value calculated from the formula [94] is 3944.8 ($cm^2 \cdot g^{-1}$). This value is more than 20 times the value of the corresponding STW wave sensor operating at the same frequency.

13.7 Physical Sensing of Liquids

13.7.1 Density Sensing

Liquid density is an important parameter for quality control in a large range of industries. In addition, knowledge of the density is essential for the determination of mass flow from volumetric flow and for the determination of viscosity by acoustic methods.

Liquid density is most commonly and easily measured by the reflection coefficient of longitudinal waves at normal incidence for a known solid in contact with the liquid to be measured. Then $R = (Z_2 - Z_1)/(Z_2 + Z_1)$ as in Chapter 7. For this method to be accurate it is important to choose a low attenuation solid with acoustic impedance not too much greater than that of the liquid. Most ultrasonic techniques used for measuring density are based on this principle. Lynnworth [102] describes a transmission variant of this idea, as well as the use of torsion oscillators and liquid waveguides for density measurement.

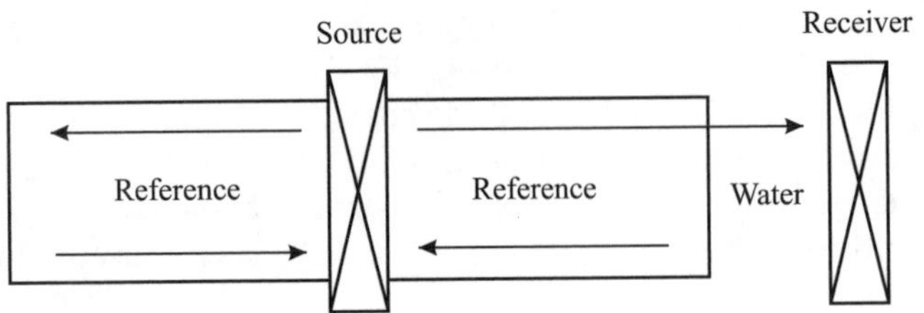

FIGURE 13.10
Configuration used for ultrasonic density sensor used in [103].

Puttmer et al. [103] describe a system, shown in Figure 13.10, based on the acoustic impedance principle, which is claimed to be of general validity for a wide range of pipe diameters, temperatures, etc. An exciting transducer is sandwiched between two identical low impedance buffer rods that are chosen according to the liquid to be studied. The left buffer rod serves to provide a level proportional to the amplitude of the excited signal and the right buffer provides the solid liquid interface whose reflection coefficient is to be measured. Then, directly,

$$R = -\frac{A_{meas}}{A_{ref}} \cdot \frac{1}{k} \tag{13.75}$$

$$Z_l = \rho_0 c_0 \frac{1+R}{1-R} \tag{13.76}$$

where k is a shape factor determined mainly by the acoustic loss in the buffer rods. The transducer on the right is used to determine V_0 by a simple time of flight. Measurements on methanol-water and water-salt solutions indicated a relative error inferior to 0.2%.

The previous device is essentially macroscopic in size as it employs standard piezotransducers and buffer rods. For many applications, such as on-line process control it is important to have a microsensor that can monitor very small quantities of liquid. The FPW sensor is ideally suited for such applications; the motion is mainly normal to the surface and hence is very sensitive to the longitudinal acoustic impedance.

The result given previously for the FPW sensor can be rewritten as

$$V_p = \sqrt{\frac{T+\beta^2 D}{M+\rho\delta}}, \quad \text{and} \quad \alpha = 0 \tag{13.77}$$

where

$$\delta = \frac{\lambda}{2\pi\sqrt{1-(V_P/V_L)^2}}$$

represents the decay length of the evanescent wave in the liquid [104].

For typical FPW plate structures $d \sim 2$ to 5 μm and ~100 μm so that the mass of an evanescent depth of fluid is of the order of the plate mass M. Hence any variation of fluid density leads to very high velocity and frequency shifts. The experimental results show agreement to within ±0.2% of known density values for aqueous solutions.

13.7.2 Viscosity Sensing

Viscosity sensing is important for a wide variety of industrial processes and is becoming increasingly useful for biosensing. Various ultrasonic shear modes provide a convenient measurement approach, as shear movements parallel to the solid-liquid interface are coupled to the fluid by the viscosity. All of the methods to be described involve a term $(\rho\eta)$ so an independent measure of the density is essential.

The APM sensor has been successfully applied to viscosity sensing [105]. The configuration used was to apply a small drop of liquid to be studied on the upper surface of the device and to measure the loss in the device due to coupling into the liquid. Since the decay distance is small ($d \sim 50$ nm at $f \sim$ 150 MHz in water), very small quantities of liquid can be used. The insertion loss involves the term $\sqrt{\rho\eta}$ and is given by

$$IL = A\sqrt{\omega\rho\eta}F(\omega\tau) \tag{13.78}$$

$$F(\omega\tau) = \left(\frac{\sqrt{1+(\omega\tau)^2}+\omega\tau}{1+(\omega\tau)^2}\right)^{\frac{1}{2}}$$

and $A \sim \omega\delta/\rho_s V_0^2$, where δ is the wave-liquid interaction length and ρ_s is the substrate density.

Relaxation effects in the liquid are described by a Maxwellian model with $\tau \equiv \eta/\mu$. The condition $\omega\tau = 1$ for $\eta = \eta_c$ separates into two distinct regimes:

1. For $\eta \ll \eta_c$, the liquid behaves as an ideal Newtonian liquid with loss $IL = A\sqrt{\omega\rho\eta}$. Since density variations are in many cases much smaller than those due to viscosity they can be accounted for by a small correction.
2. For $\eta \gg \eta_c$, the fluid molecules cannot follow the motion. The fluid behaves as a solid and the loss saturates at $IL = A\sqrt{\mu\rho}$. The behavior in the two regimes is clearly seen in the figure.

For viscosities between 1 and 50 cP, the average absolute error is 7%.

An alternative approach is to use the FPW sensor as a microviscometer. A mass of fluid M_η corresponding to the viscous penetration depth can be thought of as clamped to the plate. The effect on the attenuation of the FPW mode has been shown to be [104]

$$\alpha \cong \left(\frac{\omega}{2}\right)\frac{\delta_E^2(\lambda/2\pi)\sqrt{\eta\rho}}{T+2B+\omega^2\rho_F\delta_E/(\lambda/2\pi)^2} \tag{13.79}$$

where B and δ_E are defined in Equations 13.54 and 13.59, respectively.

This relation has been used to fit the T dependent viscosity of different solutions of DMSO. Due to the low operating frequency of the FPW device, it can be used for liquids that are much more viscous than those studied with the APM before saturation at $\omega\tau = 1$. A disadvantage is that absolute values of the attenuation are difficult to measure.

A third approach was originally suggested by Martin et al. [106] for a dual TSM configuration. One element had smooth surfaces, the second a series of corrugations defined by gold strips. The corrugations trap liquid independently of the viscosity so that the difference measurement is proportional to the density. The density can then be inferred from the single element response. As TSMs have the disadvantages of limited sensitivity, large size, pressure effects, and wetting of electrodes, this approach has recently been successfully adapted to Love mode sensors, which have the additional advantage of very high sensitivity [107].

13.7.3 Temperature Sensing

There are many other application areas where ultrasonic sensors can be used advantageously in the laboratory and in industry. It is not feasible to cover these in detail but temperature, flow, and level indication will be described briefly. A detailed account of these and other applications has been given by Lynnworth [102].

Temperature is an important parameter in all areas of instrumentation and control. Like other sensing areas there are many alternative approaches and a particular one will be chosen according to its competitive advantages. There are many cheap and reliable sensors available for routine work at ambient or near ambient conditions, where an ultrasonic-based system would be too cumbersome and expensive. Ultrasonic systems come into their own for subsurface interior temperature sensing or in hostile environments (high temperatures or corrosive media). The principle used in all ultrasonic methods is the variation of sound velocity with temperature.

Lynnworth makes the distinction between the medium being used as its own sensor and external or "foreign" sensors. Using the medium as its own sensor is based explicitly on the use of the temperature dependence of the sound velocity. Thus for gases

$$V_0 = \sqrt{\frac{\gamma RT}{M}} \tag{13.80}$$

and to take into account pressure variations using the virial coefficients α, β, etc.

$$V^2 = V_0^2(1 + \alpha p + \beta p^2 + \cdots) \tag{13.81}$$

This basic technique can also be used in high temperature plasmas, up to 800 K, by introducing the ultrasonic probe only momentarily into the plasma (~0.1 s), just long enough to carry out a sound velocity measurement. The same approach can be used in molten liquids (Al and Na) by insertion of a suitable probe such as titanium. Liquid sodium in fast breeder reactors can be probed ultrasonically noninvasively [102].

Solids can also be probed in the same way as liquids and gases with the caveat that the sample must be sufficiently large that it is certain that time bulk modes are being excited. If so, then ultrasonics can give unique information not available otherwise. In a steel mill, for example, most temperature measuring techniques will measure the surface temperature, while ultrasonics gives an average over the interior.

Foreign sensors come in a variety of forms. The earliest was the notched wire [108], which is similar in form to the thermocouple. Unlike the latter, it gives an average value over the length of the sensing region. The ultrasonic probe is simpler but requires more complicated electronics. The use of notches also allows the user the choice of several spatial measurement zones.

The wire thermometric probe is, of course, a nonresonant device, and increased resolution and sensitivity can be obtained by using resonant configurations. The classical example is the quartz thermometer, which uses the temperature coefficient of a high Q quartz cuptal resonator. Sensitivities as high as 30 ppm/°C can be obtained leading to temperature resolution of the order of 10^{-4} °C. Such devices are stable and have a high resolution, low cost, and simple technology. They are good for applications such as precision calorimetry and other precise measurements. A macroscopic version of the quartz thermometer is the tuning fork, which is a low-frequency version (~200 kHz) of the same basic principle. A high-frequency version is the SAW resonator [109]. These devices have potentially higher sensitivity and faster response than the quartz resonator. One particular case will be briefly described.

Viens and Cheeke [110] developed a highly sensitive SAW temperature sensor based on YZ-cut $LiNbO_3$, which has a high temperature coefficient of the order of 94 ppm/°C and a high coupling coefficient (k^2 = 0.048). The device used IDTs with a center frequency of 79 MHz and the resonator was formed by 300 frequency selective isolated electrodes placed on either side of the IDT, leading to a calculated reflection coefficient of 0.998 at the center frequency. This gave rise to a sharp minimum in the insertion loss of less than 10 dB at that frequency and an unloaded Q of 1350. The resonator was configured as an oscillator by use of a feedback loop with a 40 dB amplifier and band pass filter, so that the frequency shift varied linearly with the temperature over the range –30 to +150°C. The experimental sensitivity was about 80 ppm/°C, close to the predicted value.

13.7.4 Flow Sensing

There are many ways of measuring flow in liquids and gases. These include variable differential pressure across an orifice (flow nozzle, Venturi, Pitot tube, etc.), Coriolis, oscillatory method, displacement, thermal, magnetic, and ultrasonic. Ultrasonic flow meters have one important advantage over all others in that they can be noninvasive, for example, clamped on the outside of a pipe in an existing system. They also have excellent long-term stability, low power consumption, and low capital cost.

The measurement of flow by any method must take into account several characteristics of flow pattern including possible inhomogeneities caused by turbulence and flow profile. Generally, one desires an average value, which makes point sensors unsatisfactory in this regard. Ultrasonic sensors have an advantage here as the beam can be broad, and multiple-path propagation for integration and averaging is possible. There are two main ultrasonic approaches that will be mentioned: contrapropagating and Doppler.

In the contrapropagating mode, two transducers are placed on opposite sides of the conduit and displaced so that the distance between them is $L >$ diameter and they make an angle α with the flow direction. They are used alternatively as transmitter and receiver and if the transit times against and with the flow direction are t_1 and t_2, respectively, the flow velocity V_F is given by

$$V_F = \frac{L}{2\cos\alpha}\left(\frac{1}{t_2} - \frac{1}{t_1}\right) \tag{13.82}$$

since the ultrasonic waves are transported by the fluid at velocity V_F. The method gives a measure of volume flow (l/s) and a measurement of the density must be made if mass flow is required. According to [111], this ultrasonic drift measurement has the advantages of high accuracy, high linearity, rapid response, integration over the sound path, bidirectionality, and applicability to a wide range of gases and liquids.

Doppler is a traditional ultrasonic method for flow measurement, widely used in medical ultrasound where only one transducer can be accurately positioned on the patient. It is basically a reflection method in the frequency domain based on the Doppler effect, as outlined in Chapter 6. In the 1950s and 1960s, Doppler was introduced in medicine (blood flow), industry (e.g., flow of corrosive liquids in pipes), oceanography (ocean currents), and a diversified range of industries such as paper, food, and textile processing. The technology used in medical ultrasonics is now very sophisticated.

Speckle tracking is an alternative reflection mode method used in the time domain, where multiple reflections from scattering centers moving with the fluid are recorded. It is useful for very low flow that cannot be accessed by other methods. It is also useful for mapping flow profiles. Like Doppler, speckle tracking is best for monitoring liquids with known scattering centers such as liquids with entrained gas bubbles or slurries.

We will now briefly describe one application of a microsensor to flow. SAW devices can be used for this purpose by heating the device above ambient temperature [112 and 113], which is then cooled by the flow and registering a frequency shift as a consequence. This device has been improved by the use of a self-heating SAW [114] by coating a lossy material in the propagation path, the dissipated acoustic energy supplying a constant average heat input. The device then acts as above in much the same way as a hot wire anemometer. The device in [114] was made on 128° rotated Y-cut, X-propagating $LiNbO_3$. Its favorable characteristics include a high temperature coefficient (~ 70 ppm/°C), linearity from –40 to +100°C, modest thermal conductivity, and a strong coupling coefficient ($K^2 = 0.056$). The device was operated at 73.6 MHz. A dual oscillator configuration yielded a temperature compensated output with linearity at low flow rates and a sensitivity greater than 4.10^{-6}/sccm over a range of 0 to 500 sccm.

13.7.5 Level Sensing

As in the case of flow there are many approaches to liquid level sensing, including mechanical, flotation, optical magnetic, and ultrasonic techniques. Again, as for flow, the noninvasive possibilities offered by ultrasonics are advantageous, in addition to the comparatively low cost and compact devices. There are, however, some situations where ultrasonics are not appropriate, as with foams or in cases where there is mechanical mixing. The three types of ultrasonic devices that will be described briefly include air reflection, dipstick, and clamp-on noninvasive.

The air reflection type is adapted to storage containers with an accessible cover. An air transducer is fitted into the cover and an emitted ultrasonic pulse reflected at the air-liquid interface. The technique is conceptually simple but it cannot be applied universally.

There have been several different versions of an "ultrasonic dipstick" in which a rod, plate, or tube carrying ultrasonic guided waves is immersed in the liquid from above. The torsional wave structure using a thin rod [102] can measure the liquid level by using the travel time between a reference pulse and the liquid interface or the change in travel time for immersion in a liquid of known density. A correction factor must be applied for temperature changes. Another variant of the dipstick uses flexural waves [115]. Results have been reported for partially immersed duraluminum tubes (14 mm diameter and 1 mm wall thickness). The instrument works on simple travel time monitoring of the echo from the liquid interface. A sensitivity of about 1 mm for probe lengths up to 10 m has been attained. Extensional waves have also been used in measuring the amplitude of the signal transmitted between two parallel waveguides [102, 116]; it was reported in [102] that the spatial resolution was not very good. Another approach proposed using Rayleigh waves transmitted along the surface of a bar projecting into the liquid [117]. This approach proposes a reflector to reflect the leaky wave back into the bar to improve signal strength.

The clamp-on device is potentially of most interest for general industrial applications. Lynnworth and co-workers have used two approaches. One excites the A_0 Lamb mode around the outside of a large storage tank [118]. This method can detect a change in level of several percent. A second approach is the "hockey stick" delay line for monitoring shear wave reflectivity [119]. There is no dispersion indicating that bulk shear waves are involved. The device can be clamped or welded to hot pipes, so it can be used in a variety of industrial situations.

The potential of circumferential waves for level sensing of horizontal thin-walled aluminum and stainless steel tubes has been investigated [120, 121]. Aluminum tubes of 9 cm outside diameter and 0.8 mm wall were interrogated by 1.0 MHz circumferential waves for 36 fill levels. There was a monotonic increase in the arrival time of the echoes as the tube was filled. For example, the second principal echo arrived after 1.5×10^{-4} s for the empty tube and 3.2×10^{-4} s for the full tube. The results were analyzed in terms of leakage of the circumferential waves into the water with subsequent multiple echoes inside the tube. This path was demonstrated by suppression of the echoes by insertion of a coaxial empty tube inside the recipient tube.

Similar experiments in stainless steel tubes revealed large variations (up to 25%) in the travel time with fill level of water. In this case, the propagation analysis is more complex and is thought to arise from coupled guided modes in the steel-water system.

13.8 Chemical Gas Sensors

13.8.1 Introduction

The basic goal of chemical sensing is to detect, identify, and measure the concentration of chemical contaminant species in a gaseous environment. Such sensors are very important for industrial and environmental applications, which may include detection of the following:

- Toxic/polluting gases in industrial processes
- Lethal solvant vapors in factories
- Environmental effluents
- Food (e.g., fish) for freshness
- Perfumes, alcohols, etc.
- Indoor air quality

The vast majority of acoustic chemical sensors operate on the principle of applying a chemically selective coating layer on the sensor. Such coatings are typically polymers or chemical reagents. Adsorption of gaseous species changes the mechanical properties of the layer that in turn are reflected in the velocity and attenuation of the acoustic waves. The actual interaction mechanisms between the acoustic waves and the layer have already been described: mass loading, elastic and viscoelastic effects, electrical conductivity, permittivity, etc. These different interaction mechanisms can be separated by the parametric dependence on temperature, concentration, frequency, etc., as well as artifices as metallic layers to short out acoustoelectric effects, thickness effects, etc.

In real life the situation is very different, and most of the problems encountered come about due to difficulties with the technology of the chemically selective layer, some of which are listed below:

- Not chemically selective
- Irreversible
- Saturation
- Lack of adhesion
- Huge cross effects due to temperature, humidity, etc.
- Several mechanisms operative at the same time
- Irreversible and unpredictable swelling effects
- Long equilibrium time

In the next section the principal characteristics of these chemically selective layers will be presented, with examples of chemical sensing using them, followed by alternative strategies to deal with the problems outlined above.

13.8.2 Chemical Interfaces for Sensing

Free energy minimization is a simple way to describe analyte or gas phase surface interactions [81, 122]. Since there is an entropy decrease S_a on adsorption of gas molecules on a surface, there must be a lowering of the energy of the adsorbed species, which corresponds to binding energy H_a. In equilibrium the concentration of adsorbed species C_s is given in terms of the concentration in the gas phase C_a by the partition coefficient

$$K_a = \frac{C_s}{C_a} = e^{-(\Delta H_a - T\Delta S_a)/RT} \tag{13.83}$$

Clearly, for a given C_a, as the binding becomes stronger, i.e., H becomes more negative, the adsorbed concentration increases as do the sensitivity and MDM. However, a compromise must be maintained; as the binding increases, gas atoms spend a longer time on the surface leading to longer equilibrium times, which is undesirable from a sensor operation point of view. The binding energy depends on the nature of the gas-substrate interaction. There are two broad categories: physisorption and chemisorption. Physisorption is weak bonding based on van der Waals forces, which exist between all atoms and molecules. In its simplest form, this type of bonding is totally nonselective. Such bonds have a low binding energy (0.1 to 1.0 $\text{kcal}\cdot\text{mole}^{-1}$) and they can easily be broken, for example, by raising the temperature sufficiently. Since the binding is weak the equilibrium time is short (~ 10 to 12 s). Chemisorption is the opposite limit, that of strong binding (20 to 40 $\text{kcal}\cdot\text{mole}^{-1}$) which leads to long equilibrium times (10^2 to 10^{17} s) and essentially irreversible bonding. By its nature, chemisorption is much more selective than physisorption. In general, a given sensing layer will involve a mixture of both types of bonding. Some of the more common cases are described briefly below.

Self-assembled monolayers combine both physisorption and chemisorption, where a very thin self-assembled film is formed, e.g., hexadecanethiol on a gold substrate. Such films are also adapted to a quantitative study of kinetics. Porous films [123] such as the zeolite family have two very attractive features. First, it is possible to increase the effective surface area dramatically leading to an increase in sensitivity and MDM. Second, the pore size of the zeolite family can be engineered with precision, so that in principle it is possible to tailor the size of the pores to the molecular diameters of interest, thus providing some size selectivity in that only the small molecules can penetrate into the interior of the pores. A compromise must be reached for small pore sizes as the equilibrium time increases dramatically as the pore size is reduced. The structures can be regenerated by heating to a sufficiently high temperature. Since water molecules have one of the smallest dynamic diameters, there are intrinsic problems with cross effects due to humidity for this type of film. Coordination/complexation chemistry for increased sensitivity and selectivity over that provided by physosorption. Some examples are detection of NO_2, iodine, and aromatics by metal phthalocyanine films [81].

Absorption-based sensors constitute one of the most commonly used approaches. In this case, the adsorbed gas atoms diffuse into the bulk of the film. Some important examples are the detection of hydrogen by palladium and of mercury by gold [81]. But the most widespread example of this approach is the use of various polymer films; the chemistry of the polymer can be controlled by adding appropriate chemical, complexes along the carbon backbone. The physical properties of the film can be radically altered by exposure to large concentrations of analytes, which may lead to swelling and changes in elastic and viscoelastic properties [89].

One of the most systematic and scientific approaches to chemical sensing has been provided by the linear solvation energy relationship (LSER) [122]. In this model, chemical reactions are grouped into five exclusive (orthogonal) types of interaction: polarizability, dipolarity, hydrogen bond acidity-basicity, and dispersion. Corresponding parameters have been determined for a large number of analytes and polymers. In this way, it is possible to choose a polymer film with the appropriate functionality to detect a given analyte species. In principle, this approach provides a solid scientific basis to the art of chemical sensing.

13.8.3 Sensor Arrays

In the absence of true chemical selectivity for a single sensor, pattern recognition of the response of an array of sensors, the so-called electronic nose, has been developed [124]. In the earlier work [124], a number of sensors, ranging from 3 or 4 up to 32 were employed; each sensor had a different response so that a given gas mixture gives a characteristic pattern in the arrays output response. This pattern could be characterized by a neural network which had been trained in its response to known gas components. This technique has been used to analyze the composition and relative concentration of gas mixtures. Of course, this approach can be used for any type of sensor, not just acoustic sensors. The LSER model has been used to refine the pattern recognition approach [122]. If an array is made up of a small number of sensors, each one responding to one of the LSER orthogonal characteristics, then pattern recognition can be used to analyze the analyte. In this approach, it was found that the best discrimination is provided by the minimum number of sensors needed to represent the LSER model; if more than this minimum number of sensors is used then there is actually a degradation in performance.

13.8.4 Gas Chromatography with Acoustic Sensor Detection

Originally discovered by King [125] and later perfected in [126] and [127], this intriguing approach is unique in that it separates completely the chemistry (analyte identification) and the physics (signal detection), allowing an independent optimization of the two functions. Analyte identification is based on the use of a gas chromatograph, which uses the principle of difference in molecular diffusivity down a long capillary (chromatographic column) to distinguish between molecular species.

A quantity of the gas to be analyzed is admitted to a preconcentrator and then injected into the system as a sharp pulse. Different molecules diffuse down the column at different rates, so that a spectrum of the detected pulse as a function of time for a calibrated column can be used to identify the different molecular species, which solves the chemical problem. The physical detection of the output signal can be carried out in a number of ways, but clearly the principle of the ultra high sensitivity of the acoustic microbalance is pertinent here. Recent work has focused on the use of uncoated SAW [127] and FPW sensors [128]. For the results to be shown here, an uncoated SAW device at 500 MHz was used. Unlike previous chemsensors, coating would be deleterious as it would reduce response time and possibly uncalibrate the system. The SAW detector is placed on a Peltier cooler to allow rapid control of the temperature of the detector. In the current version of the instrument (Figure 13.11), the signal is normally displayed in a polar diagram (vapor print) where the radius is proportional to the amplitude (SAW frequency shift) and the polar angle represents time (clockwise variation with $t = 0$ and $t = \max$ at 12 o'clock). These polar prints give coherent image patterns that

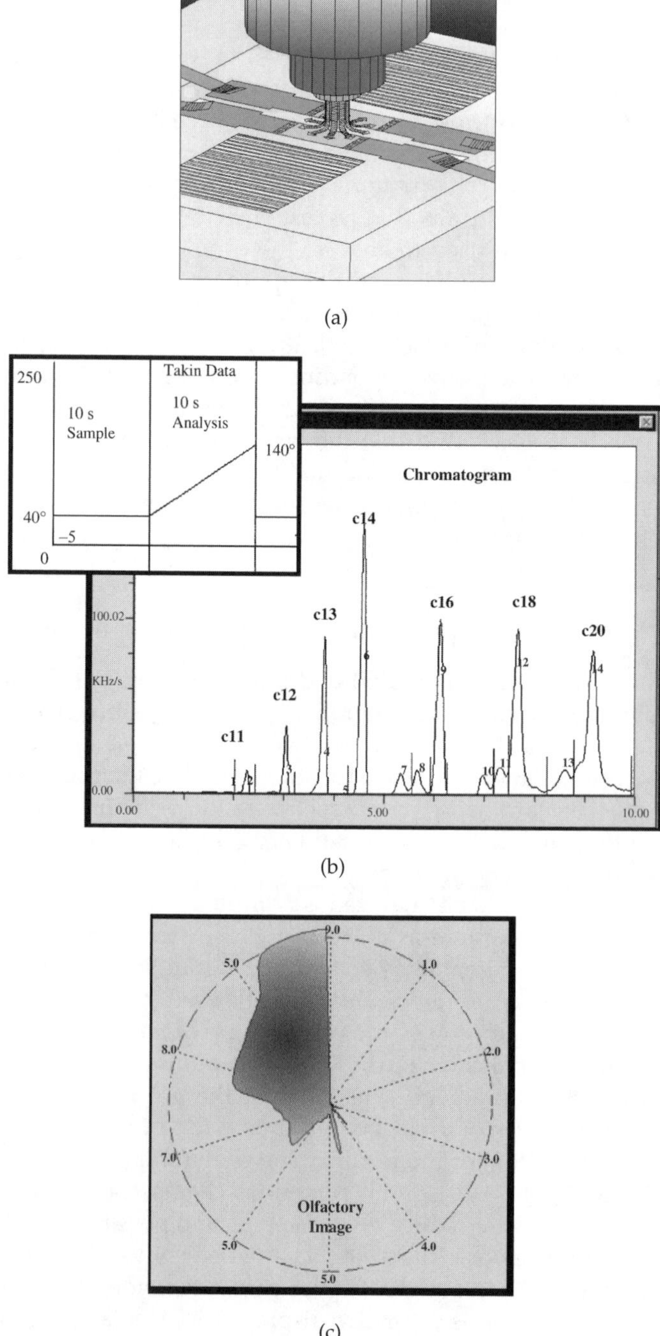

FIGURE 13.11
Gas chromatograph used as a chemical sensor with SAW detection. (a) Detection scheme. (b) Chromatogram. (c) Olfactory image. (From E. J. Staples, EstCal Corp. With permission.)

are intuitively interpretable; like most natural images, they are easy to remember and relate to, unlike the random histograms of conventional sensor array responses. Moreover, this representation is implicitly and fully consistent with the notion of orthogonal sensor response as each time gives a specific sensor response orthogonal to that at other times. This approach yields a portable instrument which is very fast (<10 s measuring time) and sensitive (better than 1 ppb of trace impurity concentration).

13.9 Biosensing

Biosensing affords a good example of liquid phase acoustic sensing. The present discussion will concentrate on immunoassay, the detection of immunoglobins, for special protein molecules also known as antibodies. These molecules are able to recognize and interact with alien proteins, antigens, by an ideally perfectly selective "lock and key" coupling. In the normal biosensing configuration, the antibody is coated directly on the surface of the sensor, and the antigen in solution couples selectively to it. Several examples of some of the previous acoustic sensor configurations will be given.

Baer et al. [94] have applied the 250 MHz "Attila" STW sensor to biosensing and detection of human immunoglobin (HIgG). An amino silane was formed on the SiO_2 surface and the antibody coupled to it. A dual path (sensor and reference arms) liquid flow system was employed and an RF interferometer was used to measure the phase difference between the two channels, engendered by the velocity shift in the sensor due to mass loading. The reference channel, identical to the other except that there was no antibody coated on it, was used to compensate for any nonspecific changes in temperature, density, or conductivity.

The measurement cycle involves the following flows in series: pure buffer solution, pure analyte flow (HIgG), and recycling agent, which regenerates the sensor and removes the antibody-HIgG complex. Absence of nonspecific spurious protein binding was verified. The system could be cycled at will. Taking into account phase noise and the mass sensitivity of the device, an MDM of 45 ng/ml was estimated. Andle et al. [129] have made extensive improvements to the APM configuration to produce an equally sensitive device. A distributed acoustic reflecting transducer design was employed to reduce the insertion loss and render the device unidirectional. The device was operated at 50.5 MHz. Experiments similar to those of Baer et al. were carried out with HIgG. They were able to detect 20 ng/ml of HIgG. The same device was used to detect double stranded DNA. Polymerase-amplified genomic DNA (200 ng/ml) from cytomegalovirus was detected. The results indicate an MDM of 40 ng/ml in this case. These examples show that SH devices are very well adapted to biosensing. The field is still in its infancy and many new applications can be anticipated in the future.

14

Acoustic Microscopy

14.1 Introduction

Acoustic microscopy involves imaging the elastic properties of surface or subsurface regions using acoustic waves as well as measuring the mechanical properties on a microscopic scale. In most of the work done so far, this has involved focusing acoustic waves by an acoustic lens that is mechanically scanned over the field of view. Following the initial work of Sokolov [130], the real start of the field was the development of the scanning acoustic microscope (SAM) by Lemons and Quate in 1973 [131]. This was essentially an extension of the traditional focused C scan ultrasonic imaging system, which is a broadband scanned ultrasonic imaging system using a spherical lens of high F number to image surface detail or defects in the interior of opaque samples.

The heart of the Lemons-Quate SAM is the acoustic lens, shown in Figure 14.1. A radio frequency (RF) tone burst, typically 50 to 500 ns wide and containing a single RF frequency in the range 10 to 1000 MHz is applied to a piezoelectric transducer fixed on the top surface of the acoustic lens body. The transducer converts the RF pulse into an ultrasonic wave with the same frequency that is emitted into the lens body. This ultrasonic wave propagates to the opposite face and impinges on the surface of a spherical cavity that has been carefully ground and polished in the lens body. The lens cavity is coupled by a liquid drop, usually water, to the sample surface, which is placed at the focal point of the spherical lens. The ultrasonic pulse is thus transmitted into the water, comes to a focus, and then is reflected back to trace out the same path in reverse. The amplitude of the reflected pulse is proportional to the difference between the acoustic properties of the sample and that of the water at the focal point, so that the amplitude gives a measure of the microscopic properties of the sample at that point. The pulse is reconverted to an RF pulse by the inverse piezoelectric effect, and this RF pulse is then fed into an RF receiver tuned to the appropriate frequency. The average amplitude of the pulse is determined, converted into a digital signal, and sent to a computer imaging system. The lens is then mechanically displaced a small distance

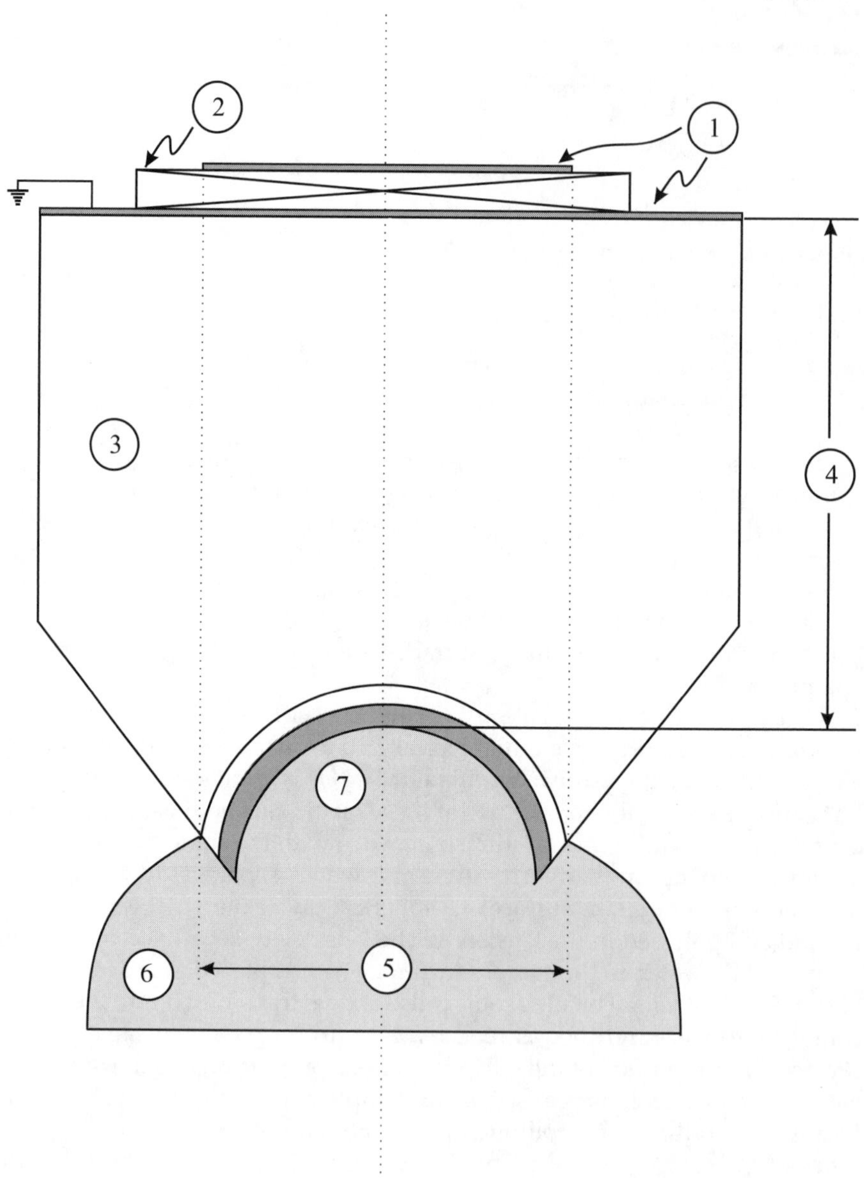

FIGURE 14.1
Spherical acoustic lens used in scanning acoustic microscope. (1) Upper electrode. (2) Transducer. (3) Lens body. (4) Lens body length. (5) Lens aperture. (6) Coupling liquid. (7) Lens diameter.

and the whole process is repeated. In order to form an image, the lens is scanned successively from point to point along a line that typically contains 500 points or pixels. Successive lines are then scanned in raster fashion, so that an image is formed in the same way as on a TV screen.

Despite its simplicity, the spherical acoustic lens is an almost perfect imaging device. All of the usual aberrations that enormously complicate the design of optical microscopes are absent from the SAM, principally because the imaging is always done on axis at a single frequency. An essential aspect is that the acoustic velocity of the lens is chosen to be much greater than that of the coupling liquid, which reduces spherical aberration to a minimum. The result is that the spatial resolution, the smallest distance between neighboring image points, is close to its ideal theoretical value, being limited by diffraction or the natural broadening of any wave focused to a point. As Lord Rayleigh showed, a point can best be described as a circle of diameter equal to the wavelength of the wave that is used for imaging. The wavelength is inversely proportional to the frequency, so to increase the resolution and decrease the size of the smallest circle, the frequency must be increased. Herein lies one of the main design considerations of acoustic lenses.

With increase in frequency, while the resolution increases proportionally, the acoustic losses increase even faster, so that eventually the reflected pulse becomes too small to measure. Hence, systematic steps must be taken to reduce losses if the goal is to maximize the resolution. From the RF source to the receiver, such steps include the following:

- Maximize peak power and minimize pulse width to separate closely spaced echoes.
- Match electrical impedance between transducer and the electronics to maximize power transfer.
- Use the most efficient and low-loss transducer possible.
- Choose lens body material that is low loss and high velocity.
- Use a highly oriented single crystal to avoid beam steering and ensure that maximum acoustic intensity reaches the cavity.
- Use a small diameter lens to reduce transmission length in the liquid.
- Use acoustic matching layers to maximize transmission into the liquid and reduce stray reflected echoes in the lens body.
- Choose a low attenuation liquid.
- Use a high sensitivity, low-noise receiver.

These conditions are easy to fulfill at 10 or 100 MHz; at 2 GHz, the upper operating frequency of the Leitz ELSAM, where the resolution is about that of the standard optical microscope, they are exceedingly difficult, and indeed relatively little work has been done in this range.

Reflection SAM is generally done in one of two imaging modes: (1) high-resolution surface imaging, where a high-frequency, high numerical aperture (NA) lens is chosen or (2) subsurface imaging, for which a sufficiently low-frequency and low NA lens is used, so that most of the ultrasonic wave penetrates into the sample. Many examples of reflection SAM imaging can be given, including biomedical imaging of soft and hard tissues, thin films, substrate materials, subsurface defects in materials and devices, stress, cracks, etc. An example of each type will be given in the next section. With increasing frequency, the most common applications are:

1. Low-frequency regime (10 to 100 MHz) is generally used for detecting defects in microelectronic chips and other subsurface damage.
2. The medium-frequency range (100 to 1000 MHz) is generally used for a wide variety of nondestructive evaluation (NDE) and biological samples, as well as quantitative microscopy (to be described).
3. High-frequency range above 1 GHz is restricted to special studies needing very high resolution. The highest resolution attained in this range is 20 nm using liquid helium as a coupling liquid.

In the early days of acoustic microscopy, it was discovered that slight defocusing is needed to obtain high-contrast images. A theoretical understanding of this phenomenon quickly leads to the realization that one could obtain quantitative information from the SAM by continuously defocusing and bringing the sample toward the lens for a fixed x, y position of the lens axis (z direction). Periodic variations of the voltage of the reflected signal are observed, the so-called $V(z)$ phenomenon [132]. Typical behavior is seen in Figure 14.2, which shows a series of oscillations of $V(z)$, with constant distance Δz between the minima. It is possible to obtain the Rayleigh surface

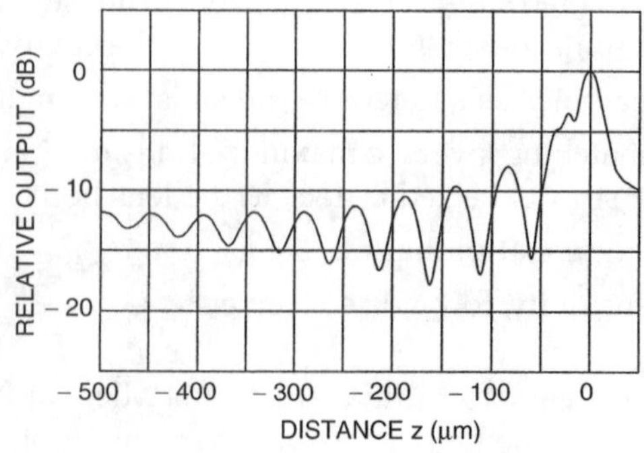

FIGURE 14.2
$V(z)$ curve for a gold film on a fused quartz substrate at 190 MHz. (From Kushibiki, J., Ishikawa, T., and Chubachi, N., *Appl. Phys. Lett.*, 57, 1967, 1990. With permission.)

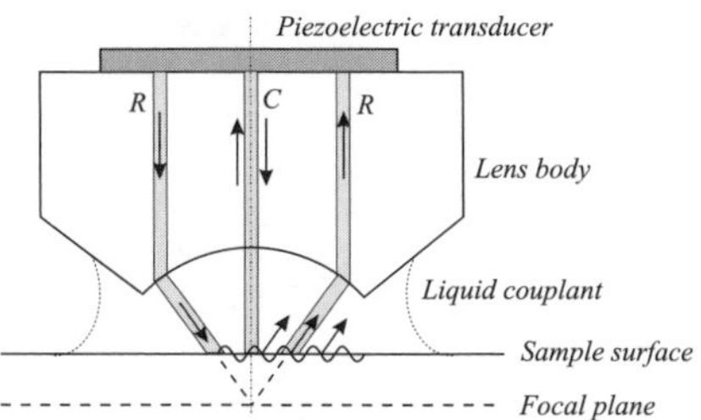

FIGURE 14.3
Simplified two-beam model to show the physical origin of $V(z)$.

wave velocity directly from Δz and this forms the basis for the quantitative applications of the SAM.

A simple explanation for the $V(z)$ effect is as follows: One can divide the acoustic wave incident on the sample into two beams, a central one (C) and an outside cone of rays (R), as shown in Figure 14.3. The central beam is directly reflected by the sample and serves as a reference. The outer conical beam arrives at the sample surface at the appropriate angle to set up Rayleigh surface waves. These are reradiated or leaked back into the liquid and eventually return to the transducer. These two components interfere constructively or destructively, depending on the lens-to-sample distance, which results in the set of interference fringes observed in $V(z)$.

The consequences of the $V(z)$ effect are many, and in fact the phenomenon is fundamentally important for all aspects of acoustic microscopy. For the spherical lens, the Rayleigh surface waves are excited in all directions and some appropriate average pertains for each point on the surface. This is important for high-contrast imaging, for example, of the grain structure of an alloy. Each grain has a particular crystallographic orientation compared to its neighbor, and so each one has a different average surface wave velocity. This leads to a different reflected signal for each grain via the $V(z)$ effect, so that some grains will give a maximum reflection and others a minimum one. The situation will be reversed for some other neighboring value of z. All of this results in the SAM having very high intrinsic contrast, so that special staining or etching techniques often used in metallography are not required. This identifies one important advantage of SAM for studying metals, alloys, and inhomogeneous samples.

One specific application of quantitative acoustic microscopy has been the development of the line focus beam (LFB) for directional measurements [133]. The spherical Lemons-Quate lens is replaced by a cylindrical lens, so that the focal point is replaced by a focal line. Of course, it is no longer possible to obtain acoustic images, but there are compensations for quantitative microscopy. The $V(z)$ phenomenon remains essentially the same with the important proviso that Rayleigh surface waves are now emitted in the direction perpendicular to the focal line. Therefore, the $V(z)$ can be related to a specific propagation on the sample surface. By rotating the lens one can

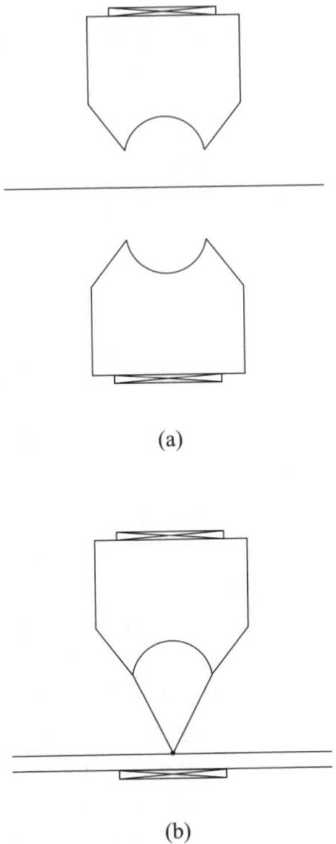

FIGURE 14.4
Transmission acoustic microscopy configurations. (a) Confocal. (b) Focal plane imaging.

measure the anisotropy in the Rayleigh wave velocity and, by an inversion procedure, the elastic constants. These effects have been studied by many workers. It has been shown that accuracies of the order of 10^{-4} in the velocities are possible providing displacements, water temperature, and frequency are measured very accurately. Examples of studies carried out with the LFB are crystal anisotropy, anisotropic films on substrates, wafer mapping, optical fibers, etc.

In addition to reflection mode acoustic microscopy, some work has also been done in transmission. In the original work of Lemons and Quate, two identical "Quate" lenses were placed face to face in a confocal configuration and a thin sample holder was positioned perpendicular to the lens axis in the focal plane, as shown in Figure 14.4. Although found to be very effective, this configuration has some serious alignment problems, especially at high frequency. It is especially suitable for the imaging of acoustically transparent or biological samples where the transmitted signal is very sensitive to variations in sample attenuation and to a lesser extent the phase. Extremely high resolution, high contrast images of red blood cells were obtained. The configuration is also well adapted to the study of living cells. A second configuration for focal plane imaging, Figure 14.4(b), was used by Germain and Cheeke [134]. Because of its geometry and its use at low frequencies there were no particular alignment problems. It was used for quantitative determination of the nonlinear acoustic properties of liquids as well as for harmonic imaging using resonant transducers.

Another different but complementary tool to the SAM is the scanning laser acoustic microscope (SLAM) [135]. In the SLAM, the sample is irradiated from the back side by a continuous uniform beam of ultrasound, which is then transmitted to the front surface. The impinging ultrasonic beam creates a surface disturbance on the front surface and an image is formed by scanning a laser beam over the front surface to image this disturbance. The SLAM is basically a near field shadow imaging system. Another feature is that it is possible to obtain simultaneous optical and acoustic images in real time (30 frames per second). Since the transmitted ultrasound intensity is affected by defects in the bulk of the sample, these can be detected by SLAM imaging. The technique has been widely used for evaluating bonding, delamination, defects in microelectronic devices, biomedical imaging, and many other applications. The real-time aspect is particularly interesting for NDE, for example to study the propagation of a crack in a material under stress. Other advantages include the possibility of detecting surface waves of extremely small amplitudes($\sim 10^{-6}$ nm/$\sqrt{\text{Hz}}$ bandwidth) and doing plane-by-plane imaging by a holographic technique. The resolution of the acoustic images is determined principally by the width of the laser beam, the depth of the defect, and the ultrasonic wavelength.

14.2 Resolution

The characteristics of an acoustic beam focused by a spherical radiator have been described in Chapter 6. For SAM, an acoustic lens of this type at the end of a buffer rod (Quate lens) is used for focusing and reception of the ultrasonic signal. This means that the point spread function (PSF) for the confocal configuration used in the SAM is sharper than that for the single lens. In fact, it is the square of the single lens PSF, and so varies as $\text{jinc}^2 X$. The transverse definition is therefore sharper for the SAM and is given by

$$d_{rc}(3\,\text{dB}) = \frac{0.37\lambda}{\sin\theta_0} \tag{14.1}$$

$$d_{rc}(\text{Rayleigh}) = \frac{0.56\lambda}{\sin\theta_0} \tag{14.2}$$

where the subscript c refers to the confocal configuration and θ_0 is the half aperture angle. The result is that the sidelobes in the SAM are much lower, 35 dB down from the focal peak as compared to 17.6 dB for the single lens. However, while the diffraction limited performance is greatly improved by the confocal configuration, it then becomes very sensitive to the presence of aberrations, which will now be regarded in detail. By the very fact of on-axis imaging at a single frequency, four out of five aberrations identified in optical microscopy are immediately eliminated: chromatic, barrel distortion, pincushion distortion, and astigmatism. That remaining, spherical aberration (SA) can easily be eliminated, in theory and in practice, in the following way. A geometrical optics approach will be adopted to determine both the paraxial focal length and an approximate formula for the aberrations, following the original approach of Lemons and Quate. One of the principal differences with the case of optics is that SA is greatly reduced in the acoustic case, which makes it possible to design SA-free spherical acoustic lenses. This is a fortunate circumstance as high-frequency acoustic lenses are very small (~20 μm diameter), where it would be difficult to grind and polish aspherical surfaces.

In Figure 14.5(a) the center of curvature is taken as the origin of the coordinate system. A reduced velocity variable is defined as

$$n = \frac{V_0}{V_1} \tag{14.3}$$

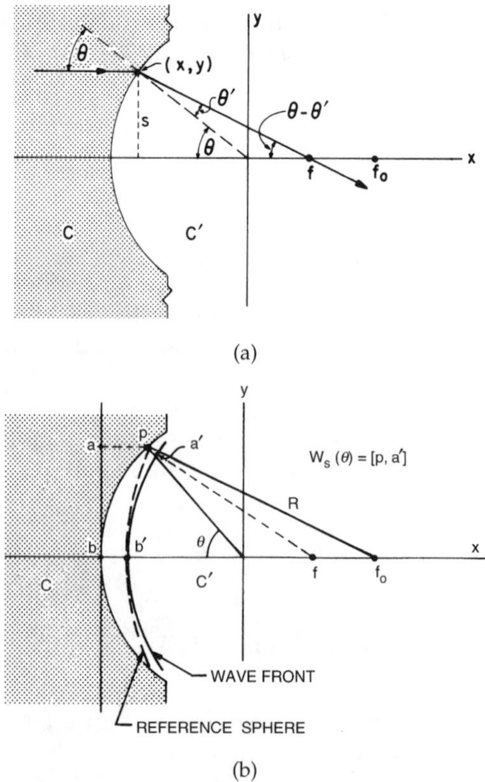

FIGURE 14.5
Geometry used for analysis of spherical aberration of acoustic lenses. (a) Ray tracing analysis. (b) Wavefront analysis. (From Lemons, R.A., Acoustic Microscopy by Mechanical Scanning, Ph.D. thesis, E.L. Ginzton Laboratory, Stanford University, Stanford, CA, 1975. With permission.)

where
V_0 = sound velocity in liquid
V_1 = sound velocity in solid

A ray incident parallel to the axis at aperture angle θ will be focused at point f, where by Snell's law $\sin\theta' = n\sin\theta$ and from simple geometry

$$f = \frac{y}{\tan(\theta - \theta')} + x$$
$$\approx y \cdot \frac{1 + \tan\theta \cdot \tan\theta'}{\tan\theta - \tan\theta'} + x \tag{14.4}$$

Passing to variables normalized to the radius of curvature ($r = 1$) and using $s = \sin\theta$

$$y = s, \quad x = -\sqrt{1-s^2}, \quad \tan\theta = \frac{s}{\sqrt{1-s^2}}, \quad \tan\theta' = \frac{ns}{\sqrt{1-n^2s^2}}$$

giving the focal distance at angle θ

$$f = \frac{n}{\sqrt{1-n^2s^2} - n\sqrt{1-s^2}} \tag{14.5}$$

For $s \ll 1$ this expression yields the usual expression for the paraxial focus

$$f_0 = \frac{n}{1-n} \tag{14.6}$$

The deviation of the position of $f(\theta)$ with respect to f_0, $f_0 - f(\theta)$, corresponds to SA.
Equation 14.5 shows explicitly that the SA

- Increases with angle θ
- Increases with reduced velocity n

It is the variation with n that is the most important for SAM. The key point is that the SA can be made very small indeed if n is minimized, and it is here that the acoustical situation becomes very favorable compared to its optical counterpart. In optics there is only a small (~30%) variation of relative refractive index, so that SA is intrinsically quite large. However, in acoustics there is typically a big difference between sound velocities of liquids and solids and in fact the original choice of Lemons and Quate of sapphire ($V_1 \sim 11.1 \times 10^5$ cm·s^{-1}) and water ($V_0 \sim 1.5 \times 10^5$ cm·s^{-1}) corresponding to $n = 0.135$ was almost optimal from the point of view of SA.

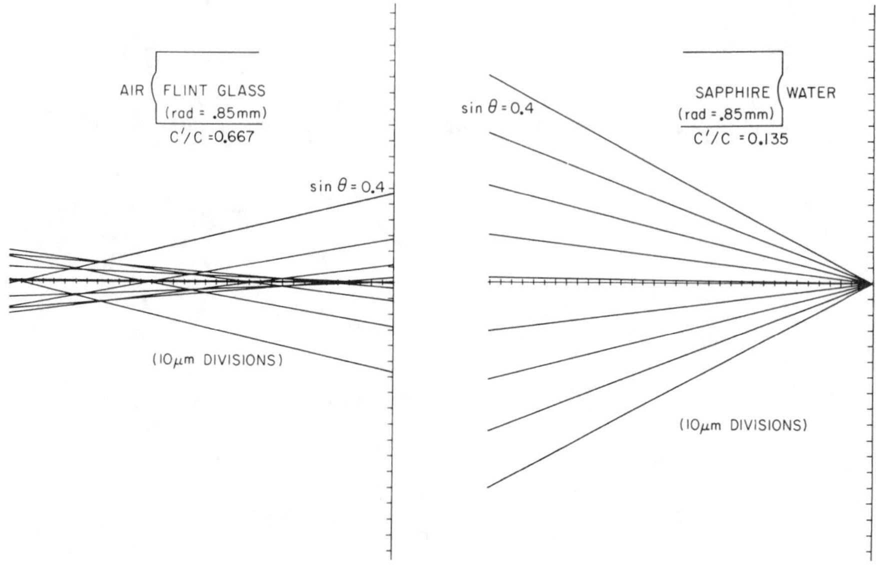

FIGURE 14.6
Ray tracing comparison of the performance of a single surface lens in an optical system (left) and an acoustic system (right). (From Lemons, R.A., Acoustic Microscopy by Mechanical Scanning, Ph.D. thesis, E. L. Ginzton Laboratory, Stanford University, Stanford, CA, 1975. With permission.)

Lemons and Quate have shown the difference between optical and acoustical SA by using ray tracing comparisons for $n = 0.667$ for the optical case and $n = 0.135$ for the acoustical case, as shown in Figure 14.6. The broadening of the focal point due to SA and hence the limit of resolutions due to SA is usually described by the circle of least confusion, the minimum diameter of the bundle of rays around the focus. For this calculation, the aberration limit for the optical case is about 40 μm and only of the order of 0.5 μm for the acoustic case.

An important point is that as SA is a geometrical effect, it scales with the size of the lens and so decreases as the lens is made smaller. The lens radius used in the above simulation was about 0.8 mm, which is much bigger than an actual acoustic lens that would be used for an acoustic wavelength in water of 0.5 μm. Thus, in the latter case, the SA would be even more negligible than it was estimated to be above. The conclusion is that for a sapphire-water system the SA will be negligible at all frequencies.

Additional insight can be afforded by approaching the problem from a wave point of perspective. The ideal case of no aberration corresponds to a spherical wave converging on the focus. The presence of SA causes a deformation of the wavefront at the exit pupil of the lens, so that the effect can be described as a phase deformation $kW(r, \theta, \varphi)$ of the actual wavefront compared to the spherical reference. Combined with the amplitude distribution over the exit pupil $p(r, \theta, \varphi)$, this gives the generalized pupil function

$$P(r, \theta, \varphi) = p(r, \theta, \varphi)\exp(jkW(r, \theta, \varphi)) \tag{14.7}$$

The SA, $W(r, \theta, \varphi)$, can be calculated with the aid of Figure 14.5(b). A plane wave A-A′ is shown incident on the lens and the corresponding converging wavefront B-B′ in the exit pupil of the lens is shown, together with a reference sphere that crosses the lens at reference angle θ. Thus, if there were no aberration, the point B′ would be found at P; in other words, the SA can be expressed by $W_s(\theta)$ = [P, B′]. This distance can be found by application of Fermat's principle, as the time of propagation from A-B must be the same as A′-B′. Formally, this means

$$\frac{1 + f_0 - R}{V_0} = \frac{1 - \cos\theta}{V_1} + \frac{W(\theta)}{V_0}$$

which yields

$$W(\theta) = 1 + f_0 - R - n(1 - \cos\theta) \tag{14.8}$$

We know $f_0 = \frac{n}{1-n}$ and finally, after a Taylor's expansion

$$W(\theta, n) = 2n^2(1 - n)\left(\sin^4\frac{\theta}{2} + 2n(1 - n)\sin^6\frac{\theta}{2}\right) \tag{14.9}$$

The first term gives the primary SA and has a form similar to that given by the Seidel theory of aberrations. For small values of n this gives $W \sim n^2$, which shows explicitly that SA is negligible for sufficiently small n.

As a specific example, Lemons [136] shows that for a sapphire-water lens with $r = 0.4$ mm, $W(50°, 0.135) \sim 0.4\ \mu$m. For a 500-MHz wave, which would be a typical frequency for such a lens, this is about an eighth of a wavelength.

For surface imaging, where maximum resolution is normally desirable, the NA is normally made as large as possible. As described before, increasing the resolution can be most directly accomplished by raising the frequency, and the acoustic attenuation in the liquid then becomes the main parameter. A resolution coefficient has been defined to compare the best resolution that can be obtained for various coupling liquids [137], taking into account the focal length and the attenuation in the liquid. This resolution coefficient R_c is defined as [138]

$$R_c = \sqrt{V_0^3 \alpha_0} \tag{14.10}$$

where $\alpha = \alpha_0 f^2$ for a given liquid. In general, one has to go to cryogenic liquids to obtain significant improvement over water. Relevant acoustic parameters for solids, including those used for lens fabrication, are given in Table 5.1 and Appendix B.

Various strategies can be employed to increase the resolution, depending on the experimental conditions. The following points can be made:

1. For the vast majority of applications at not-too-high frequencies, water is the simplest and almost optimal choice.
2. The liquid metals gallium and mercury have attractive acoustic properties, but they are difficult to work with, and this fact has greatly reduced their use in practical applications.
3. Significant gains can be achieved by heating the water to 60°C or higher.
4. High-pressure gases such as argon are in principle attractive; however, the acoustic impedance difference between sample and gas means that topography dominates the image properties.
5. Cryogenic liquids can be used to advantage because of their low attenuation and velocity [139]; however, the acoustic impedance mismatch is so great that the reflectivity is almost 100% everywhere on the sample surface, so that topography again dominates. Also, this is not a practical route for most industrial applications.
6. Nonlinear enhancement of the resolution can be used to advantage. The high acoustic intensities at the focus mean that harmonic generation is very pronounced in this region. Rugar [140] showed that the threshold power for significant generation of the second harmonic is given by

$$P_0 = \frac{4s_d L(NA)^2}{f^2} \tag{14.11}$$

where s_d = fractional depletion of the fundamental

$$L = \frac{\rho_0 V_0^5}{16\pi^3 \beta_L^2}$$

where β_L is the fluid nonlinear coupling constant.

It is known that an increase in resolution by $\sqrt{2}$ is obtained by generation of the second harmonic at the focus, and Rugar [140] showed that in reflection microscopy this enhancement is maintained even though the second harmonic is subsequently converted down to the fundamental. This work was extended by Germain and Cheeke [141], who showed experimentally that a similar resolution enhancement by $\sqrt{n}$ occurred for higher harmonics

n and that significant resolution improvement could be obtained by detecting them directly at the focal plane in a transmission configuration. They showed that this mode of operation is particularly advantageous for samples in solution and biological samples.

14.3 Acoustic Lens Design

Lens design for the SAM is a pinnacle of achievement in ultrasonic science and engineering. All of the essential ingredients of ultrasonic propagation in solids and liquids are assembled and an optimum solution must be found by compromise and ingenious choice of materials and design. The difficulties increase exponentially with frequency and it is somewhat miraculous that resolution in the 10-nm range has been attained under special laboratory conditions.

Our goal here is much more modest and it is mainly to see how the building blocks of Chapters 2 through 8 can be applied to the problem. We describe first the common components that make up a typical acoustic microscope lens, as shown in Figure 14.1. We then give a critical discussion on how these parameters are chosen for the case of a lens to be used either for surface imaging or quantitative $V(z)$ measurements; lens design considerations for subsurface or interior imaging will be given in Section 14.4.

The starting point in Figure 14.1 is the piezoelectric transducer, which in full detail would be described by the Mason equivalent circuit. The transducer assembly includes the two metallic electrodes. These can be neglected in analysis for frequencies less than 100 MHz; they should be included above that frequency and they must be included for frequencies at 1 GHz or above. Matching of the transducer assembly to the RF source can be accomplished crudely with stub tuners but much more effectively by carefully designed series and parallel inductances to tune out the transducer capacitance. The diameter of the top electrode defines the active region of the transducer; this should be slightly larger than the diameter of the lens cavity to assure roughly uniform illumination of the lens. The coupling factor (K^2) of the piezoelectric material should be as high as possible to ensure a high dynamic range, which is essential for good image contrast. The design of the transducer assembly will generally be one of two general approaches. If very sharp DC pulses are used then the electronics and transducer assembly must have broadband characteristics. Alternatively, very-high-frequency systems for high resolution and $V(z)$ applications will intrinsically involve narrow band highly resonant transducer assemblies.

At first glance, the lens body is a supporting structure but in fact it is much more than that and there are a number of key design issues. The role of the lens body is to act as a propagation medium between transducer, lens cavity, and back. A primary requirement for the lens body material is that its velocity be much higher than that of the liquid used so that spherical aberration is reduced to acceptable levels. The material must have very low attenuation to minimize insertion loss. At low frequencies, it can be isotropic or polycrystalline; fused quartz and aluminum work well below 100 MHz. At sufficiently high frequencies, well above 100 MHz, it must be an insulating crystal oriented along a pure longitudinal wave symmetry direction. In the GHz range, precision alignment of top electrode, crystal axis, and lens center are critical. If the alignment is not perfect, the acoustic beam will be displaced and then will only partially illuminate the lens. For example, for a 2-mm-long sapphire lens body, a 1° lens body axis misalignment will give an 11-μm beam displacement at the lens, which would be intolerable for a 20-μm diameter acoustic lens for the GHz range. Apart from precision alignment, choice of lens body material, e.g., using cubic YIG instead of trigonal Al_2O_3, can reduce the beam displacement. The length of the lens body is a key parameter. It is generally accepted practice that this be 1 to 3 or 4 Fresnel lengths in order to achieve uniform illumination of the lens. Below a Fresnel length of one, we enter the nonuniform regime of the near field; much above 3 or 4 Fresnel lengths, most of the beam power will miss the lens. Chou et al. [142] have calculated the point spread function for acoustic lenses with Fresnel parameters $S = 0.5$, 1.0, and 10.0 and find that $S = 1.0$ is optimal for reducing the sidelobes at the focal plane. Another influence of the lens body is that ideally there would be no echoes in it, as these often overlap and obscure the focal plane echo to be detected. To some extent this can be avoided by a careful choice of dimensions and materials so that the lens echo falls between two lens echoes. The best solution is to use matching layers on the lens surface which provide a double benefit: they significantly reduce the undesirable lens echo and significantly increase the focal plane echo to be detected. Undesirable reflections can also be reduced by roughening the outer surfaces of the lens body.

The heart of the acoustic microscope is evidently the lens cavity itself. This cavity is produced by carefully grinding and polishing a spherical recess in the face opposite the transducer. The lens cavity radius is an important parameter for several reasons:

1. It determines the maximum pulse width as the reflection from the front face of the lens and from the sample at the focal plane must be clearly time resolved.
2. This pulse width determines the maximum receiver bandwidth and hence the receiver noise figure.
3. The pulse width also determines the axial resolution or depth of field for the case of subsurface imaging. Apart from resolution considerations, the choice of the NA follows directly from the lens diameter. For surface imaging, it is critical that the NA be sufficiently large to include the specimen Rayleigh angle, which is an essential element of the contrast mechanism.

4. Most important, the lens diameter must be sufficiently small so that loss in the liquid between lens surface and sample is within acceptable limits.

The choice of the liquid has been covered in the discussion on resolution; we assume from here on that water has been chosen. This leads to the water-lens interface as the last major design issue. Because of all of the other constraints already mentioned, at this stage we usually end up with a big difference in acoustic impedance between lens and liquid. Since this interface has to be traversed once in each direction, at sufficiently high frequencies the insertion loss of the lens will become unacceptably large. Acoustic matching layers is the indicated solution for what is already a highly constrained problem.

From Equation 7.21, we have

$$R_p = \frac{Z_2^2 - Z_1 Z_0}{Z_2^2 + Z_1 Z_0} \tag{14.12}$$

where

R_p = reflection coefficient for waves incident from the lens body of acoustic impedance Z_1
Z_2 = acoustic impedance of the layer material
Z_0 = acoustic impedance of the liquid

with the condition $d = \lambda_2/4$ where d is the layer thickness.

The desired condition is $R_p \equiv 0$, which is achieved by $Z_2 = \sqrt{Z_1 Z_3}$, the standard quarter wavelength matching condition. For a given solid-liquid combination, choice of potentially suitable matching layers is given by consulting Figure 5.2. Of course, it is unlikely to find a material that exactly fits the bill but it does enable the lens designer to optimize the layer material chosen or, perhaps, rethink the solid-liquid combination chosen.

For such combinations as sapphire-water, borosilicate glass or carbon layers are quite suitable. Rugar [143] has found that amorphous carbon films on sapphire give almost perfect matching into water at 1 GHz on planar surfaces. In practice, high-frequency matching layers for acoustic lenses becomes rather technical. The sputtered or evaporated films produce a $\cos\theta$ thickness profile, which reduces the transmission efficiency. Also, in cases of extreme mismatch, two matching layers may be needed. A detailed discussion is given by Rugar [143]. For our purposes it is important that above 500 MHz matching layers are virtually essential, and reduce the insertion loss by tens of dB. At low frequencies (below 100 MHz), it is nontrivial to control the required matching layer thickness but if careful lens design is carried out matching layers are not really needed in this regime.

In the ideal case lens design for surface imaging involves the choice of the required resolution, which is approximately 0.7λ for an F 0.75 lens. This determines the frequency. Once an acceptable coupling fluid has been chosen, the maximum lens diameter is determined, assuming a maximum allowable loss (say 60 dB) in the liquid. The lens body material is chosen to respect $n \ll 1$ and minimum attenuation. The lens body length is then chosen to optimize the illumination of the lens. Further steps to maximize the signal-to-noise ratio include the following:

1. Choice of high performance transducers such as bonded lithium niobate or PZT plates below 150 MHz, sputtered ZnO or AlN at higher frequencies
2. Impedance matching of the transducers to the RF source
3. Choice of a suitable matching layer for the lens
4. Choice of low-noise, high-sensitivity electronics

For order of magnitude considerations at 1 GHz, if we assume a receiver dynamic range of the order of 120 dB, then this might be divided up as follows: 30 dB for various losses in the lens, 60 dB for losses in the liquid, and 30 dB available to provide sufficient image contrast.

If these principles are followed, then we arrive at the order of magnitude choices presented in Table 14.1 for two quite different cases: 30 MHz and 1 GHz. Of course, in a given situation there might be special constraints, e.g., available instrumentation or lenses, so the approach might have to be somewhat flexible. High-resolution surface imaging is assumed above; quantitative $V(z)$ measurements would require a similar approach but with other considerations.

TABLE 14.1

Typical Design Parameters for Low Frequency and High Frequency Acoustic Lenses for Scanning Acoustic Microscopy

Property	30 MHz	1 GHz
Transducer		
• Material	Lithium niobate	ZnO (1.4 μm)
• Diameter	6 mm	200 μm
• Electrodes	Au-Cr	Au-Cr (0.2 μm)
Lens body		
• Material	Fused quartz	Sapphire
• Orientation	NA	C axis
• Diameter	10 mm	6 mm
• Length	10 mm	4 mm
Lens cavity		
• Diameter	5 mm	100 μm
• Opening	4 mm	
• Matching layer	NA	Borosilicate glass
Liquid		
• Type	Water	Water
• Temperature	20°C	60°C

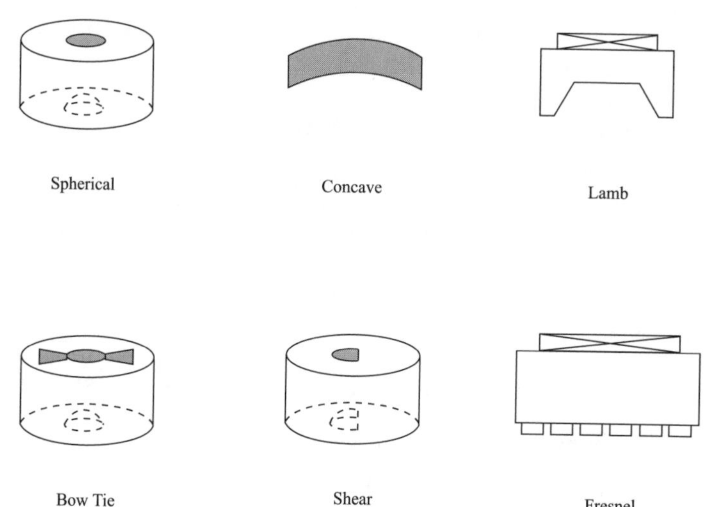

FIGURE 14.7
Different types of acoustic lenses used in scanning acoustic microscopy.

Although spherical and cylindrical lenses are by far the most used in SAM, other lens designs have been proposed at various times for particular applications. These are displayed in Figure 14.7 and described very briefly below:

1. Standard spherical lens, already described in detail. The main variables are the lens diameter and the aperture.
2. Concave lenses (spherical emitters). Several variants have been developed. Liang et al. [144] carried out many low-frequency experiments with these devices at 3 MHz. At higher frequencies, various piezoelectric films were posed in the spherical cavity to focus at the center of curvature. These devices include ZnO films [145] and PVDF [146]. The latter has a low intrinsic insertion loss as it is well matched to water.
3. Lamb wave lens [147]. As described in Section 15.3.1, the higher-order Lamb modes can be excited at characteristic angles given by $\sin\theta = V_0/V_P$ for different values of plate phase velocity V_P. Atalar has developed a lens where the central part emits at normal incidence and the outside part at a constant angle θ. If the lens is placed in a defocused position above the plate to be studied as the frequency is swept, various Lamb modes become excited as the above condition becomes satisfied. It becomes useful to create a $V(f)$ curve in analogy to the $V(z)$ curve.
4. Bow-tie transducer. Davids and coworkers [148] developed this transducer in an attempt to produce a spherical lens with directional characteristics.

5. Shear wave lens. Khuri-Yakub et al. [149] attempted to use shear wave transducers illuminating a spherical lens to create directionality as in [148] by mode conversion at the lens interface. A characteristic $\cos\theta$ variation was confirmed experimentally.
6. Fresnel lens. Yawada and coworkers used a Fresnel-type lens [150] that allowed simultaneous optical and acoustic images of the same local region of the sample.

14.4 Contrast Mechanisms and Quantitative Measurements

14.4.1 *V*(*z*) Theory

A typical $V(z)$ curve is shown in Figure 14.2; by convention, negative z corresponds to a decreased lens-to-sample distance. The two main interfering beams are shown in Figure 14.3, and those that appear to come from the focal point interfere at the piezoelectric transducer, which is sensitive to the phase. By simple geometry from Figure 14.3, the relative phase difference between the two beams is

$$\phi_G - \phi_R = -2kz(1 - \cos\theta_R) + \pi \tag{14.13}$$

where θ_R is the Rayleigh angle, defined as $\sin\theta_R = V_0/V_R$. Clearly the interference condition depends on Z, giving rise to the series of minima seen in Figure 14.2. The period of the oscillations is

$$\Delta z = \frac{2\pi}{2k(1 - \cos\theta_R)} \tag{14.14}$$

so that measurement of Δz for a given f and V_0 gives θ_R, hence V_R, for the sample at this position. Similar considerations give for the attenuation

$$\Delta\alpha = 2z(\alpha_0 \sec\theta_R - \alpha_R \tan\theta_R) \tag{14.15}$$

However, the attenuation is much more difficult to obtain accurately, and most of the work has been done on measurement of V_R.

While the simple two-beam model is useful for understanding the physics of $V(z)$, many simplifications have been made. A more rigorous mathematical treatment of the phenomenon is provided by scalar wave theory [151], which is used to describe the refraction of all acoustic waves over the lens aperture into the liquid. For a given z, the result is

$$V(z) = \int_0^{\pi/2} P(\theta)R(\theta)e^{-2zk\cos\theta}\sin\theta\cos\theta\, d\theta \tag{14.16}$$

where $P(\theta)$ is the pupil function that characterizes the lens transmission properties, which depend on the geometry and the lens material parameters, and $R(\theta)$ is the amplitude reflectance function. By redefining variables such that $u = kz$, $t = (1/\pi)\cos\theta$, and $Q(t) = P(t)R(t)$, we find

$$V(u) = \int_0^{1/\pi} Q(t)e^{-i2\pi ut}dt \tag{14.17}$$

so that $V(u)$ and $Q(t)$ are a Fourier transform pair for a lens with a known pupil function. Thus the measurement of the full $V(z)$ curve over the full range of z should lead in principle to a determination of $R(\theta)$, which will be given below. Analogous treatment can be given for transmission, although the applications have been much less numerous. The formulation is

$$A(z) = \int_0^{\pi/2} P(\theta)T(\theta)e^{-i(z-d)k\cos\theta}\sin\theta\cos\theta\, d\theta \tag{14.18}$$

where $P(\theta)$ is the lens function for the two lenses and $T(\theta)$ is the transmission function for a layer of thickness d for incident and refracted angles θ.

In addition to the wave theory, a ray model more complete than the simplified version already mentioned has also been developed [152]. It is an interesting complement to the wave theory, as various modes such as surface-skimming bulk waves may be put explicitly into the model, as described in detail in [138].

14.4.2 Reflectance Function from Fourier Inversion

Inversion of the wave theory gives:

$$R_t(t) = \int_{-\infty}^{\infty} \frac{V(u)}{V_0}e^{-i2\pi ut}du \tag{14.19}$$

so that measurement of $V(u)$ can give $R(\theta)$. As mentioned by Briggs [138], there are several precautions to be observed with this formula: one can only obtain $R(\theta)$ for the range of angles included within the lens opening; the full curve $V(u)$ is needed, as truncation can cause errors; the results are sensitive to attenuation associated with fluid loading, especially at high frequencies; and $V(u)$ is a complex function, so measurement of the amplitude and phase are needed. The first measurements were carried out by Liang et al. [144], shown in Figure 14.8 for water-fused silica interfaces at 10 MHz. A lead sample, for which no Rayleigh waves are excited in this case, was used as a reference to obtain the pupil function. The most spectacular result was an observation of an expected phase change of 2π, at the Rayleigh angle, which allowed accurate determination of v_R. A dip in the amplitude is also seen at ϑ_R; this is usually due to damping of the Rayleigh wave, but care must be taken as such dips could also be due to anisotropy and/or truncation of the data.

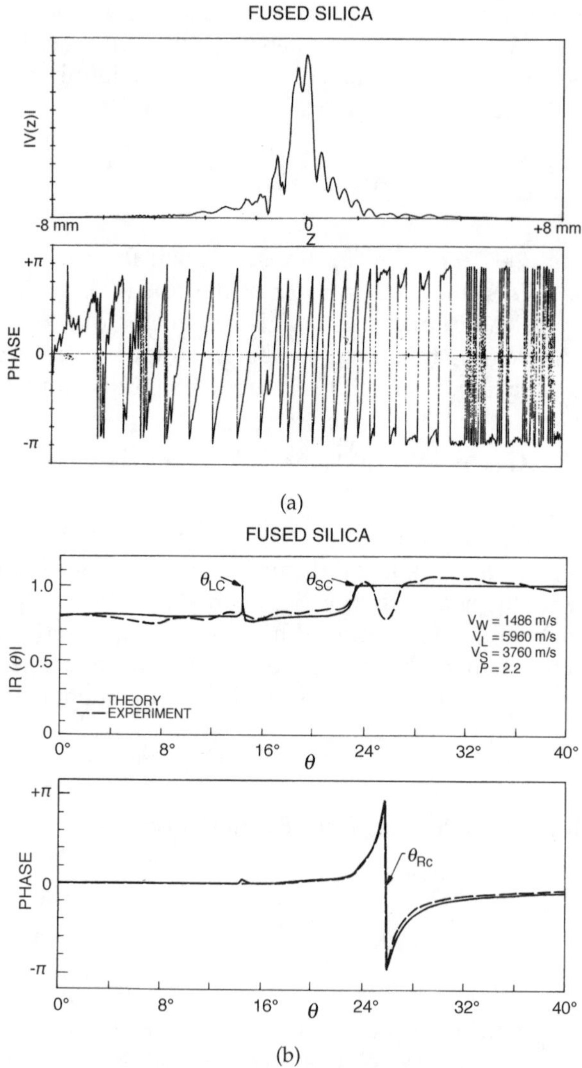

FIGURE 14.8
(a) Experimental $V(z)$ of water-fused silica interface at 10.17 MHz. (b) Comparison of the theoretical and experimental reflectance function for a water-fused silica interface. (From Liang, K.K., Kino, G.S., and Khuri-Yakub, B.T., *IEEE Trans. Sonics Ultrasonics*, SU-33, 213, 1985. © IEEE. With permission.)

14.4.3 Line Focus Beam

Developed by Kushibiki and Chubachi [133], the line focus beam (LFB) technique exploits Rayleigh waves emitted perpendicular to the focal line of a cylindrical lens. The generally accepted analysis uses a ray approach that can be summarized as follows. The reflected signal can be written as

$$V = V_G + V_R \tag{14.20}$$

where V_R is the Rayleigh wave contribution and V_G is due to the sum of all other scattered waves.

For square law detection

$$|V|^2 = |V_G|^2 + |V_R|^2 + 2|V_G||V_R|\cos\theta \tag{14.21}$$

where θ is the phase angle between V_G and V_R and all terms are z dependent. The measuring system is calibrated using a lead sample; to a good approximation, $V_L = V_G$. Two assumptions are then made to complete the analysis for the LFB:

1. $|V_R| \ll |V_G|$, which reduces to

$$|V| - |V_L| = |V_R|\cos\phi$$

2. The phase depends linearly on z, leading to

$$\phi = -2kz(1 - \cos\theta_R) + \pi$$

Neglecting attenuation, it is found that the spatial frequency of the Fourier transform of $V(z)$ is centered at

$$\xi_0 = \frac{2\pi}{\Delta z} = 2k(1 - \cos\theta_R)$$

Taking attenuation into account, the final results are expressed as

$$V_R = V_0\left\{1 - \left(1 - \frac{V_0\xi_0}{4\pi f}\right)\right\} \tag{14.22}$$

$$\alpha_N = \frac{\alpha\cos\theta_R + 2\alpha_0}{2k_R\sin\theta_R} \tag{14.23}$$

Several experimental precautions are needed to obtain very high accuracy for V_R and α_N with the LFB; steps include use of goniometers for tilt alignment, careful temperature control of the water drop, and careful measurement of the lead reference calibration curve. Likewise, there are several steps in the data analysis necessary to get accurate data reduction for Fourier analysis, including filtering and subtracting out V_L by an iterative procedure. The accuracy of the LFB can be written in terms of precision in temperature, distance, and frequency measurements as

$$\frac{\delta V_R}{V_R} = \sqrt{\left\{(0.0011\,\delta T)^2 + \left(0.464\frac{\delta f}{f}\right)^2 + \left(0.464\frac{\delta\Delta z}{\Delta z}\right)^2\right\}} \tag{14.24}$$

from which it can be deduced that for a relative accuracy of 10^{-3} in $\Delta V/V$, ΔT is needed to ±0.9°C, $\Delta f/f$ to 0.2%, and $\Delta z/z$ to 0.2%. For a relative accuracy of 10^{-4}, ten times greater precision is needed for each parameter. Full details are given in [133] for determinations of $\Delta V/V$ and α_N over 30 different materials. Accuracies of 10^{-4} for $\Delta V/V$ and 2% for α_N are claimed.

14.4.4 Subsurface (Interior) Imaging

Subsurface imaging is one of the unique capabilities of SAM because unlike light many other forms of radiation ultrasonic waves can penetrate optically opaque media. They can thus be used to detect and image defects and other subsurface structure.

Surface imaging and metrology ($V(z)$) applications mentioned previously used wide aperture lenses, to maximize resolution and to excite Rayleigh waves, respectively. In subsurface imaging, the requirements are different, in fact, complementary. In this case, one aims to maximize the acoustic energy that penetrates into the substrate. In that context, it is undesirable to generate Rayleigh waves at the surface, as this would simply subtract acoustic energy out of the main beam. For this, and for other reasons to be seen, it is preferable to use a narrow beam, at least one narrow enough to avoid Rayleigh wave generation.

A number of other factors limit the subsurface imaging effectiveness. The signal levels are small due to the large acoustic mismatch between sample and coupling liquid. One solution would be to increase the pulse width, but there is a conflicting requirement of keeping narrow pulses to ensure good temporal resolution for resolving subsurface structure. Pulse compression techniques are one solution to resolve this problem. Another factor is that the resolution is degraded compared to that obtained for surface imaging as the sound velocity inside the sample is typically much higher than that in the coupling liquid. For a given frequency, this leads to longer wavelengths and lower resolution. The resolution is also degraded by the higher spherical aberration introduced at the interface, as shown in Figure 14.9.

There are a number of solutions to the above difficulties. The use of a reduced aperture lens, say 30°, not only avoids Rayleigh wave generation but also reduces spherical aberrations. Use of high-density, low-attenuation liquid metals such as gallium and mercury helps to increase the transmitted acoustic intensity. Finally, use of the shear mode for subsurface imaging is advantageous as for many materials the shear wave velocity is close to that of the coupling liquid. With these subterfuges, most of what was lost in signal level and resolution has now been regained.

Yet another approach to improve performance in subsurface imaging is to use aspherical lenses. Pino et al. [153] have used a geometrical optics approach in designing an aspherical lens to attain a diffraction limited focus inside the solid. They used Fermat's principle to equalize the transit time from the exit pupil to the focus. This leads to an aspherical surface with a smaller curvature on the outside of the lens compared to that in the axial region.

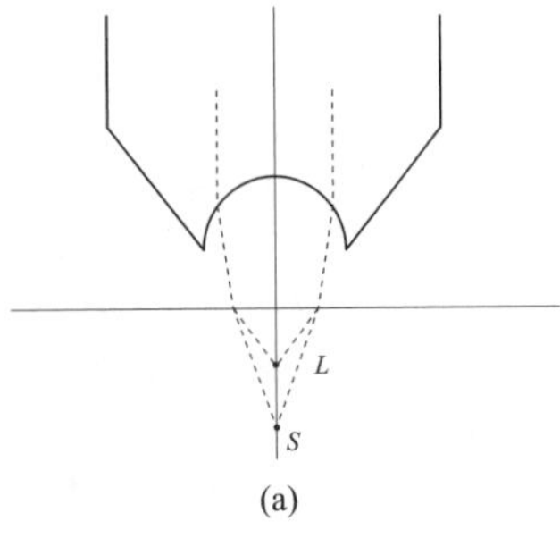

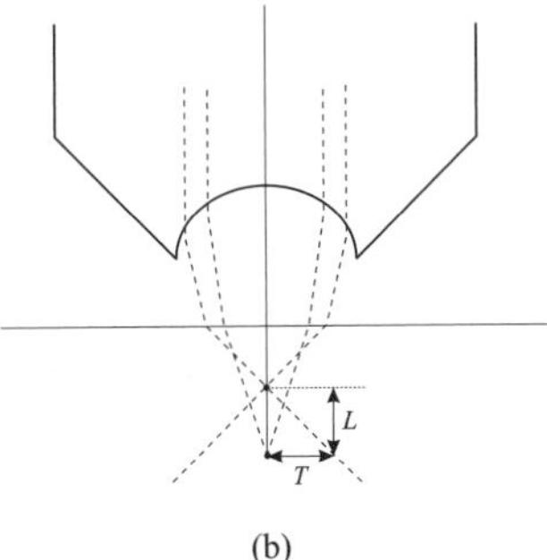

FIGURE 14.9
Subsurface imaging in acoustic microscopy. (a) Longitudinal and shear focal regions. (b) Longitudinal (L) and transverse (T) spherical aberration associated with either one of the focal regions shown in (a).

The calculations were confirmed experimentally by the observation of diffraction limited imaging inside the solid with no increase in the sidelobes.

A final point is that high resolution is not an overriding concern in subsurface imaging. The main application is in NDE and in most cases one wishes to observe if there is a defect and not to study it in refined detail.

14.5 Applications of Acoustic Microscopy

14.5.1 Biological Samples

Ultrasound imaging and quantitative study of biological tissue have several characteristic differences from similar studies on materials. There is no flat, well-defined reflecting surface, and biological tissues are generally more homogeneous in their structures, typically with high attenuation and sound velocity in the range of that of water. Since the shear modulus is low and shear viscous damping is high, we are only concerned with longitudinal waves.

As a consequence, while the technology is generally the same as for materials, there are important differences. Transmission mode imaging or through transmission substrate reflection is much more frequently used although the analog of reflection SAM, ultrasonic backscatter microscopy (UBM), has been used in some work. Traditionally, the frequency range for medical imagery has been below 10 MHz although in some of the work to be described here this has been extended toward the 30 to 100 MHz range. As in NDE, ceramic transducers with their high-coupling coefficients are frequently employed although polyvinylidene fluoride (PVDF) and copolymers find relatively more frequent use than in NDE because of their good impedance match to water. It should be noted that medical imaging has several imaging modes, namely A scan (amplitude/time trace as on the oscilloscope), B scan (section normal to the sample surface), and the usual C scan used for imaging materials.

Acoustic microscopy in the 10 to 100 MHz range can be either *in vivo* or *in vitro* [285, 286]. One of the common imaging applications in this range is for dermatological diagnosis. A wide bandwidth and sufficiently high frequency of the transducer and electronics are essential to obtain sufficient axial and lateral resolution. Typically, the transducer is placed at the end of a lever and mechanically scanned by a DC motor, with acoustic coupling supplied by an ultrasonic gel. B scan is used to identify the various layers and interfaces of normal skin (epidermis, dermis, and hypodermis) and muscle. One of the main applications is imaging of pathological skin in order to determine the size and depth of tumors, a complement to other techniques for determining malignancy. Inflammatory diseases such as psoriasis plaques can also be monitored by B scan. Most of the commercial units operate near 20 MHz. Recent work at 50 MHz shows that the depth of exploration is limited to about 4 mm at this frequency.

Another much studied area is that of opthalmological applications. At low frequencies (<15 MHz) commercial instruments are routinely used to measure dimensions of internal structures of the eye and to detect structures hidden by the eye lens. More recently, there have been developments of high-frequency (30 to 100 MHz) biomicroscopes, which are useful for imaging small structures a few millimeters below the surface, for imaging the cornea for thickness, for state of corneal grafts, and for detecting cysts and tumors. This high-frequency work provides new, unique information on eye structures and is a promising area of development.

Intravascular ultrasonic imaging is another important area in medical applications, where the main problem is detection of hardening of the arteries, or atherosclerosis. *In vitro* studies have been carried out to establish a correlation between ultrasonic images at about 50 MHz and histology. The agreement is excellent for detection of arterial wall thickening due to plaque in most arteries, and good calculation is also obtained for the more elastic carotid artery. *In vivo* ultrasonic imaging is under development, while *in vivo* ultrasound is already useful for diagnosis and monitoring during surgery. The prime advantages of acoustic imaging are good resolution, contrast, and real-time imaging.

All of the very-high-frequency (>500 MHz) work has been done by SAM. Developments have proceeded more slowly than originally anticipated, in part because of the difficulties in image interpretation, but also because it is not a well-known technique so that it has not been easy to make connections with traditional cell biology. Several studies have been made of cells in culture, notably fibroblasts. SAM can be used as a tool to measure (1) topography, with the aid of the observed interference fringes; (2) attenuation, which is, however, difficult to interpret because of model dependence and assumptions on homogeneity; and (3) reflectivity, which suffers from similar ambiguities. A key issue for image interpretation is the model used for the acoustic properties of the cytoplasm. An effective medium approach for the sound velocity in the saline/fibril system can be used, as for porous systems, for example. There are, at present, too many uncertainties in the acoustic parameters to provide a basis for interpretation of SAM images. Work is ongoing using all of the techniques of acoustic microscopy to elucidate the mechanical properties of cytoplasm. Important work is under way to study cell dynamics by SAM. Interference reflectometry has been used to visualize the elasticity distribution in cells. A subtraction scheme for images taken at different times has been used to image cell motility and relate this to changes in elasticity, topography, or attenuation. This is a promising tool in its ability to detect all motile responses to applied stimuli.

14.5.2 Films and Substrates

Achenbach and coworkers [154] have used the LFB to determine the elastic constants of isotropic materials in bulk, plate, or thin film configurations at a single frequency. The heart of the method is an inversion procedure in which best estimates of elastic constants are put into a theoretical model for $V(z)$ to calculate velocities and amplitudes of leaky waves, which are then compared with those determined experimentally by LFB. The difference, or deviation D, is used to adjust the input elastic constants, and the process is repeated until convergence by least squares is obtained. Good agreement, of the order of 1%, has been obtained for velocities for glass and aluminum in bulk form, glass plates, titanium films on gold, and a gold film on glass. The advantage of the method is that it only requires a single frequency measurement by LFB.

Anisotropic films on anisotropic substrates [154] have been studied as an extension of the inversion method for isotropic systems. The wave model is used as the starting point for calculating $V(z)$. The reflection coefficient is calculated for the anisotropic case by a matrix method, where layers are represented transfer matrices, which are multiplied together to give the reflection coefficient. The measured and calculated $V(z)$ give good overall agreement for various isotropic/anisotropic combinations, such as TiN films on MgO substrates. As in the previous section, the actual inversion procedure for determining elastic constants is carried out by comparing surface acoustic wave (SAW) velocities extracted from the experimental $V(z)$ curves with those calculated by finding the roots of the Christoffel equation. In making the comparison for anisotropic materials, the distinction must be made between regular SAW and pseudo-SAW; for the latter, V_R is greater than for out-of-plane transverse waves. These two components correspond to qualitatively different regions of the angular velocity variation for SAW.

For a given specimen, the most reliable inversion technique is to compare SAW velocities obtained experimentally and those calculated theoretically by an iteration process with minimization by a Simplex method over a wide range of frequencies. A less acceptable alternative is to measure a few frequencies or several specimens with different thicknesses at a single frequency, but the latter approach is based on the dubious assumption that the properties of films of different thicknesses remain the same.

Considerable work has been done on film thickness measurements by SAM. For medium-thickness films (1 to 20 μm), ultrasonic microspectrometry (UMSM) [155] has been demonstrated to be an effective real-time online device. This technique works for films such that the layer transverse wave velocity is smaller than that of the substrate. Dispersion calculations show that the two lowest-frequency modes are the Rayleigh mode and the Sezawa mode, the latter having a low-frequency cutoff when it leaks into the substrate. It has been demonstrated in practical conditions in UMSM that, when the frequency is scanned, a dip occurs at cutoff, which enables a determination of the film thickness. For very thin films ($d < 1$ μm), the thickness can be determined easily in the laboratory by a $V(z)$ measurement, given an appropriate knowledge of the film parameters.

Film adhesion is another important problem which, in principle, is ideally suitable for study by the SAM. There have been a number of studies, which have been well summarized in [156]. The basic idea is to use a high NA acoustic lens, so as to excite Rayleigh and Lamb waves in the multilayer system. One can then compare experimentally measured dispersion curves to those predicted by the theory for various states of interfacial contact: perfect, intermediate, or loss of contact. It was found that in the two limiting cases there was excellent agreement between theory and experiment, and that known imperfect interfaces fell between the two. Finally, it was found that surface-skimming compressional waves (SSCW) were even more sensitive than generalized Lamb waves to the interface conditions.

14.5.3 NDE of Materials

Subsurface imaging is carried out by focusing an acoustic lens below the surface. The attenuation in the solid, which may be very high (composites) or very low (single crystals), will be a major factor in limiting the maximum imaging depth. The confocal nature of subsurface acoustic imaging is such that it is possible to obtain plane-by-plane image slices; a demonstration for composites is given in [157].

The presence of stress in materials can be measured by acoustic microscopy by the effect of stress on sound velocity via the third-order elastic constants. For surface and near surface stress, the SAM is a useful tool for detecting the presence of applied and residual stress, with reasonably high spatial resolution depending on the approach that is used, by use of Rayleigh or SSCW detection. Applied stress leads to the acoustoelastic effect, the change in velocity due to an applied stress field. There is an advantage to using the LFB instead of more conventional SAW technology because of the flexibility of liquid coupling and the directionality and the 1 or 2 mm spatial resolution provided. A demonstration of the technique has been given by Lee et al. [158]

for 6061-T6 aluminum using Rayleigh waves and SSCW, and for polymethyl methacrylate (PMMA) using SSCW. Samples were cut in a dog bone shape and placed in a uniaxial loader, with strain gauges attached to the surface. Calibration was carried out using a uniform load and measuring velocity parallel and perpendicular to the loading direction as a function of strain and by measuring $v(\theta)$ for several fixed values of strain. This procedure gives the two principal acoustoelastic constants for the material, which allows subsequent measurements of unknown nonuniform stress fields. In both cases, good agreement was obtained with finite element calculations.

Near-surface residual stress can also be measured using the Rayleigh wave velocity. The study by Liang et al. [159] used time-resolved phase measurements of the Rayleigh waves using a spherical lens. Excellent agreement was obtained for the spatial variation of residual stress by comparison with actual Vickers hardness measurements. Again, the acoustic technique would require a calibration procedure for a given material.

It has also been shown that bulk stress in solids can be imaged using the acoustic microscope [160]. The technique is based on measuring acoustic birefringence under applied stress. Shear modes created by mode conversion can be imaged; those propagating through the stressed region have a decreased amplitude compared to those that traverse stress-free regions. By comparing the two, one can measure and image the variation in a stress field throughout the volume of the sample. Longitudinal waves give complementary results, that is, maximum amplitude where the shear mode has minimum amplitude. Possible applications include residual stress detection and crack-induced stress in ceramics and composites.

Qualitative and quantitative assessment of crack forms, dimensions, and growth rates in materials is important for NDE, particularly in determining the estimated lifetime of industrial components. SAM imaging adapts to this problem well particularly because of its subsurface ability. One characteristic of SAM images of cracks is the strong fringing observed with spacing of $\lambda_R/2$, which clearly demonstrates that Rayleigh waves are involved. This conclusion is also confirmed by detailed theoretical analysis [138].

The smallest cracks that can be detected by SAM are determined by acoustic considerations for the minimum width [138]. Since Rayleigh waves need to propagate in a continuous fashion and they involve a strong, shear component, the viscous penetration depth determines the smallest crack width at a given frequency. This length varies as $\sqrt{1/f}$, and for water at 1 GHz it is about 18 nm. The minimum length is determined mainly by ultrasonic time of flight considerations because short pulse techniques are mainly used to determine this dimension. For example, a pulse width of about 8 ns is needed to detect a crack 100 μm long. The time of flight diffraction technique (TOFD) has been used to identify various possible paths from the acoustic lens to the crack and then by use of a ray model to identify the observed rebound echoes by transit time. The model was validated in plastic materials and then applied to the measurement of actual cracks in aluminum-lithium alloys down to a depth of 220 μm. The same technique was used to measure crack growth under elastic loading in aluminum alloys, and good agreement was obtained with subsequent destructive inspection. Crack detection is thoroughly explored in [161].

14.5.4 NDE of Devices

This section is concerned with two complementary areas of the application of acoustic microscopy to the NDE of microelectronic and optical devices. The first is the important industrial area of microelectronic packaging of single chips, stacked chips, multichip modules, and stacked modules. The need here is for low-cost, high-speed detection of packaging defects such as leaks, voids, delaminations, etc., and their visualization. The principles involved are based on those of subsurface imaging of defects and acoustic studies of defects as discussed above. Ideally, these tests will be carried out online in real time. The second is more laboratory-level research and development to characterize the homogeneity of microelectronic chips and optical fibers, which is achieved by measuring the spatial variation of the acoustic parameters.

The application of SAM and SLAM to microelectronic packaging has been fully covered in [162] and [163], with many examples of acoustic and other images. In [163], particular emphasis is put on the complementary nature of SAM, SLAM, x-rays, and optical and destructive analysis. One of the important areas is in the ceramic packaging of chips, where one of the chief issues is leaks in the lid sealing. Entry of moisture and other contaminants leads to corrosion or change in electrical properties. SAM at 50 MHz was shown to be a useful technique for lid seal inspection, giving depth-specific information and void detection for both solder and glass-seal devices. Shear-wave imaging was shown in [162] to give good resolution for void detection up to 2 mm depth. Failure in plastic-packaged devices was found to be due largely to differential contraction, and SAM was found to be useful for detecting internal cracking and delamination, and to be very complementary to x-ray inspection.

Die-attach, the bond between a semiconducting chip and the substrate, is another area where SAM and SLAM have proven very useful. Bond integrity is important to provide good thermal, electrical, and mechanical contact, which are all essential for proper device operation. Voids, cracking, and poor adhesion are among the main problems, and it is shown by numerous images in [162] and [163] that these can be detected by SAM and SLAM. SAM is good for work in the reflection mode and can give unique information on the disbond. Other special applications in microelectronic packaging include detection of voids at tape automated bonding (TAB) interfaces, poor adhesion at soldered joints, and detection of delaminated leads. These detailed studies clearly show that SAM and SLAM are now indispensable diagnostic tools for microelectronic packaging.

Two other microscopic monitoring tools of device components and materials should be mentioned. Kushibiki et al. [164] have done extensive studies of wafer mapping using the LFB. For example, studies were carried out on a 36° *Y*-cut $LiTaO_3$ wafer to be used for shear horizontal (SH) SAW. Rayleigh-type

SAW waves were excited along the x axis, as this direction was found to be most sensitive to chemical composition and elastic inhomogeneities. Experiments were carried out as a two-dimensional mapping of 6 × 6-mm squares over a 76-mm diameter wafer. The results showed that by measuring velocity variations it is possible to carry out physical and chemical quality control as follows: (1) V_{LSAW} was proportional to the Curie temperature varying as 0.52 m/s per °C, (2) variations of 0.03 Li_2O mol% could be detected, and (3) residual multidomains produced during poling were detected by elastic inhomogeneities. A similar study was carried out over the section of cladded optical fibers [165], where different sections were doped with GeO_2, F, and B_2O_3 to produce a controlled variation in refractive index. The LFB was used to compare the profile of V_{LSAW} with that of the refractive index. Very good agreement was obtained indicating the potential of the LFB as a characterization tool for optical fibers and preforms.

14.6 Perspectives

Conventional acoustic microscopy is now a mature subject. Its use in the microelectronics industry as an NDE tool is becoming more frequent; there is still a need for faster, ideally real-time imaging in this area. The LFB technique is finding increasing application as a research tool. The high-frequency SAM is used mainly for specialized applications and its future may well be in the biological area. As for future development, it seems likely that this lies with the application of atomic force microscopy to acoustic imaging.

In conventional (far-field) acoustic microscopy, it is axiomatic that the spatial resolution is limited by the wavelength. However, this condition can be circumvented by using the principle of near-field imaging, in which a probe or pinhole is placed very close to the surface. If the size and distance, d, of the probe are much less than the wavelength, the resolution is limited by d and not λ. This principle is valid for any type of wave and was first demonstrated by Ash and Nicholls [166] for electromagnetic waves and Zieniuk and Latuszek [167] for ultrasonic waves. The development of the atomic microscope has led to several variance of a near-field acoustic microscope.

Takata et al. [168] used a vibrating tip provided by a scanning tunneling microscope, whereby the tip generated strains in the sample, which were detected by a piezoelectric transducer coupled to the sample. The detected signal depended on the tip-sample interaction and the ultrasonic propagation from the tip to the transducer. Cretin et al. [169] have developed microdeformation microscopy, again based on a vibrating tip that is mechanically scanned across the face, which in this case creates microdeformations in the surface. In transmission mode, a cantilever beam terminated with a diamond or sapphire tip is vibrated at frequencies from 20 to 200 kHz. The microdeformations induce strain in the sample, which is detected by a piezoelectric transducer fixed onto the opposing face. Experiments on silicon wafers and

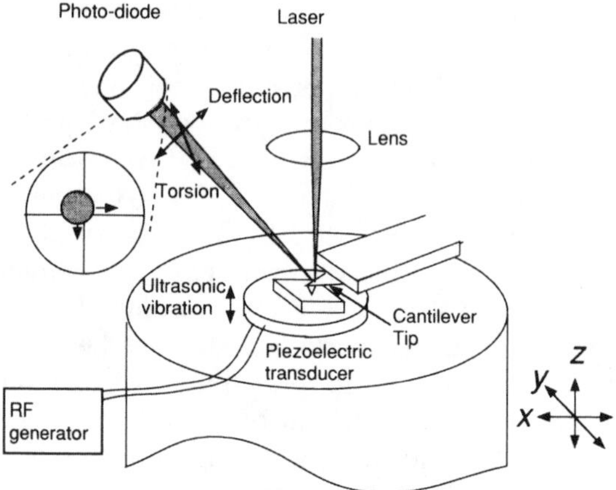

FIGURE 14.10
Schematic illustration of an AFM/UFM. A thickness mode PZT transducer is bonded to the sample stage to excite vibrations of 1 to 10 MHz. (From Yamanaka, K., New approaches in acoustic microscopy for noncontact measurement and ultra high resolution, in *Advances in Acoustic Microscopy*, Vol. 1, Briggs, A., Ed., Plenum Press, New York, 1995, chap. 8. With permission.)

polycrystalline stainless steel showed that the image contrast is related to grain orientation. In the reflection mode, the cantilever is fixed to a piezoelectric transducer; results complementary to transmission are observed.

All of the most recent work is based on the use of atomic force microscope as the detector of vibrations set up by ultrasonic waves applied to the sample [170]. This has the advantage that one can control the frequency, mode direction and amplitude of the applied wave. Most of the work has been done for vertical surface displacements and this will be discussed first. A typical experimental setup is shown in Figure 14.10. The system is integrated with a commercial AFM, and the cantilever displacement is measured optically. Low-frequency scanning for the AFM mode is done in the range 1 to 20 kHz. The sample to be studied is placed on an appropriate piezoelectric transducer. One big advantage of this geometry is that one can define the mode, amplitude, and frequency of the driving ultrasonic wave.

In the low-frequency limit, the ultrasonic frequency f is much less than the cantilever resonance frequency f_0; this is called the force modulation mode (FMM). The peak-to-peak cantilever deflection amplitude is given by [170]

$$V = 2z_c\frac{a/z_c}{1+(k/s)} \tag{14.25}$$

where a = sample vibration amplitude, k = cantilever spring constant, s = tip-sample contact stiffness, z_c, cantilever deflection due to static repulsive force, and $K = k/s$. It is clear that V depends little on K for $K \ll 1$, so we expect little intrinsic contrast for imaging in this regimen. This regimen is also characterized by the absence of tip-sample indentation.

The much more interesting limit, called ultrasonic force microscopy (UFM), corresponds to $f \gg f_0$. At low amplitude, $a < (k/s)z_c$, the average force per cycle on the tip is zero, and so the tip stays on the sample surface. At sufficiently high amplitude a, even though the cantilever cannot follow the ultrasonic vibration, the tip comes away from the sample surface during part of the cycle, as the average tip-sample repulsive force is nonzero. Above the threshold amplitude a_0, the cantilever deflection due to the ultrasound is [170]

$$z_a = z_c\left[\frac{k}{s} + \frac{a}{z_c} + 2\frac{ka}{sz_c} - 2\sqrt{\frac{ka}{sz_c}\left(\frac{k}{s}+1\right)\left(\frac{a}{z_c}+1\right)}\right] \quad a > (k/s)/z_c \qquad (14.26)$$

The procedure followed was to amplitude-modulate (triangular) the ultrasonic beam at a frequency below f_0 and measure the cantilever deflection. From Equation 14.26, the instrument performance is governed by three factors: normalized cantilever deflection, normalized ultrasonic amplitude a/z, and normalized cantilever stiffness $K = k/s$. Contrary to the FMM regimen, in the UFM regimen the deflection signal depends strongly on K, so the intrinsic contrast is expected to be high. Thresholds are observed for various values of k/s, and in principle, dynamic elastic effects can be determined from them. The full theory of Hertzian contacts shows that the force curve $F(d)$ is very nonlinear, which is determined in the detection mechanism. The additional cantilever deflection for different repulsive forces and different values of effective elastic constants can be calculated as a function of vibration amplitude; as before, the added deflection is very sensitive to these variations for UFM and not for FMM leading to the theoretical prediction of good contrast in the former case and not in the latter. Several examples are presented in [170] for UFM vertical mode imaging, principally of defects in highly oriented pyrolytic graphite (HOPG) and structure of a floppy disc surface. This work confirms sensitivity of the technique to subsurface elasticity variations and the good image contrast of UFM. Typical fields of view are 400 nm by 500 nm, with ultrasonic frequencies of the order of 5 MHz.

A smaller amount of work has also been done on lateral displacements using the UFM by suitable choice of piezoelectric transducer. For the UFM, the lateral mode AFM (LM-AFM) is based on measurement of the torsional vibration of the cantilever, which is dominated by friction. This principle is used to image the frictional force distribution by amplitude measurement, while the phase gives the energy dissipation. The image is free from topographical effects, which are automatically subtracted in real time. There is some indication that the technique is sensitive to subsurface shear modulus variations.

Atomic force acoustic microscopy (AFAM) is a related study on vertical mode imaging [171]. Again, an ultrasonic transducer in the low MHz range sets up surface vibrations, which are coupled to the tip as described by the mass-spring model. These vibrations excite flexural waves in the cantilever in the MHz range, which can be detected by a very fast knife-edge detector. Resonance of the cantilever surface system can easily be calculated in a mass-spring model and were measured up to mode $n = 9$ by impulse excitation, knife-edge detection, and Fourier transform. The nonlinearity of the tip-sample force is used to explain the ultrasonic coupling to the cantilever. At sufficiently high amplitude, the mean displacement is shifted by nonlinearity, and at very high amplitude, cantilever frequencies other than that used for the ultrasonics are excited. Examples are given of the use of the AFAM for imaging as a function of amplitude; strong variations in contrast are observed, which are interpreted as reflecting the variation in the sound transmission due to the tip-sample interaction forces.

The configuration used in scanning local acceleration microscopy is again basically similar to that used for the UFM and the AFAM [172], at frequencies much higher than the tip cantilever resonant frequency. Imaging is done in CW at low amplitude in the so-called contact mode. The basic result is that at sufficiently high frequencies the output signal is determined by the cantilever acceleration and not the static force acting on it. The cantilever stays in contact with the sample and the sample stiffness can be mapped as it enters in the interaction stiffness. As before, it has been shown experimentally that the contrast is better than with the AFM, and, moreover, that a contrast variation with frequency predicted by the theory is observed. A "diode" and "subharmonic" mode are identified at higher amplitude, and their interpretation is ongoing.

15

Nondestructive Evaluation (NDE) of Materials

15.1 Introduction

NDE is a huge and diverse field. Regarding experimental methodology it includes not only ultrasonics but also a wide range of complementary techniques such as x-rays, optical, thermal, electrical, and magnetic measurements. We restrict ourselves to ultrasonics and even then the scope is wide; we may look for defects in existing structures, measure elastic constants and other material parameters in process technology, carry out thickness measurements on thin-walled vessels or in layered structures, etc. We will concentrate on describing those approaches that depend on the propagation properties of ultrasonic waves, principally the use of guided waves in supporting structures such as surfaces, plates, layers, etc. The aim is to discuss the principles by which such guided waves interact with defects and how they can be useful for determining elastic properties. The chapter will not be a compendium or systematic review of all known results, nor a complete academic treatise on acoustic wave defect interaction. Moreover, any objective such as the latter would be exceedingly complex, as defects can be clothed in an almost infinite variety of guises.

The general approach to ultrasonic NDE has evolved considerably over time and from a historical perspective we can identify several main approaches to the subject:

1. Classical NDE

   The field has followed two separate but related paths. In the first, defect detection, we are looking at *in situ*, field inspection techniques for the detection of various defects or material failures that occur with time and/or wear in existing structures. One can think of countless examples: cracks in rails of railway lines, stress corrosion cracks in pipelines, lack of adhesion at certain areas of a protective coating to its substrate, etc. Detection and characterization of these defects may include determining their existence, making an acoustic image of them, and obtaining quantitative information on them. A second avenue of NDE involves determination of the intrinsic material properties themselves, such as ultrasonic attenuation measurements to determine grain size distribution in alloys, determination of the elastic constants of an anisotropic ceramic coating, etc.

2. NDE for material processing

   This is a more modern approach, in many cases an ideal yet to be attained. In the so-called intelligent processing or manufacturing operation, sensors are placed at strategic points in the process technology diagram so that the physical and chemical properties are monitored during production. If these differ from their required values the process is modified accordingly in real time to correct the situation. Thus the appropriate preventive steps are taken during the manufacturing process to eliminate or reduce the probability of failure of the material during its working life. The success of this approach depends on the development of a full process model that establishes the important processing variables.

3. Intelligent materials

   Many large and complex industrial structures such as aircraft fuselages, bridges, etc. incorporate networks of embedded sensors. This is particularly feasible in composite materials where the sensor probes can be incorporated during fabrication. Such sensors give continuous monitoring of the state of the material during service, and by detecting signs of incipient failure by the detection of small cracks, stress, etc. they perform a valuable NDE function.

4. Modern signal processing

   Ultrasonic inspection has evolved considerably from the original direct pulse echo approaches. Techniques have been developed to recover signals buried in the noise. New approaches, such as neural network signal processing, can be used to help in flaw identification and determination of the ultrasonic propagation path in complex geometries.

Formal NDE classification procedures make a clear distinction between forward (inductive and predictive) and inverse (deductive) approaches. In the forward problem, we set up a model based on established physical laws that enable us to calculate measurables (for example, the ultrasonic velocity of a certain acoustic mode) from known model parameters (elastic constants). So the forward problem corresponds implicitly to the approach taken by a typical textbook in ultrasonics; if we know the elastic constants then by the standard theory (Christoffel equation) we can calculate the (measurable) sound velocity for a mode of specified polarization in any direction in the crystal.

The approach is inductive in that if we do this for a given set of parameters we can generalize the result to predict the measurables for all similar directions, materials, etc.

Although textbooks implicitly adopt the forward approach, practical NDE implicitly takes the opposite tack, the inverse approach. NDE is in fact very much like detective work. With a given set of tools (ultrasonic techniques), we determine experimentally the measurables (ultrasonic velocity and attenuation) and use the resulting data set to propose a suitable model that will enable us to infer or deduce the model parameters (i.e., to catch the criminal, the elastic constants). The general problem is, of course, very difficult, and like the criminal counterpart, there is no general solution. Specific approaches must be devised to account for the particularities of a given problem, and to a general extent the development/outcome is dependent upon how many other parameters are known by other means. Thus the general NDE inverse problem is unsolved, as the general NDE field variables are neither known nor constrained. More often, there are more unknowns than knowns, so the problem is in general underdetermined.

An example of a typical ultrasonic NDE problem has been described by Gordon and Tittmann [173]. Suppose that we want to deduce the hardness gradient in a steel rail from measurements of ultrasonic velocity and attenuation. This really involves two separate problems:

1. An inverse problem to infer the model parameters from the ultrasonic measurements.
2. A forward problem to determine hardness from the model parameters. Clearly, there may not be, probably will not be, complete intersection between the parameters involved in the two problems. It is likely that there are some parameters that are needed to predict the hardness that cannot be deduced from the ultrasonic data and that may not be available from other sources. Other concerns are sensitivity issues related to how sensitive the calculated values are to the parameters and questions of uniqueness of a given solution.

There are many possible approaches to the inverse problem. One classification scheme distinguishes between direct and indirect approaches. In the direct approach, a given model (for example, a one-dimensional Born model for scattering) is used to describe the scattering of an ultrasonic wave, as described by Chaloner and Bond [174]. We describe briefly an indirect method used by the same authors, based on the Monte Carlo search method, which is rather more general in applicability and makes few *a priori* assumptions on the nature of the defect. As a first step, the information sought from the inversion must be parameterized; for example, three possible parameters might be flaw radius, density, and longitudinal sound velocity. A four-dimensional parameter space containing all possible solutions is then set up.

A random point is chosen to start a Monte Carlo search. Synthetic scattering data from this point is then compared to the experimental data. The degree of fit by least squares is recorded and the search continues. In the "Hedgehog" modification described in [174], for points where the fit is good, all neighboring points are examined to optimize the process.

After this is done, a new random starting point is chosen and the same procedure is repeated. Although this approach can be computationally cumbersome, it is very flexible and can be used with any desired forward model for the defect scattering.

Concentrating on the inverse problem, we are then faced with the question of how to obtain ultrasonic data on the field sample in question. In what follows, in the spirit of Chapter 12, we assume that contact piezoelectric transducers will be employed. A number of issues have to be addressed in order to make the ultrasonic measurement, including surface state, surface shape, coupling medium, temperature, and sample geometry. These will be covered briefly in order to give an idea of the practical constraints involved.

Let us consider as an example an irregularly shaped sample placed in a field inspection setting. Before attaching the transducer to it, one should clean off all the relevant surface region of any dirt, loose scale, sand, loosely adhering layers of paint, coatings, rust, etc., which could cause air gaps and prevent the ultrasonic wave from penetrating into the material. Strongly adhering layers are not generally a problem and they may inadvertently provide some degree of impedance matching. If the surface can be mechanically formed then a uniformly curved or plane surface is desirable. Surface roughness should be of the order of a tenth of an acoustic wavelength or less. Otherwise, moderate roughness is not a problem and may help to retain the coupling fluid between transducer and surface. Regarding surface curvature a convex shape is preferable to concave, with the additional positive result that it will probably produce a narrower beam. A specially shaped coupling block could be made to ensure transmission for concave surfaces.

An appropriate acoustic coupling agent is an important part of the process. Its nature is not critical in the low MHz range but it becomes increasingly difficult to make an acceptable acoustic bond if higher frequencies and/or lower temperatures are involved. For low MHz ultrasonics at room temperature, most oils and greases will work. Since they are generally high attenuation, low-impedance media they should be thin. For this reason it is not recommended to fill a very rough inspection surface with copious quantities of couplant. Commercial couplants are available; otherwise, silicone oil and vacuum grease are quite satisfactory. If the sample is to be cooled below room temperature, it is advisable to remove the water from vacuum grease by heating and pumping. Salicylic acid phenol ester (SALOL) is a useful couplant for shear waves where a solid bond is needed. SALOL has a low melting point, only slightly higher than room temperature, so that it can be liquefied to place the transducer then left to solidify to form the desired solid bond.

High temperatures pose a special set of problems. One solution is to transmit the ultrasonic pulse by a water squirter, which can be used up to 300°C.

Since splashes may degrade the ultrasonic reflection, squirter techniques are best used in transmission. Dry coupling is also possible for higher temperatures; various special high temperature pastes have also been developed. For temperatures above 200 or 300°C, water-cooled delay lines can be used, as described in Section 15.7.

Sample shape provides a separate set of potential problems as this can distort the ultrasonic signal or introduce spurious echoes, such as those due to end faces, etc. If the defect is near an edge of the sample, then the transducer emplacement becomes critical. If the transducer is also placed near the edge in an attempt to get a direct echo from the defect, diffraction of the ultrasonic beam and subsequent reflection from the edge will distort the signal. It is preferable to place the transducer far from specimen edges and irradiate the suspected defect at an angle. Other complications arise due to mode conversion into transverse or surface waves. The resulting parasitic echoes can be identified by manual ray tracing or by the use of commercially available software packages.

Some of the above problems can be alleviated by using noncontact methods. One popular approach is to use a powerful laser beam to generate the ultrasonic pulse and an air-coupled ultrasonic transducer to receive the signal. In analogy with the liquid squirter, Hutchins et al. [175] have recently developed a gas jet for coupling to the sample. Electromagnetic transducers (EMATs) are an effective noncontact method for generating and detecting shear waves in metals. The most universal and undoubtedly the most commonly used noncontact method is the water immersion tank where the sample is placed in a water bath and irradiated by an ultrasonic beam in the water whose position and angle of incidence can be controlled independently. Many such systems are available commercially.

15.2 Surfaces

Material surface coatings, layers, or surface modifications are an important part of modern technology. Clearly NDE of such structures to determine their quality is important. Type I NDE, defect detection, is best carried out with Rayleigh waves, which are ideally suited to surface inspection problems. Scattering of Rayleigh waves by defects in these structures, cracks, voids, delaminations, etc. is analogous to the bulk wave situation, so that a defect echo could be detected by pulse echo studies. For frequencies in the range 1 to 10 MHz defects can be detected in a surface region of 0.5 to 5.0 mm. Scanning acoustic microscopy (SAM) is an excellent approach for very small defects, which can be imaged as well as quantified by $V(z)$. It has been shown that cracks as small as 20 nm can be detected by this method. Of course, defects that intersect the surface can be observed optically and in many cases this is sufficient.

15.2.1 Principles of Rayleigh Wave NDE

SAW is also ideally suited to Type II NDE of surface structure, the determination of elastic constants, for example, in the reconstruction of the gradient of near-surface elastic properties. As the frequency is varied, different depths of the surface region can be sampled by the SAW signal. A good example for the forward problem is the prediction of the SAW velocity as a function of frequency for an inhomogeneous sample, as discussed by Gordon and Tittman [173]. Three models were considered:

1. Mixture rule

   It is assumed that the SAW is influenced by a weighted summation of contributions from layers near the surface region. Additional assumptions are linear weighting, penetration depth proportional to one wavelength scaled by a constant C, and isotropic lossless layers. For m layers, the SAW velocity is then given by

$$V_{\mathrm{SAW}}(\lambda) = \frac{\sum_{i=1}^{n} V_{R_i} d_i}{c\lambda} + \frac{V_{R_{n+1}}(c\lambda - \sum_{i=1}^{n} d_i)}{c\lambda} \tag{15.1}$$

   where

$$\sum_{i}^{n} d_n < \lambda \leq d_{n+1},\ n = 1, 2, m$$

$$d_i = \text{thickness of layer } i$$

2. Perturbation

   In this approach it is assumed that the variations in parameters were mutually proportional, specifically that the density and elastic constants vary with time according to the same functional form $F(z)$.

3. Thomson-Haskell method

   This is a propagator matrix operator adapted from geophysics. Stresses and displacements between two points are related by a series of matrix multiplications. By application of the free surface boundary conditions, the eigenvalue problem for Rayleigh waves in the layered structure can be solved.

These three approaches were applied to two quite different systems where the variations were known. The first was a hardened sample 1043 steel that was known to have constant density and Poisson's ratio with depth. The second was a Ni/Cu/Ni structure of constant density. In both cases, the elastic constants varied in a subsurface region 1 to 3 mm in depth. It was found that the rule of mixtures gave a very good approximation to the known profile while the perturbation method was the least satisfactory. The Thomson-Haskell method gives an exact solution and would be a good general approach if the number of independent variables could be limited.

15.2.2 Generation of Rayleigh Waves for NDE

Of the methods already encountered for generating Rayleigh waves, it should be said that the IDT does not provide a practical approach to field NDE. Critical angle reflectivity (CAR) is a very useful laboratory tool for Type II NDE, but again it is not a practical way of looking for defects. However, there are a number of other practical ways for generating low-frequency Rayleigh waves (≤30 MHz) and these are summarized in Figure 15.1. Bulk wave transducers (Y-cut quartz) can be placed flat on the surface with oil or grease coupling. Evidently, most of the energy is lost as a radiated transverse bulk wave, but a usable Rayleigh wave can still be generated. Also shown in Figure 15.1(a), a longitudinal wave transducer making contact with an edge can generate Rayleigh waves. The wedge configuration shown in Figure 15.1(b) is one of the most popular methods. A longitudinal wave is launched into a plastic block (longitudinal wave velocity V_{LP}) cut at the appropriate angle ($\sin\theta = V_{LP}/V_R$) to excite Rayleigh waves in the substrate. The principle used is exactly that of phase matching (conservation of parallel wave vector) used in critical angle experiments.

The basic idea of the IDT can be retained to make a flexible, portable device, shown in Figure 15.2(c). A comb structure is built on an aluminum plate with an array of parallel grooves; the device is then excited ultrasonically by a longitudinal transducer placed on the back side of the structure. The device is then pressed against the sample to be tested and Rayleigh waves are generated when the frequency is adjusted so that the distance between neighboring teeth is equal to the Rayleigh wavelength ($\lambda_R = 2a$). For various reasons, all of these techniques are restricted to frequencies well below 50 MHz.

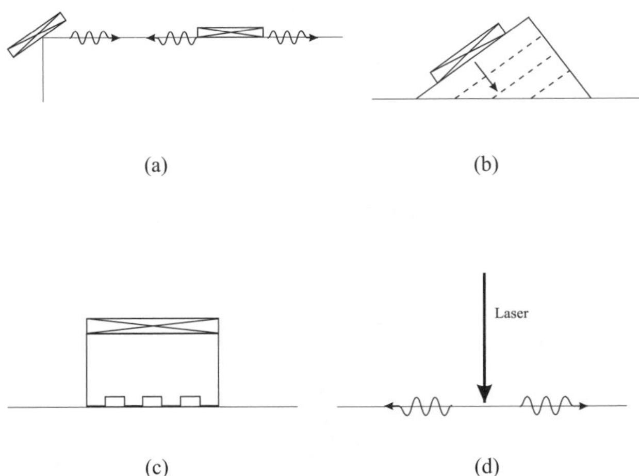

FIGURE 15.1
Practical methods for excitation of Rayleigh waves for NDE. (a) Edge-bonded longitudinal bulk wave transducer. (b) Wedge excitation at angle $\sin\theta_R = V_W/V_R$ where V_W is the longitudinal wave velocity of the wedge material. (c) Periodic comb-like excitor. (d) Laser generation.

In any case, industrial quality surfaces would scatter and attenuate Rayleigh waves at significantly higher frequencies.

15.2.3 Critical Angle Reflectivity (CAR)

The physics of CAR is at the basis of much of modern NDE of surface regions by ultrasonics and is an essential component of acoustic microscopy. As originally developed by Mayer [176] and Rollins [177], it is a simple technique, requiring only one free surface of the sample, as opposed to the more stringent conditions of two flat parallel surfaces of a large sample required by standard ultrasonic techniques. A simple CAR goniometer is shown in Figure 15.2. The system is immersed in a fluid bath, usually water. Transmitting and receiving transducers are designed so that a reasonably well-collimated beam is incident on the surface at angle θ. The reflected beam is detected at the same angle. Sweeping through the values of θ from 0 to 90 enables tracing of a reflectivity curve; a typical one was sketched in Figure 8.6. Key features of the curve include critical angles for longitudinal, shear, and

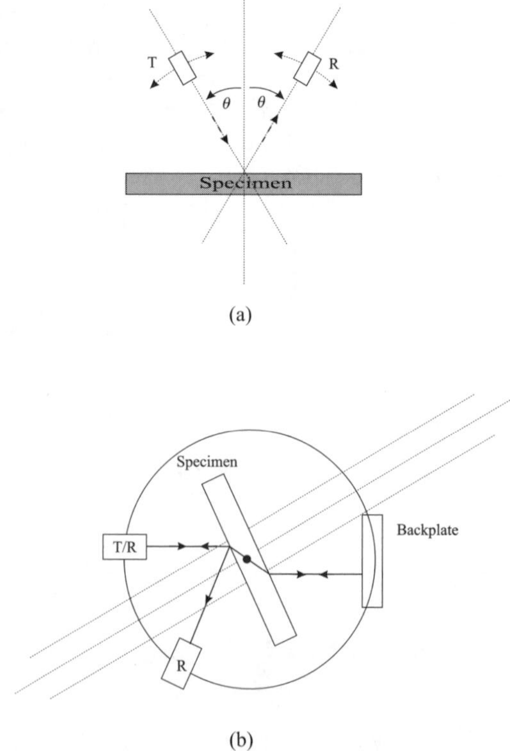

FIGURE 15.2
Goniometer for reflectivity and transmission experiments. (a) Reflection. (b) Combined reflection and transmission configuration based on [179].

Rayleigh waves, from which the corresponding sound velocities can be obtained by $\sin\Theta = V/V_i$. The origin of these has already been explained; when the x component of the incident wave vector matches that of a wave in the solid travelling parallel to the surface, there will be an anomaly in the reflection coefficient (RC). Determination of the reflection coefficient at normal incidence can then be used to determine the density. Thus one simple series of noncontact measurements can be used to determine the principal acoustic parameters of the solid. Potentially, the most useful feature of Figure 8.6 is the minimum at the Rayleigh angle observed for actual samples, but which is not predicted for ideal lossless solids. For samples with zero material attenuation, the Rayleigh wave is purely evanescent, i.e., energy is stored but not propagated in it. In this case, theoretical and experimental reflectivity curves show total reflection in this region. If the sample has nonzero material attenuation, then energy is dissipated from the Rayleigh wave, giving rise to the dip at the Rayleigh angle shown in the figure. This effect has been investigated quantitatively by Becker and Richardson [43] and the variation of dip depth and width with shear wave attenuation is shown in the figure.

It is seen that the RC has a minimum for a certain value of shear wave attenuation per wavelength l_s, where $l_s = \alpha\lambda$ is the shear wave loss parameter, after which the RC increases and the Rayleigh dip broadens out. Thus the sharp resonance observed at Θ_R for small attenuation is smeared out as the loss increases, which is expected physically.

The main conclusions from this work are as follows:

- The shear wave parameters are the most important in determining the Rayleigh angle (shear velocity) and nature of the Rayleigh dip (attenuation).
- There is a critical value of l_s and a frequency for which the reflection coefficient is zero, which can be related to material parameters.
- The FWHM is constant up to the critical value of l_s, then it increases linearly with l_s.
- The phase of the RC goes from 0 to 2π for zero shear attenuation and from $\pi/2$ to $-\pi/2$ at the critical value of l_s.

In addition to the above, Rollins [178] showed that the size, shape, and position of the Rayleigh minimum had a strong dependence on frequency, roughness, presence of coatings, strain, and crystal anisotropy. The frequency dependence can be easily understood from the Schoch displacement considerations of Chapter 8. Likewise, since all sound velocities depend on crystal orientation, it is quite normal that the phase velocity of Rayleigh waves should also depend on the propagation direction. The basic configuration of CAR is of interest here as it corresponds exactly to the textbook treatment of ultrasonic reflectivity and Rayleigh wave generation. In its modern applications, the technique is used indirectly in acoustic microscopy, as has been seen, and in a modified form for the study of Lamb waves and layered systems, which will be developed in the following sections.

15.3 Plates

15.3.1 Leaky Lamb Waves: Dispersion Curves

The CAR technique can be extended directly to the study of Lamb waves in plates using the same type of goniometer. Such measurements can be used to measure the dispersion relations of plates and layered structures and they can also be extended to the study of defects (e.g., lack of adhesion between layers, point defects). Moreover, this geometry is particularly well suited to the generation of guided interface waves. Guided waves are an ideal probe for carrying out NDE of layered structures as they are very sensitive to the conditions at the interface, the study of layered structures. Since they are dispersive and multimodal in nature, a large number of data points can be generated for different frequencies, as compared to the data available from bulk waves. Also, the different modes are directly accessible from reflectivity measurements at different angles.

The plate is immersed in water and is insonified by tone bursts at a single frequency by a transducer placed at a distance greater than a diameter but less than two Fresnel lengths for optimal operation. Specular reflection and transmission and leaky waves are obtained. For a given angle of incidence, the wave in the fluid has an x component $k_x = k_f \sin\theta$. When this is equal to the wave number of a Lamb mode along the plate, this mode becomes excited by phase matching. This plate wave then immediately leaks into the liquid as in Chapter 8 for Rayleigh waves. As in the latter case, there is creation of a null zone due to destructive interference between the specularly reflected wave and the reradiated guided wave, followed by an extended decay region of the leaky waves. A similar phenomenon occurs for the transmitted component.

A practical goniometer of a universal nature has been developed by Rochlin and Wang [179], as shown in Figure 15.2(b), where it is possible to use a reflector behind the sample so as to detect reflected and transmitted beams. The theoretical expression for the reflection and transmission coefficient have been presented in a convenient form by Chimenti and coworkers [180] as

$$R = \frac{AS - Y^2}{(S + iY)(A - iY)} \tag{15.2}$$

$$T = \frac{iY(S + A)}{(S + iY)(A - iY)} \tag{15.3}$$

where A, S, and Y are complicated functions of the material parameters given in [180].

The zeros of the RC correspond roughly to transverse resonances in the plate, so we expect the RC as a function of the frequency to exhibit a series of sharp minima, corresponding to the excitation of successive Lamb modes in the plate. This is in fact observed, as is shown in Figure 15.3 for an aluminum plate immersed in water at an angle of 20°. In an actual experiment, the full frequency range required for the transducers is determined by calculating the wavelengths corresponding to the characteristic lengths of the problem, namely, the thicknesses of layer and substrate. This range can be covered by a set of appropriate broadband transducers. A series of incidence angles (every degree or half degree) are assigned between the longitudinal and transverse critical angles. The reflected amplitude spectrum for a given angle, determined by FFT, is used to determine the set of minima corresponding to the generation of Lamb waves at the corresponding values of *fd*. The phase velocity V_P corresponding to this angle is given by $V_P = V_0/\sin\theta$. In this way, by sweeping through various fixed angles one can build up the full dispersion curve, as shown in Figure 15.4.

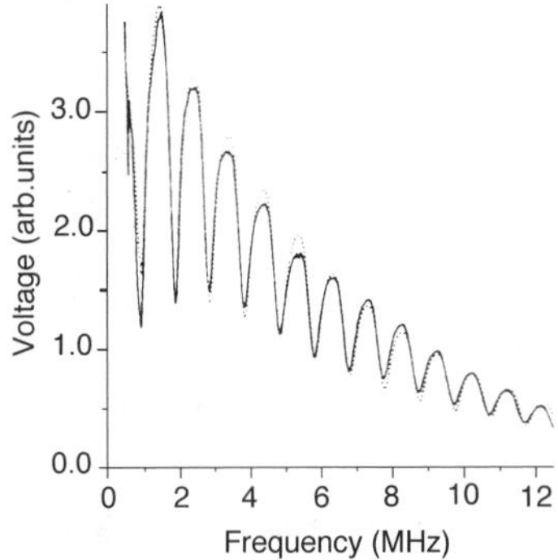

FIGURE 15.3
Leaky Lamb wave spectra for a smooth surface of 2024 aluminum plate at 20° incidence. Solid curve: deconvoluted measurement results; dashed curve: theoretical prediction. (From Lobkis, O.I. and Chimenti, D.E., *J. Acoust. Soc. Am.*, 102, 150, 1997. With permission.)

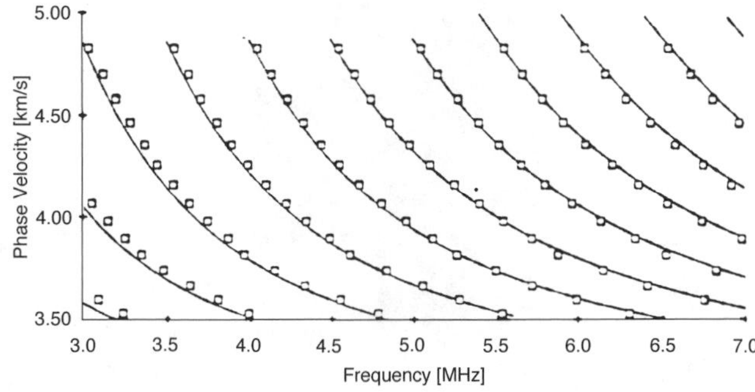

FIGURE 15.4
Simulated Lamb wave dispersion curves for an uncoated sample of aluminum 4 mm thick compared with experimental data. (From Xu, P.-C., Lindenschmidt, K.E., and Meguid, S.A., *J. Acoust. Soc. Am.*, 94, 2954, 1993. With permission.)

15.3.2 NDE Using Leaky Lamb Waves (LLW)

Chimenti [180] has described in detail how the LLW configuration can be used as an inspection method for detecting and imaging defects in plates. The sample is set up for insonification at an angle θ as shown in Figure 15.2, and an RF tone burst is emitted. The frequency of the tone burst is swept over the available bandwidth, which should be sufficiently wide to encompass a large number of Lamb modes. An FFT of the data is taken to produce a pseudo-frequency spectrum that mimics the behavior of an RF pulse in the time domain. If there is a defect in the specimen, this produces the equivalent of a defect echo in the pseudo-spectrum. Thus a measure of the change in the spectrum for this particular data point gives information on the presence of the defect. If the sending and receiving transducers are scanned over the plate then a C scan image of the plate with defect can be produced. Chimenti showed that use of the median pseudo-frequency, as the quantity to be measured, was the most satisfactory of several possible approaches. This frequency, $f_{1/2}$, is defined by

$$\int_0^{f_{1/2}} R(f, x)df = \frac{1}{2}\int_0^{f_{max}} R(f, x)df \tag{15.4}$$

For a given point this procedure then gives a single number, $f_{1/2}$, to describe the defect. The process can be repeated at each point in the full x-y plane and the result displayed as a C scan image. An example is shown in Figure 15.5

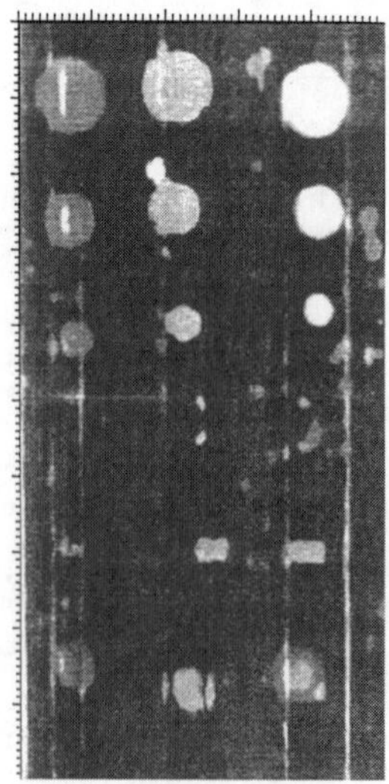

FIGURE 15.5
Leaky Lamb wave scan of a composite plate containing simulated defects. Images were formed by the median frequency representation. Defects closest to the sound entry field are on the left. (From Chimenti, D.E., Ultrasonic plate wave evaluation of composite plates, in *Proc. 1993 Ultrasonics Symp.*, Levy, M. and McAvoy, B.R., Eds., IEEE, New York, 1993, 863. ©IEEE. With permission.)

for a number of simulated defects in a composite plate. The method is an attractive alternative to conventional C scan imaging.

15.4 Layered Structures

15.4.1 Inversion Procedures

We are principally concerned here with single or multilayer coatings on substrates that can be assumed to be fully characterized (thickness, density, and sound velocities V_L and V_S are assumed known). Such coatings are extremely important industrially as they may provide mechanical, chemical, or electrical protection to prevent abrasion, wear, corrosion, etc. In this case, the general NDE problem is considerably more complicated than that for the single homogeneous plate. Type I NDE, defect detection, can be approached in a similar way to that for plates, but there is a new and perhaps dominant feature for the layered system, namely the question of adhesion of the layers between themselves and with the substrate. This is treated as a separate issue below. Type II NDE, dimensions and elastic constants, can be approached through the inverse problem. However, thickness gauging is generally seen as an important problem in its own right, and again, this will be treated as a separate issue in Section 15.4.5.

Because of their guided mode character, LLWs adapt well to the study of multilayer structures. Several inverse approaches have been used in the past, and one of them, the Simplex algorithm [183], will be described briefly here. The Simplex algorithm has the advantage that it always converges. The procedure for the LLW case is as follows:

1. Obtain LLW data experimentally, as outlined above.
2. Calculate the RC and the dispersion function. If there is a theoretical relationship between RC and material properties, as above, then this could be used in principle. However, there are difficulties in mode identification as well as lack of a precise functional form near the reflection minima. A more useful approach is to use the dispersion equation for the free plate in the form

$$G(f, V_p) = 0 \tag{15.5}$$

   It has been shown [180] that in most cases the dispersion curve for the free and water-loaded plate is very nearly equal and it simplifies the procedure enormously to use that for the free plate.
3. Inversion procedure

   For a given multilayered isotropic plate specimen, the properties h (thickness), ρ (density), V_L (longitudinal wave velocity), and V_S

(shear wave velocity) are to be determined from the dispersion equation

$$G(f_i, V_i, h, \rho, V_L, V_S) = 0 \quad i = 1,\ldots,x \tag{15.6}$$

where (f_i, V_i) are the sets of the LLW data points.
The optimization procedure consists in minimizing the sum of squares

$$SSR = \sum_{i=1}^{n} W_i G_i^* G_i$$

where G_i^* is the complex conjugate of G_i and W_i is a weighting function for the data points. Numerous examples on aluminum epoxy, aluminum plates, and graphite/epoxy composites showed that the Simplex algorithm recovered exactly the known material parameters, even when starting with initial values with a 50% error with respect to the true values.

Despite the overall power of the technique, there are a number of problems in the inversion of LLW data on multilayer structures. These include the following:

1. Coupling effects. The minimum of eight parameters (thickness, density, longitudinal, and shear wave velocity of the layer and substrate) for a single isotropic layer on an isotropic substrate are generally coupled together, leading to lengthy and complex computations. It is desirable to decouple these parameters as much as possible, and this cannot be done in a straightforward inversion approach.
2. Irregular or wrong data can derail the convergence of the inversion procedure. However, this procedure is so complex and nontransparent that it is not easy to identify which subset of the data is reliable.
3. Mode identification is essential for the full inversion approach. However, the actual data may only cover a limited frequency range on an uncharacterized sample and some data may be missing.

These difficulties and others can be avoided by a recent, simpler approach, modal frequency spacing, which will be described in the next two sections.

15.4.2 Modal Frequency Spacing (MFS) Method

The basic idea of the MFS method is to simplify the inversion procedure as much as possible by decoupling the parameters. The method was developed by Xu et al. [182], who showed that simple relations could be obtained between the elastic constants of the plate to be inspected and the shifts between frequencies of the leaky Lamb waves. As a simplification, the plate and the coating will be considered as isotropic. The method has been used for transversely isotropic coatings on isotropic substrates and could in principle be extended to more anisotropic systems. In the MFS work, the thickness of coating and substrate and the density of the substrate were assumed to be known. The goal was then to determine the longitudinal and transverse sound velocities in the substrate and coating and the density of the coating. Again, the effect of the fluid used for inspection on the dispersion characteristics of the bare and coated plate has been neglected as the effect is known to be small.

The MFS method starts with an analysis of the bare plate ($h_c = 0$). The dispersion relations are then given by the Rayleigh-Lamb equations as

$$\frac{\tanh\left[\nu_1\left(\frac{h_s}{2}\right)\right]}{\tanh\left[\nu_2\left(\frac{h_s}{2}\right)\right]} = \frac{(\nu_2^2 + k^2)^2}{4k^2\nu_1\nu_2} \tag{15.7}$$

for symmetric modes and

$$\frac{\tanh\left[\nu_1\left(\frac{h_s}{2}\right)\right]}{\tanh\left[\nu_2\left(\frac{h_s}{2}\right)\right]} = \frac{4k^2\nu_1\nu_2}{(\nu_2^2 + k^2)^2} \tag{15.8}$$

for antisymmetric modes, where

$$k = \frac{2\pi f}{V_P}, \quad k_\alpha = \frac{2\pi f}{V_\alpha}$$

$$V_P = \frac{V_0}{\sin\theta}$$

$$\nu_\alpha = \sqrt{k^2 - k_\alpha^2} \qquad k \ge k_\alpha$$

$$= -i\sqrt{k_\alpha^2 - k^2} \qquad k < k_\alpha$$

and h_s (h_c) = substrate (coating) thickness and $\alpha = L, S$. As the frequency increases, the symmetric and antisymmetric modes are usually excited alternately, as was seen in Chapter 9.

A first objective is to decouple V_L and V_S in Equations 15.7 and 15.8 so that each can be related independently to a set of experimental parameters. V_L can be decoupled by having the wave from the fluid at normal incidence, where only longitudinal waves are excited in the plate. In this case, instead of Equations 15.7 and 15.8, we have a simple cutoff frequency equation for longitudinal waves in the plate

$$\sin\left(\frac{2\pi f h_s}{V_L}\right) = 0 \tag{15.9}$$

so that V_L can be determined by the frequency difference between adjacent modes Δf

$$\Delta f = \frac{V_L}{2h_s} \tag{15.10}$$

Decoupling of V_S can be accomplished at high frequency for oblique incidence. If the latter is chosen so that

$$V_S < V_P < V_L$$

then Equations 15.7 and 15.8 can be rewritten as

$$\frac{\tanh\left(\pi f h_s \sqrt{\frac{1}{V_P^2} - \frac{1}{V_L^2}}\right)}{\tanh\left(\pi f h_s \sqrt{\frac{1}{V_T^2} - \frac{1}{V_P^2}}\right)} = \frac{1}{A} \tag{15.11}$$

for symmetric modes and

$$\frac{\tanh\left(\pi f h_S \sqrt{\frac{1}{V_P^2} - \frac{1}{V_L^2}}\right)}{\tanh\left(\pi f h_S \sqrt{\frac{1}{V_T^2} - \frac{1}{V_P^2}}\right)} = -A \tag{15.12}$$

for antisymmetric modes, where

$$A = \frac{4\sqrt{\left(\frac{1}{V_P^2} - \frac{1}{V_L^2}\right)\left(\frac{1}{V_S^2} - \frac{1}{V_P^2}\right)}}{\left(\frac{1}{V_P^2} - \frac{1}{V_S^2}\right)^2 V_P^2}$$

For high frequencies, i.e.,

$$\left| \pi f h_s \sqrt{\frac{1}{V_P^2} - \frac{1}{V_L^2}} \right| \gg 1 \tag{15.13}$$

Equations 15.11 and 15.12 become, respectively,

$$\tan\left(\pi f h_s \sqrt{\frac{1}{V_s^2} - \frac{1}{V_P^2}}\right) = A \tag{15.14}$$

and

$$\tan\left(\pi f h_s \sqrt{\frac{1}{V_s^2} - \frac{1}{V_P^2}}\right) = -\frac{1}{A} \tag{15.15}$$

with respective roots

$$\phi_1 = \tan^{-1} A \pm n\pi, \quad n = 0, 1, 2$$

$$\phi_1 = \tan^{-1}\left(-\frac{1}{A}\right) \pm 2\pi n$$

so that adjacent symmetric and antisymmetric modes have a spacing of π. From trigonometry $\tan(\phi_1 - \phi_2)$ is infinity, so that

$$\phi_1 - \phi_2 = \frac{\pi}{2} + 2\pi n \tag{15.16}$$

Finally, using Equations 15.14, 15.15, and 15.16, we have the desired result, that the frequency spacing Δf between adjacent symmetric and antisymmetric modes is given by

$$\Delta f = \frac{1}{2h_s}\left(\frac{1}{V_S^2} - \frac{1}{V_P^2}\right)^{-\frac{1}{2}} \tag{15.17}$$

Thus, from Equations 15.10 and 15.17, using the experimental values of the appropriate Δf, h_s, and V_P, V_L and V_S can be determined directly.

For the coated plate, the cutoff frequency equation at normal incidence becomes

$$\tan\left(\frac{2\pi f h_s}{V_L}\right) + \frac{\hat{V}_L \hat{\rho}}{V_L \rho} \tan\left(\frac{2\pi f h_c}{\hat{V}_L}\right) = 0 \tag{15.18}$$

where the two terms correspond to standing wave resonances in the substrate and coating, respectively. The roots of Equation 15.18 can be determined graphically from the intersections of the two families of curves as shown in Figure 15.6(b). In this approach the MFS is taken to be a constant and an average value is used. At oblique incidence the theory is more complex. In practice, the velocities of coating and substrate were weighted by their relative thickness by the relation

$$h_s V_\alpha + h_c \hat{V}_\alpha = (h_s + h_c) V_\alpha^n \tag{15.19}$$

where V_α^n is the nominal phase velocity of an equivalent uniform plate that includes the presence of the coating.

In applications, the following procedure was adopted:

1. For the bare substrate, measure Δf at normal incidence, Δf and V_P at various angles θ. This gives average values of V_L and V_S of the bare substrate.

2. Use these values as the starting point of a trial-and-error inversion to find the parameters V_L and V_S that best fit the experimental dispersion curves.
3. For coated samples, the substrate is now fully characterized and h_s, ρ_s, V_L, and V_S are fixed. Step 1 is repeated for the coated sample to give average values of V_L^n and V_S^n.
4. Equation 15.19 is used with the values of V_L^n and V_S^n from step 3 to determine starting values of $\hat{V}_L$ and $\hat{V}_S$.
5. Step 2 is repeated and $\hat{V}_\alpha$ are varied to get a best fit with the experimental diffusion curves.
6. $\hat{\rho}$ is adjusted to optimize the fit, which is not very sensitive to this parameter.

Excellent agreement was obtained with experiment for three different plasma sprayed coatings on titanium and aluminum samples.

15.4.3 Modified Modal Frequency Spacing (MMFS) Method

The basic approach [184] is the same for MFS except that the explicit variation of the MFS with frequency is taken into account. This fact is used to determine simple relations that can be used to determine all of the elastic constants from the measured MFS. In this case, experimental cutoff frequency data for the separate longitudinal and transverse-like modes must be obtained.

The same coating on substrate configuration as shown in Figure 15.6(a) is considered. The sample is immersed in a water bath and a longitudinal wave at normal incidence is incoming from the fluid. As before, the boundary value problem leads to a cutoff frequency equation for the longitudinal waves

$$\tan(k_L h_s) + \frac{\hat{\rho}\hat{V}_L}{\rho V_L}\tan(\hat{k}_L h_c) = 0 \qquad (15.20)$$

if the effect of the surrounding liquid is ignored. The roots of Equation 15.20 are the cutoff frequencies and are given by the intersection points in Figure 15.6(b). The parameter MFS = $\Delta f_n = f_n - f_{n-1}$ is no longer a constant for the coated plate but varies periodically, as shown in Figure 15.6(c). As shown in the figures, there are two characteristic regions. In the regular region the MFS is roughly constant and changes smoothly. The region where the MFS changes abruptly is called the transition region. By using Equation 15.20 and looking at the behavior of the second term, we have the following results:

1. In the regular region, the second term is very small, i.e.,

$$\frac{\omega h_c}{\hat{V}_L} = m\pi \qquad (15.21)$$

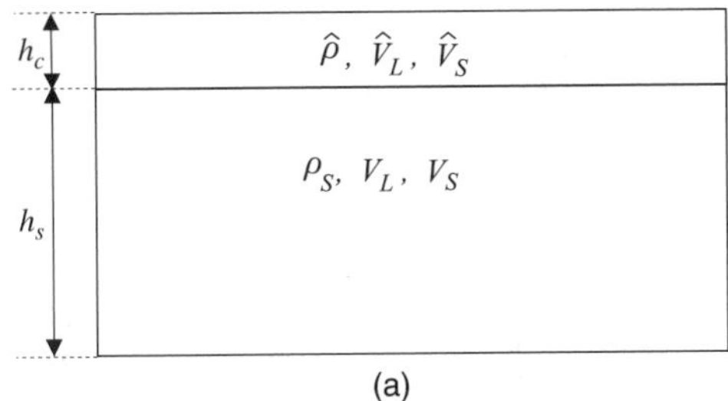

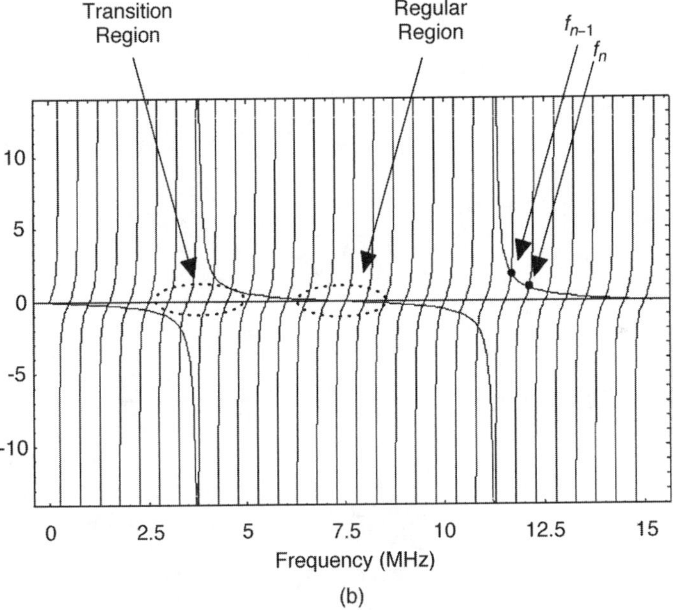

FIGURE 15.6
Geometry and analysis for MMFS method. (a) Definition of parameters for a coated plate. (b) Cutoff frequencies are obtained as the roots of Equation 15.20. (From Wang, Z. et al., Material characterization using leaky Lamb waves, in *Proc. 1994 IEEE Ultrasonics Symp.*, Levy, M., Schneider, S.C., and McAvoy, B.R., Eds., IEEE, New York, 1994, 1227. © IEEE. With permission.)

and the MFS at the zone center is given by

$$\Delta f_M = \frac{\Delta f_0}{1 + \frac{\hat{\rho} h_c}{\rho_s h_s}} \tag{15.22}$$

where

$$\Delta f_0 = \frac{V_L}{2h_s}$$

is the MFS of the uncoated plate.

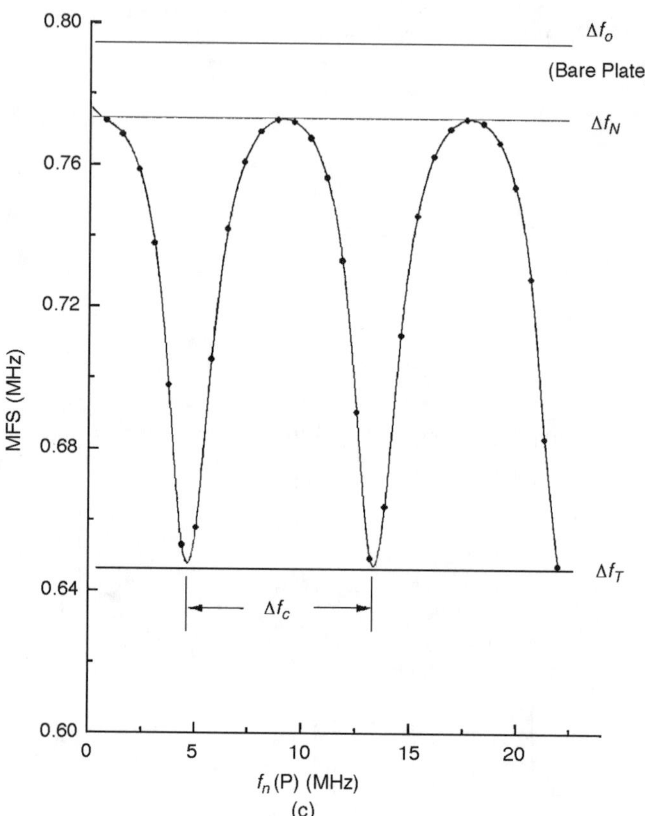

FIGURE 15.6 (Continued)
Geometry and analysis for MMFS method. (c) Distribution of MFS with frequency and identification of frequency parameters. (From Wang, Z. et al., Material characterization using leaky Lamb waves, in *Proc. 1994 IEEE Ultrasonics Symp.*, Levy, M., Schneider, S.C., and McAvoy, B.R., Eds., IEEE, New York, 1994, 1227. © IEEE. With permission.)

2. In the transition region, the second term goes to infinity

$$\frac{\omega h_c}{\hat{V}_L} = \left(m + \frac{1}{2}\right)\pi \tag{15.23}$$

and at the minimum of the transition region

$$\Delta f_T = \frac{\Delta f_0}{1 + \frac{\rho_s V_L^2}{\hat{\rho}\hat{V}_L^2}\frac{h_c}{h_s}} \tag{15.24}$$

3. Zone spacing

$$\Delta f_c = \frac{\hat{V}_L}{2h_c} \tag{15.25}$$

The physics of the composite resonator comes out in the form of Equations 15.21 through 15.25. The fine scale variations in MFS (Δf_M and Δf_T) are to first order governed by Δf_0 multiplied by a correction factor. Roughly speaking, each modal resonance corresponds to putting an extra standing wave wavelength in the substrate. Likewise, the overall frequency modulation of the MFS curve corresponds to successive resonances in the coating. In fact, Equation 15.25 shows that the experimental determination of Δf_c gives direct valuable information on the acoustic properties of the coating. The acoustic impedance ratio of coating to substrate,

$$A = \frac{\hat{\rho}\hat{V}_L}{\rho_s V_L} \tag{15.26}$$

is also an important parameter. It is shown in [184] that for $A < 1$ Equations 15.22 and 15.44 can be written as

$$\Delta f_M \sim \left[\frac{1}{\Delta f_0} + \frac{A}{\Delta f_c}\right]^{-1} \tag{15.27}$$

$$\Delta f_T \sim \left[\frac{1}{\Delta f_0} + \frac{1}{A\Delta f_c}\right]^{-1} \tag{15.28}$$

The important feature of the MMFS approach is that Equations 15.25 through 15.27 give simple algebraic approximate formulae whereby experimental measurement of Δf_0, Δf_c, Δf_M, and Δf_T leads to a direct determination of V_L, $\hat{V}_L$, $\hat{\rho}/\rho_s$, and h_c/h_s. Thus what was previously an inverse problem has turned into a forward calculation involving the resolution of a few simple linear equations. It has been shown by extensive simulations that the approximation procedure used here leads to calculation errors that are typically of the order of 1 or 2%. In order to obtain acceptable accuracy, the number N of modal frequencies within a period should be of the order of ten or more, where

$$N = \frac{\Delta f_c}{\Delta f_0} + 1 \tag{15.29}$$

The previous partial wave analysis was carried out for longitudinal waves. An identical approach can be used for SV and SH shear modes as discussed elsewhere [185]. The shear wave approach will evidently yield V_S and $\hat{V}_S$ as well as the other parameters. Thus, if all of these partial waves can be generated at normal incidence, all of the acoustic parameters of the coating and plate can be determined.

15.5 Adhesion

The quality of adhesion between coatings and substrates or between two layers is an important industrial problem. Because of the ability of ultrasonic waves to penetrate opaque media with possibility to set up guided waves, ultrasonics is one of the most promising techniques. There has been extensive work on ultrasonic studies of adhesion over the last 20 years. Earlier work has been reviewed in [186]. Recent work for isotropic and transversely isotropic layers has been reviewed in [187] and [188]. Work since then has been focused mainly on anisotropic media and cylindrical surfaces. The different layers involved are identified in Figure 15.7.

There are several types of adhesion problems [187]. Complete disbonds, voids, or porosity in the adhesive layer can generally be addressed by subsurface imaging, for example, in acoustic microscopy. A second aspect, poor cohesion due to a weak adhesive layer, has been addressed by several techniques. However, the problem of a weak interface layer (interlayer) poses many challenges and this aspect will be addressed here. An overriding consideration throughout will be that shear at the interface is clearly the key point, so that the guided or other modes used to probe the interface will

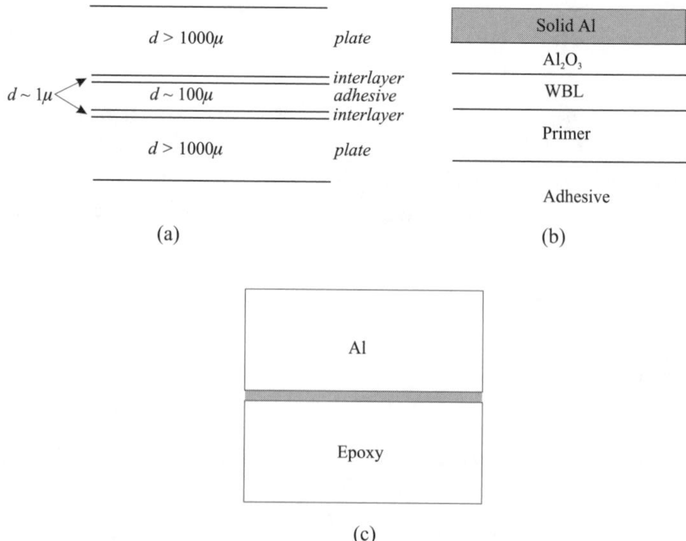

FIGURE 15.7
Models and configurations used for adhesive joints. (a) Simple model of an adhesive joint. The outer regions are the bulk adherends. The central adhesive layer is 100 μm thick. The two interlayers are about 1 μm thick. (b) The aluminum-adhesive joint consists of an anisotropic layer of Al_2O_3, a weak boundary layer (WBL), and a primer. (c) Experimental configuration for reflectivity measurements on an aluminum-epoxy interface.

have to have a strong shear component. A simple spring model of the structure to be probed will now be described.

The idealized model presented here [189] clearly shows the physics involved and provides a basis for theoretical interpretation of studies of the problem by Lamb wave interrogation. The ideal interface is described by the usual boundary conditions for two solids in contact with the x axis parallel to the interface and the z axis perpendicular to it. Any weakening of local rigidity or contact is described by a spring model, with normal and tangential stiffness constants K_n and K_t at the interface of a two-layer isotropic composite. In this model, the boundary conditions can be written as

$$T_{zz1} = T_{zz2} \tag{15.30}$$

$$T_{xz1} = T_{xz2} \tag{15.31}$$

$$K_t(u_{x1} - u_{x2}) = T_{xz1} \tag{15.32}$$

$$K_n(u_{z1} - u_{z2}) = T_{zz1} \tag{15.33}$$

where T_{zz} and u_z are stress and displacement components normal to the interface and T_{xz} and u_x are the shear stress and displacement along the interface.

For an ideal interface, two limiting cases can be identified in terms of values of the spring constants. For a "rigid" boundary, the case usually assumed for solid-solid interface problems, $K_n \to \infty$ and $K_t \to \infty$. This leads to the standard boundary conditions, $u_{x1} = u_{x2}$ and $u_{z1} = u_{z2}$, as the stresses at the interface must be finite. The opposing limit assumes "slip" between the two bodies at the interface. In this case the normal stress and displacement are continuous, as usual, so that again $K_n \to \infty$. However, the shear stress now vanishes at the interface as there is no binding contact between the media. Hence, the shear stresses vanish and the shear displacements are discontinuous, which can be obtained by setting

$$K_t \to 0$$

Stress and displacement can be expressed in terms of displacement scalar and vector potentials in the usual way. Following an approach similar to that for Lamb waves, we can express these functions as

$$\phi = [Ae^{-\varepsilon kz} + Be^{\varepsilon kz}]\exp j(\omega t - kx) \tag{15.34}$$

$$\psi = [Ce^{-\delta kz} + De^{\delta kz}]\exp j(\omega t - kx) \tag{15.35}$$

Hence

$$u_x = \frac{\partial\phi}{\partial x} + \frac{\partial\psi}{\partial z} \tag{15.36}$$

$$u_z = \frac{\partial\phi}{\partial z} - \frac{\partial\psi}{\partial x} \tag{15.37}$$

$$T_{zz} = \mu\left[\left(\frac{V_L^2}{V_S^2} - 2\right)\frac{\partial^2\phi}{\partial x^2} + \frac{V_L^2}{V_S^2}\frac{\partial^2\phi}{\partial z^2} - 2\frac{\partial^2\psi}{\partial x\partial z}\right] \tag{15.38}$$

$$T_{xz} = \mu\left[2\frac{\partial^2\phi}{\partial x\partial z} + \frac{\partial^2\psi}{\partial z^2} - \frac{\partial^2\psi}{\partial x^2}\right] \tag{15.39}$$

Substituting the form of the potentials into Equations 15.36 through 15.39, we obtain an 8×8 dispersion equation for the Lamb modes. The results are calculated for an aluminum-copper interface for a rigid and a slip interface. The rigid interface solution resembles that for a Lamb wave in a plate. However, the solution for the slip interface is quite different; the S_0 mode becomes a doublet with limiting low-frequency velocities

$$V_{01} = \sqrt{\frac{2}{1-\sigma_1}}V_{S1} \tag{15.40}$$

$$V_{02} = \sqrt{\frac{2}{1-\sigma_2}}V_{S2} \tag{15.41}$$

where σ_i are the Poisson's ratios of the two media. Thus measuring these limiting low-frequency Lamb wave velocities can, in principle, give an indication of the state of the interface layer. The model also shows the importance of shear stresses parallel to the interface, which will be investigated in some practical cases in the next section.

Cawley [187] has extensively reviewed ultrasonic inspection of adhesive joints. A promising method was found to be that of detecting the zeros in the reflection coefficient of shear waves incident from the adhesive layer. Such measurements can be performed with the goniometer shown in Figure 15.2, which has the advantage of being well suited to measuring very thin samples as well as being a very simple system. Alternatively, angular measurements on the interface of bulk samples can be carried out, as was done in [187]. In either case, the incidence angle is chosen to be larger than the longitudinal wave critical angle, so only shear waves are reflected at the interface. The state of the interlayer has a strong effect on the location and sharpness of the reflection zeros as a function of frequency. Unfortunately, as Cawley points out, the "zero" frequencies are more sensitive to the properties of the adhesive layer than those of the interlayer. Despite the difficulties in precise measurement of amplitude, Cawley concludes that the study of the modulus of the RC is a more fruitful approach.

Representative results for simulations of the RC for an interface with a porous inside layer were carried out for shear waves incident at 32° on an aluminum epoxy composite. The results clearly show that the RC is very sensitive to the thickness and sound velocity of the interlayer, while remaining virtually insensitive to a significant variation of velocity in the adhesive layer. This work was followed up by a detailed experimental study of anodized aluminum-epoxy interfaces, again for reflection of shear waves at 32° incidence. The anodized layers can be modeled as a transversely isotropic structure in which the elastic constants can be predicted as a function of porosity. The results for a 50-μm oxide layer were consistent with a porosity in the range 58 to 70%. Oxide layers down to 10-μm thickness should be detectable with this technique. The porosity of the layer determines the minimum detectable oxide thickness for the following reason. There is a large acoustic impedance mismatch between aluminum and epoxy, and increasing porosity decreases the impedance contrast between the two media, which ultimately establishes the limits of the technique. The conclusion of these studies is that ultrasonic reflectivity is a useful tool for the quantitative characterization of the interlayer in adhesive joints.

15.6 Thickness Gauging

Thickness determination of thin-walled vessels, sheets, coatings on substrates, etc. has traditionally been one of the most widespread ultrasonic techniques and this capability is provided in many commercially available instruments. There are two general approaches: time and frequency domain. Time domain studies are conceptually the simplest. A sharp ultrasonic pulse or tone burst is propagated in the sample and the time between two consecutive echoes is measured with precision. An alternative approach in the frequency domain is based on varying the frequency and looking for the fundamental resonance in the wall or layer. Both types of methods are described in [190].

This section is devoted to the description of several modern methods based on the use of guided waves. A first group is based on determination of reflectivity/transmission curves, and the second exploits the existence of cut-off frequencies in layered systems. A final example gives a demonstration of the applicability of the perturbation principle to describe layered structures.

1. Wideband acoustic microscopy

   Lee and Tsai [191] used a wideband scanning acoustic microscope (50 to 175 MHz) focused on a composite sample formed by a layer of thickness d_2 on a substrate. Sputtered pyrex films on sapphire and photoresist films on glass were studied. The acoustic beam could be focused on the surface of either the composite or the bare substrate. Labeling water, layer, and substrate as media 1, 2, and 3, respectively, we have:

a. Amplitude reflection coefficient at the water substrate interface

$$R_{13} = \frac{Z_3 - Z_1}{Z_3 + Z_1} = |R_{13}|e^{j\varphi_{13}} \tag{15.42}$$

b. Input impedance of the film-substrate composite

$$Z_{L1} = Z_2\left(\frac{Z_3 \cos k_2 d_2 + jZ_2 \sin k_2 d_2}{Z_2 \cos k_2 d_2 + jZ_3 \sin k_2 d_2}\right) \tag{15.43}$$

c. Complex RC at the water composite interface

$$R_{1(2)3} = \frac{Z_{L1} - Z_1}{Z_{L1} + Z_1} = |R_{1(2)3}|e^{j\varphi_{1(2)3}} \tag{15.44}$$

d. Phase difference between acoustic waves reflected from the composite and the substrate alone

$$\Delta\phi = 2k_1 d_2 + \varphi_{1(2)3} - \varphi_{13} \tag{15.45}$$

As the frequency is varied over the bandwidth, the RC reaches a minimum at the resonance frequency f_R where $d_2 = \lambda_R/4$. From Equation 15.45, measurement of the differential phase at resonance leads to a determination of d. In fact, Lee and Tsai [191] show that the best approach is fit the full RC as a function of frequency to Equations 15.42 through 15.45, which yields values of V_2, d_2, and ρ_2. For the frequency range used in this work, films of thickness 3 to 30 μm could be measured. Submicron films could be studied using this technique with frequencies above 600 MHz. Another advantage of the SAM technique is the high spatial resolution that can be attained.

2. Low-frequency normal incidence inspection

Low-frequency reflection and transmission at normal incidence is an off resonance technique which should be applicable to a wide range of configurations, including self-supporting foils and films [192]. The basic idea is to irradiate a thin layer situated between two identical substrates of acoustic impedance $Z_S = \rho_S V_S$. For simplification, lossless materials are considered. The through amplitude transmission coefficient for a layer of thickness h and acoustic impedance $Z = \rho V_l$ is

$$T = \frac{2}{2\cos kh + j\left(\frac{Z_S}{Z} + \frac{Z}{Z_S}\right)\sin kh} \tag{15.46}$$

where

$$k = \frac{\omega}{V_l}$$

The energy through transmission coefficient $t = |T|^2$ and finally

$$\left|\frac{R}{T}\right| = \sqrt{\frac{1-t}{t}} = \frac{\sin kh}{2}\left|\frac{Z_S}{Z} - \frac{Z}{Z_S}\right| \tag{15.47}$$

For a very thin layer such that $kh \ll 1$

$$\left|\frac{R}{T}\right| \approx \frac{\pi h}{V_L}\left|\frac{Z_S}{Z} - \frac{Z}{Z_S}\right| \tag{15.48}$$

This relation holds for longitudinal and shear waves at normal incidence. The linear behavior with frequency allows important information to be gathered from the slope. Two limiting cases are considered. For a high impedance layer $Z \gg Z_S$, Equation 15.47 can be rewritten as

$$\left|\frac{R}{T}\right| \approx \frac{\pi}{Z_S}\rho h f \tag{15.49}$$

This relation does not involve the bulk wave velocity in the layer, so if the density is known the thickness can be determined or conversely. This limit is particularly useful for such cases as immersion tank characterization of foils or studies of polymers, paper, etc. in air.

The opposite limit $Z \gg Z_S$ will be appropriate, for example, to describe an adhesive joint between metal plates. In this case, Equation 15.47 can be written as

$$\left|\frac{R}{T}\right| \approx \frac{\pi Z_S h f}{\rho V_l^2} \tag{15.50}$$

Thus the slope is given by $\pi K_{SS}h/c$ where $c = \rho V_l^2$ is the elastic modulus of the layer. This result can be used for adhesive characterization as the larger specific compliance h/c is known to be related to the state of cure and the joint quality. For this case, longitudinal and transverse waves can be used.

15.6.1 Mode-Cutoff-Based Approaches

These approaches use the basic characteristics of guided waves. They enjoy all of the usual advantages of guided waves for NDE, they are very sensitive, and they are adaptable to microscopic and macroscopic situations. The first of these, UMSM [155], was developed as a potential online NDE technique with high spatial resolution. It is effectively a miniaturized version of an RC goniometer and either planar or focused beams can be used. The method applies to the case of a lossless or low loss layer having a shear wave velocity lower than that of the substrate. It has previously been shown in this case that the fundamental mode in the layer is the Rayleigh mode and the next highest one is the Sezawa mode. As the frequency is lowered the latter has a cutoff at the point where the phase velocity equals the shear wave velocity of the substrate. Below this the Sezawa mode leaks into the substrate and becomes evanescent; in this region, it is called a pseudo-Sezawa mode.

If the Sezawa mode is excited by an incident wave from the fluid then this cutoff can be detected by a dip in the reflected coefficient at the critical frequency; in effect, the energy that is lost from the incident beam is coupled directly into the substrate by the intermediary of the layer, as schematized in Figure 15.8. Since the effect occurs at a critical value of fd, the thickness d can be inferred immediately from a knowledge of the cutoff frequency f_c. In practice, the phenomenon is observed in a UMSM goniometer. Operating over a frequency range 30 to 150 MHz, the goniometer is set at the angle corresponding to the usual leaky wave condition, in this case for Sezawa cutoff phase velocity V_c, at $\sin\theta = V_w/V_c$. The frequency is then scanned and the cutoff condition is easily identified by a dip in the RC at the appropriate frequency, as in Figure 15.8(b).

The system can be made very sensitive by the use of accurate micropositioners and temperature compensation of the water, leading to estimated stability and accuracy of ±2% and ±1%, respectively. High-speed resolution can be obtained using acoustic lenses, enabling values of the order of 200 μm to be obtained. The UMSM has been designed for rapid online measurement of film thickness in the range 1 to 20 μm for the 10 to 200 MHz frequency range. Submicron thicknesses can be measured by the LFB technique described in the next section.

Another approach is to use the leaky Sezawa modes measured by the LFB [193]. The physical principle involved is the same as for layer thickness determination by UMSM, except that now the leaky Sezawa mode is detected directly with the LFB. Above cutoff, the leaky Sezawa wave leaks only into

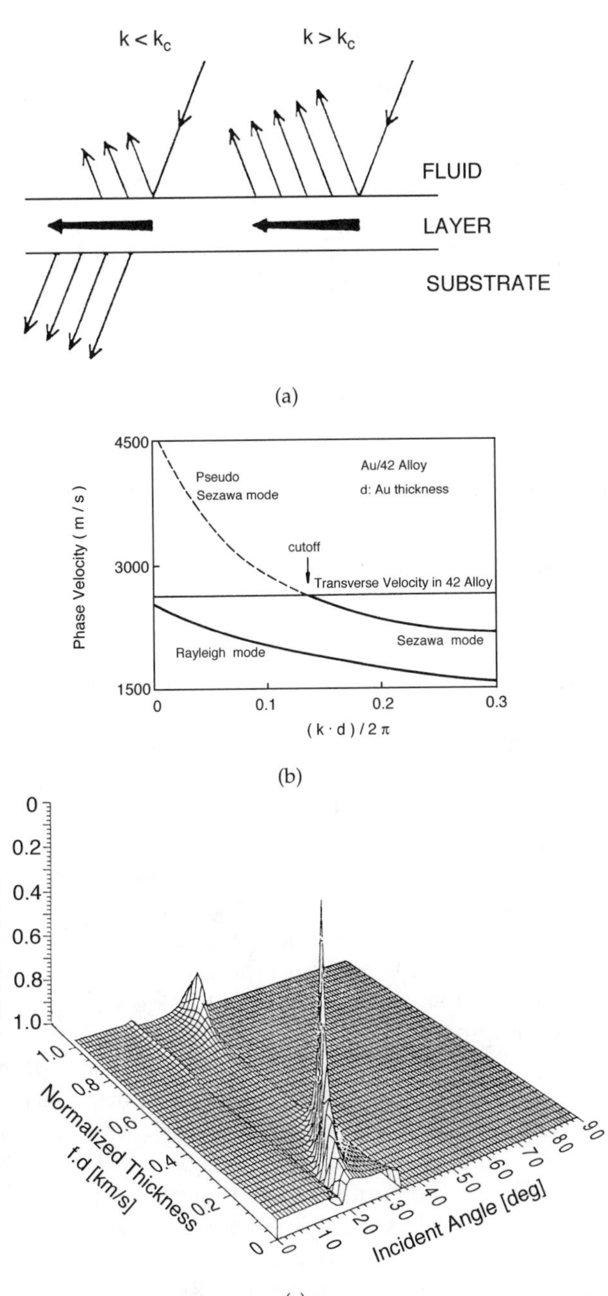

FIGURE 15.8

Ultrasonic microspectrometer. (a) Wave propagation conditions for (i) Leaky pseudo-Sezawa wave is excited for $k < k_c$ and (ii) Sezawa mode is excited for $k > k_c$. (b) Dispersion curve and cutoff condition for Sezawa modes for a gold layer on a 42-alloy substrate. (c) Reflection coefficient calculated as a function of θ and fd for case (b) above. (From Tsukahara, Y. et al., *IEEE Trans. UFFC*, 36, 326, 1989. ©IEEE. With permission.)

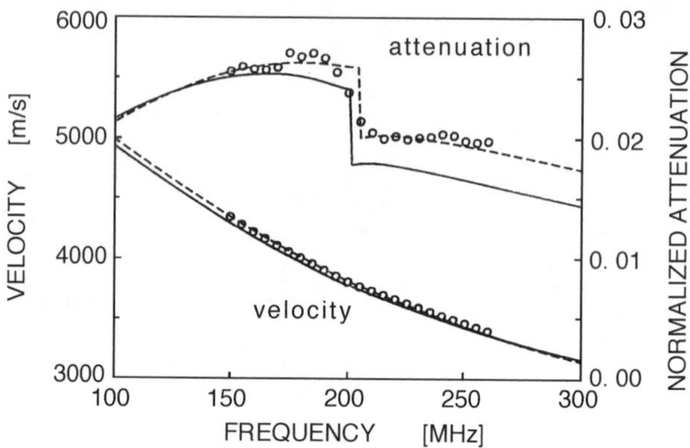

FIGURE 15.9
Frequency dependence of measured and calculated propagation characteristics of leaky Sezawa and pseudo-Sezawa wave modes for a gold film on a fused quartz substrate. The solid lines are calculated with the bulk constants of gold, while the dotted lines are computer fitted. (From Kushibiki, J., Ishikawa, T., and Chubachi, N., *Appl. Phys. Lett.*, 57, 1967, 1990. With permission.)

the water. Below it, the pseudo-Sezawa wave leaks into the substrate and the water, leading to a jump in attenuation at the cutoff frequency. The velocity and attenuation were measured for a gold film on a fused quartz substrate as a function of frequency as shown in Figure 15.9. c_{44} and ρ were used in the fitting, while the thickness d is obtained directly from the cutoff condition. Thus all three quantities could be obtained by a measurement of the leaky mode as a function of frequency. It is interesting to note [194] that at much lower frequencies the leaky Rayleigh wave could be well separated from the pseudo-Sezawa wave in the experimental $V(z)$ curve, so that c_{11} could also be obtained, thus enabling the direct determination of all four material constants in a single experiment. It is, of course, assumed that all of the corresponding parameters for the substrate are known.

The cutoff principle can also be used directly on the higher-order Lamb modes of a plate or pipe; this approach should be particularly useful for the noninvasive detection of inaccessible layers of corroded material on the inside surface of a pipe. The principle of detection [195] is easily appreciated by an examination of the group velocity curves for Lamb waves in an aluminum plate, as was shown in Chapter 9.

The higher-order modes all have a cutoff frequency at specific values of fd. Thus a wave generated at a frequency above cutoff would propagate down the plate, but one generated below cutoff would be reflected, as in Figure 15.10. Comparing cutoff frequencies for corroded and noncorroded samples would then provide a measure of the corrosion layer thickness d. The method has been tested on laboratory samples of aluminum with an accuracy of about 5%.

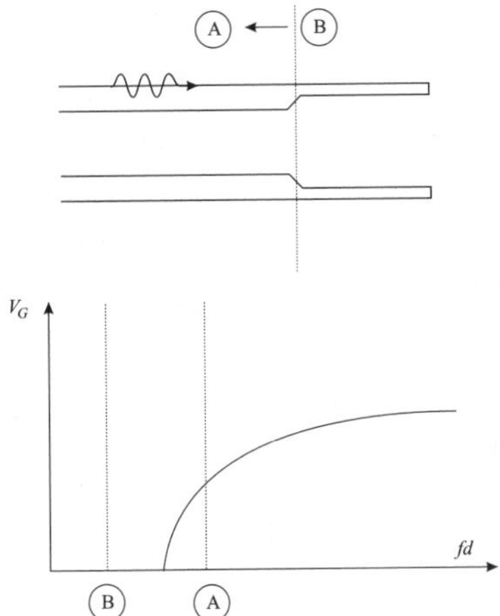

FIGURE 15.10
Reflection of Lamb waves near the cutoff condition due to pipe wall thickness reduction caused by corrosion.

15.7 Clad Buffer Rods

Cladded acoustic fiber delay lines were first developed by Boyd et al. [72] as an alternative to bulk wave polygon lines and surface wave wraparound delay lines discussed earlier. The principle involved is based on that used in optical fibers; the acoustic fiber consisted of a low-velocity (e.g., glassy) core and a higher-velocity cladding to confine the acoustic energy to the core and reduce spurious losses due to surface effects, mechanical supports, etc., as well as to eliminate crosstalk. Depending upon the transduction mechanism employed, torsional or radial axial modes can be excited in the fiber. Such long delay lines have also been used for acoustic imaging [196] when a spherical cavity is ground in one end face.

More recently, cladded buffer rods have been developed for various specialized applications in NDE [197]. The basic principle is the same in that the cladding is used to suppress spurious structure created at the surface due to diffraction and mode conversion. In this case, bulk longitudinal or transverse waves are transmitted in a low-amplitude core. Since one of the main applications is NDE at high temperatures, core materials such as Al, Zr, and fine-grained steels have been used. Thermal spray techniques have been used to deposit claddings that may be up to several millimeters thick, enough to support the cladding function as well as to permit machining of the outer surface for providing screw threads, etc. These long, high-quality buffer rods have potential for application in many industrial processes carried out in hostile or challenging environmental conditions. Several examples are given below:

1. High-temperature NDE. Many large-scale industrial processes are carried out at elevated temperatures, e.g., 700°C for aluminum die casting, 200 to 400°C for polymer extrusion, and 1500°C for molten glass and steel. Conventional ultrasonic transducers can be used at the very most up to 350 to 400°C so new solutions must be found. Clad buffer rods fill these requirements and have been used, for example, at the interface between molten Mg and an $MgCl_2$ salt at around 700°C. A 1-m-long rod was used with an air cooling device at the top end to cool the transducer and its RF connection. For aluminum melts, the buffer rods had to be chosen with care to avoid corrosion. For a stainless steel cladding in an aluminum melt at 960°C, the measurement had to be done in less than 30 min to avoid these effects.
2. Thickness measurements at high temperatures. An important problem is that of corrosion on the inner surfaces of pipes and containers carrying molten metals or corrosive chemicals at high temperature. To do this, the clad buffer rod can be put in contact at normal incidence with the outer surface of the pipe and multiple echoes in the pipe wall can be observed. If temperature effects are taken into account, an accurate measurement of the in-service wall thickness can be carried out.
3. Online monitoring during polymer extrusion. The buffer rod can be fitted into the wall of the extruder, its extremity positioned flush with the cavity surface. The study showed that accurate measurements could be made in real time of the thickness of polymer melt extruded at an angular speed of 5 rpm at constant conditions of 220°C and 540 psi. This allows real-time monitoring of the composition of polymer blends and other properties of the mixture.

16

Special Topics

The effects discussed in this chapter are twofold in nature. One group (multiple scattering, time reversal, and air-coupled ultrasonics) are part of traditional ultrasonics, but recent advances have given them topical interest. The other group (picosecond ultrasonics and resonant ultrasound spectroscopy) are relatively new techniques that promise to enlarge the scope of ultrasonic studies of materials and devices. Their inclusion here is justified by the high probability that they will be of lasting interest in future work on ultrasonics.

16.1 Multiple Scattering

Multiple scattering of acoustic waves occurs in a variety of situations and materials. It is generally associated with propagation in inhomogeneous media. Several important application areas include oceanography and oil exploration (bubbly, gassy liquids, already mentioned, and fluid-bearing sediments), and various engineering applications (e.g., fluidizers and filtration systems). We will briefly summarize results from two areas. The first of these, fluid-saturated porous media, has been extensively studied in connection with oil exploration. The emphasis here is on a complete description of the acoustic modes in such systems. The second example is more fundamental in nature and relates to the meaning of the group velocity in multiple scattering media.

Fluid-saturated porous media are correctly described by the Biot theory [198], which applies to those situations where the fluid and solid media are continuous and interpenetrating. In particular, the case of isolated, empty pores is excluded from the discussion and will be mentioned briefly at the end.

A basic assumption is that the pore and grain sizes are small compared to the wavelength and that each volume element could be described by an average local displacement. Biot then solved the equations of motion to determine the velocities of the shear (V_S), fast compressional (V_F), and slow compressional (V_{SL}) in the medium in terms of the porosity ϕ:

$$V_S = \sqrt{\frac{N}{(1-\phi)\rho_S + \left(1-\frac{1}{\alpha}\right)\phi\rho_f}} \tag{16.1}$$

$$V_F = \sqrt{\frac{K_b + \frac{4}{3}N}{(1-\phi)\rho_S + \left(1-\frac{1}{\alpha}\right)\phi\rho_f}} \tag{16.2}$$

$$V_{SL} = \frac{V_0}{\sqrt{\alpha}} \tag{16.3}$$

where

N = shear modulus of the skeletal frame and of the composite
K_b = bulk modulus of the skeletal frame
ρ_s, ρ_f are solid and fluid densities
α is a parameter proportional to the induced mass of the skeletal frame in the fluid, to be explained below

In the equations of motion, Biot introduced a density ρ and a density matrix ρ_{ij} satisfying the following relations

$$\rho = (1-\phi)\rho_S + \phi\rho_f \tag{16.4}$$

$$\rho_{11} + \rho_{12} = (1-\phi)\rho_S \tag{16.5}$$

$$\rho_{22} + \rho_{12} = \phi\rho_f \tag{16.6}$$

$$\rho_{22} \equiv \alpha\phi\rho_f \tag{16.7}$$

It then follows that

$$\rho_{12} = -(\alpha - 1)\phi\rho_f$$

which represents the inertial drag that the fluid exerts on the solid. α was shown to be a purely geometrical quantity, for example, for spheres [199]

$$\alpha = \frac{1}{2}(\phi^{-1} + 1) \tag{16.8}$$

For negligible viscosity, attenuation effects can be ignored. Also, the quoted results in Equations 16.1 to 16.3 are valid for K_b, $N \gg K_f$. In this case, the fast compressional wave corresponds to solid and fluid moving in-phase and the slow compressional wave to out-of-phase movement between solid and fluid. It is this slow compressional wave that is the characteristic feature of porous media of this type. In fact, it turns out [200] that the slow wave is a generalization of the well-known mode of fourth sound in superfluid helium contained in a porous superleak at low temperatures.

Plona [201] was the first to observe the slow compressional wave experimentally. Experiments were carried out at 2.25 MHz on a water-saturated system of glass beads (approximately 0.25 mm diameter) about 15 to 20 mm thick. Oblique incidence allowed mode conversion to take place and hence the possibility of ultrasound propagation by shear and/or compressional waves in the sample. For a porosity of 25%, sound velocities of 4.18, 2.50, and 1.00 km/s were identified as fast compressional, shear, and slow compressional, respectively. A detailed, quantitative comparison with the Biot theory allowed Berryman [199] to confirm this identification. Experiments at very low frequencies (~1 to 10 Hz) by Chandler [202] showed that in this limit the propagation passes from propagative to diffusive. The attenuation of acoustic waves in these systems was studied by Stoll and Bryan [203] and by Stoll [204]. The so-called self-consistent theory [199] was a quantitative effective medium theory for the case where the pores are isolated, rather than continuous.

Ultrasonic propagation in porous media is a traditional area of interest in the field of strongly scattered ultrasonic waves. A related subject of long-standing interest is the question of group velocity in strongly scattering media. The issue here is to elucidate the coherence of ballistic ultrasonic pulses that propagate through strongly scattering media. Strong scattering leads to high attenuation and dispersion as well as unphysical values of the group velocity. Indeed, Sommerfeld and Brillouin have questioned the existence of a true group velocity in this case. These speculations have lead to a large body of work on multiple scattering, diffusion, and acoustic localization [205].

The question of group velocity and coherence of ultrasonic pulses in ballistic propagation through strongly resonant scattering material was elucidated by Page et al. [206]. They studied ultrasonic propagation from 1 to 5 MHz through a suspension of monodispersive glass beads (radii 0.25 to 0.50 mm) in water to form samples 2 to 5 mm thick with a glass bead volume of about 63%. The samples were placed in a water bath and 25-mm diameter piezoelectric transducers were used in a transmission configuration. The scattered sound was effectively cancelled by random phase fluctuations, so that an unscattered ballistic wave was propagated through the system; this wave was found to be completely spatially and temporally coherent. Special techniques were used to measure the phase and group velocities as a function of ka. Significant dispersion was found, and surprisingly, both velocities were found to be inferior to those of the constituent media as well as inferior to the Stoneley wave velocity at a water-glass interface. The theoretical calculation of the phase and group velocities was carried out by determining the Green's function of the wave equation and hence the spectral function, given by the the negative imaginary part of the Green's function. The peaks of the spectral function were used to calculate the phase velocity in the suspension, which gave excellent agreement with experiment. The group velocity was calculated by numerical differentiation of the dispersion curve, and again excellent agreement with experiment was obtained.

The basic feature to be explained is that a pulse maintaining full spatial and temporal coherence can travel through a dispersive medium with such slow velocities. The physical picture emerges showing that with strong scattering each particle is subjected to the scattered waves from the other particles. This leads to an effective renormalization of the medium due to strong resonant scattering such that the medium takes on the properties of the scatterers and the resonances vanish. The group velocity directly senses this effective renormalization. These results are far more general than just for the case of acoustic waves, so that similar effects should be seen in light waves and microwaves, for example.

16.2 Time Reversal Mirrors (TRM)

Time reversal has received much attention in physics and is perhaps best known for its role in the famous question of the arrow of time. The microscopic laws of physics are invariant with respect to time reversal, that is, for a given microscopic process the solutions of the equation of motion at time t can also be generated for time $-t$, as second-order differential equations are involved. The paradox is, of course, that this conclusion is not true for macroscopic thermodynamic processes, which are irreversible and dissipative in nature, leading to time evolving in one direction and never in the reverse. The situation is summarized in a famous cartoon in which a man throws a bomb into a pile of debris and the destroyed house in question reconstructs, which of course never happens in nature. This paradox of microscopic reversibility and macroscopic irreversibility has been resolved in a convincing fashion by the use of Boltzmann's original concept of entropy [207].

Because of the specific properties of acoustic waves, it is possible to achieve macroscopic time reversal acoustically. The subject has been vigorously developed by M. Fink and coworkers [208]. Under the conditions of adiabatic processes the pressure field p in a heterogeneous medium of density $\rho(r)$ and compressibility $\kappa(r)$ can be described by the wave equation

$$\kappa(\vec{r})\frac{\partial^2 p}{\partial t^2} = \nabla^2\left(\frac{p}{\rho(\vec{r})}\right) \tag{16.9}$$

which is time invariant due to the second-order time derivatives. We consider the emission of an acoustic pressure wave from a point source, the wave subsequently having its trajectory modified due to multiple scattering, refraction, etc. If we can somehow reverse the waveform at some time t (in a time reversal cavity) then there is a complicated waveform $p(r, -t)$ that will

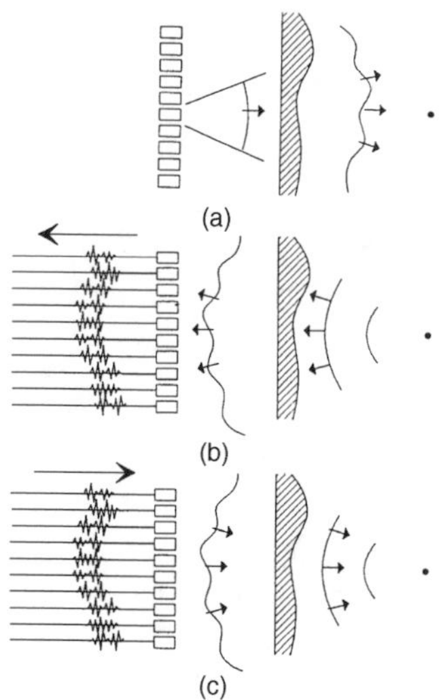

FIGURE 16.1
TRM focusing through inhomogeneous media requires three steps. (a) The first step consists of transmitting a wavefront through the inhomogeneous medium from the array to the target. The target generates a backscattered pressure field that propagates through the inhomogeneous medium and is distorted. (b) The second step is the recording step: the backscattered pressure field is recorded by the transducer array. (c) In the last step the transducer array generates on its surface the time-reversed field. This pressure field propagates through the aberrating medium and focuses on the target. (From Fink, M., *IEEE Trans. UFFC*, 39,555, 1992. ©IEEE. With permission.)

then synchronously reconverge onto the original source. For various reasons, a time reversal cavity cannot easily be constructed and it is more common to use planar time reversal mirrors (TRM), which are described below. It should be noted here that apart from time invariance, spatial reciprocity between source and receiver should also be satisfied.

With the aid of Figure 16.1 we describe the process of time reversal focusing in the transmit mode using a TRM. In the first step, an ultrasonic wavefront is emitted by the array. It travels through an unspecified inhomogeneous medium and is hence deformed in some arbitrary way. The wavefront impinges on the point target that re-emits part of it as spherical waves. This spherical wave front is again distorted by the medium. The second step consists in recording this backscattered pressure wave by the array. In the third step, the recorded signals are re-emitted in reverse order (last in, first out) and the inhomogeneous wavefront, being now perfectly matched to the medium, converges to a focus on the target. A similar type of reasoning can be applied to the receiver mode. There are some conditions on the process. Single scattering events in the medium (first Born approximation) can be compensated exactly by the TRM. For strong scattering where multiple scattering occurs, the measurement interval must be sufficiently long to receive all of the multiply scattered waves. Full details are given in [208].

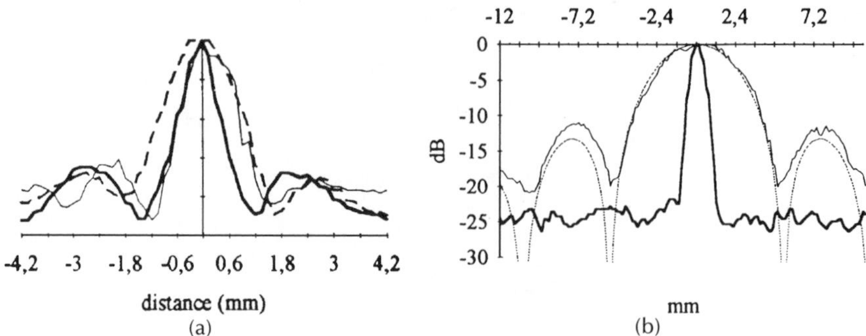

FIGURE 16.2
(a) Directivity pattern of the pressure field received by s in homogeneous medium (dashed line) through medium I (thick line) and through medium II (thin line). (b) Directivity patterns of the TRM through 2000 steel rods (thick line) and in water (thin line). The theoretical sinc function is represented by the dashed line. (From Derode, A. et al., *Phys. Rev. Lett.*, 75, 4206, 1995. With permission.)

Time reversal can be used to improve performance of focused acoustic beams and of acoustic imaging. They are applicable in all areas of acoustics, particularly in cases where strong scattering reduces the effectiveness of conventional techniques. Several examples are given below [209]:

1. Multiple scattering. A water tank experiment was carried out with an array of 96 piezoelectric transducers [210]. A 3-MHz pulse was emitted from a small source and then passed through a "forest" of about 2000 steel rods, leading to strong multiple scattering. The ultrasonic signal received by any one transducer in the array was a long incoherent echo train extending over hundreds of microseconds. Time reversal was then carried out by the array and a single sharp signal was then detected at the source by a hydrophone. The detected signal was of the order of 1 μsec in width. What is perhaps even more striking is that the width of this focal line was about six times smaller than that pertaining to a direct focusing experiment when the rods were removed. It was shown that this improvement in spatial resolution by multiple scattering was due to the fact that the whole multiple scattering medium acts as coherent focusing source with high aperture, hence the enhanced focusing performance seen in Figure 16.2.
2. Waveguide. This is another laboratory demonstration where multiple scattering is provided by a water channel bounded by steel and air interfaces. A 99-element array was placed downstream to pick up the multiple echos from the guide walls, spread out over about 100 μsec when detected with one of the array transducers. Again, time reversal led to observation of a single sharp pulse at the source position. This experiment has relevance to acoustic underwater communication in oceanography and the results have been

extended to actual measurements in the ocean for a channel 120 m below the ocean surface and 7 km long.

3. Kidney stones. This is a direct application of time reversal, but its application is complicated by the fact that the stone moves as the patient breathes. Once the most reflective part of the stone can be tracked in real time, the power is increased to the level needed to shatter the stone. Other medical applications include hypothermia for destruction of diseased tissue, including prostate cancer and applications to the brain.
4. NDE for detection of small defects in solids, which may be heterogeneous, anisotropioc, or have a complicated shape. Defects as small as 0.4 mm in 250-mm titanium billets have been detected.
5. Detection of surface roughness by displacement of the TRM before re-emission. RMS height and surface height autocorrelation function can be determined. Possible applications include arterial wall properties *in vivo*, mapping of the sea floor, and determination of interface roughness of solid joints.

16.3 Picosecond Ultrasonics

Conventional laser ultrasonic techniques have been used for a number of years as one of the preferred methods where a noncontact approach is required, often due to hostile environments that prevent the use of other techniques [211]. We briefly describe the major components of a typical laser ultrasonic system as background material for the more recent development of picosecond ultrasonics.

Pulsed lasers can be used as sources for a laser ultrasonic system. Three general generation mechanisms are employed. Thermoelastic generation occurs due to absorption of light at the surface or in the first few nanometers below it. The heated region causes thermoelastic expansion and the launching of a low-level acoustic wave. Higher levels can be obtained by use of constrained thermoelastic generation, produced by the effect of a glass slide, oxide layers, etc., on the surface. In this case, the laser beam is absorbed in a region well below the surface. Finally, if the laser power density exceeds the appropriate threshold, ablation can occur, leading to very high-amplitude ultrasonic waves. To some extent, this mechanism is destructive and may be excluded for certain applications.

Optical detection of ultrasonic waves is generally carried out by interferometric or other means for measuring surface displacement, such as Michelson or Fabry-Perot interferometry. These systems have the disadvantage of being expensive and lack sensitivity compared to conventional methods such as piezoelectric detection. For both generation and detection the surface properties of the sample are critical; in some respects, one can compare the input and probe laser beams to the coaxial cable of a conventional system and the sample surface plays the role of the piezoelectric transducer. Surface absorptivity and reflectivity are two important material parameters. Thus the choice of a laser ultrasonic system vs. a conventional pulse echo system will revolve around the importance of a large number of parameters and constraints, including cost, sensitivity, and need for contact or noncontact.

Picosecond ultrasonics is a development of laser ultrasonics with the objective of probing the acoustic properties of microstructures. With conventional ultrasonics, the best one could obtain are 1-ns pulse widths, which gives minimum spatial resolution of the order of 5 μm, which is too thick to be useful for most microstructures. On the other hand, with laser pulse widths of the order of 10^{-14} s or less, the acoustic properties of very thin structures could in principle be probed. The first results were reported by Thomsen et al. [212] in 1986 and since then work has been reported by several other groups [213 and 214]. The technique is a fairly direct extension of conventional laser ultrasonics. The typical sample to be studied is a thin film deposited on a substrate. An evaporated transducer film is deposited on the sample. This transducer, usually a metal film, absorbs the light from the laser and consequently heats up. A thermal stress pulse is emitted into the sample as the transducer relaxes. The form of the stress pulse is determined by the ratio of the acoustic impedances of the transducer and sample. Desired properties of the transducing film include a high optical absorption coefficient and a high sound velocity in order to produce an intense, short stress pulse. The film is excited by a pulsed laser, typically in the range 0.5 to 5 ps on a 20-μm diameter spot, producing a temperature rise of a few degrees Kelvin and strains of the order of 10^{-4} to 10^{-5}.

Detection is usually carried out optically to retain the flexibility of the noncontact approach and to detect such short pulses. In the original scheme, changes in reflectivity in the transducer were detected, due to changes in optical constants caused by the ultrasonic wave. The probe pulse was split off from the main pulse and delayed by the appropriate time. It has been shown that the change in reflectivity of the transducer is proportional to the average strain in it induced by the ultrasonic pulse. Since the early work, interferometric methods have also been used, which enable independent determinations of phase and amplitude to be made. Another development has been the generation of pure transverse modes in the thin samples. This has been attained by mode conversion in an isotropic film deposited on an isotropic substrate [215] from a longitudinal acoustic pulse initially generated in the film.

The overall performance of picosecond ultrasonic techniques for the study of thin films is very impressive. In the time domain, several echoes have been observed in 70-nm silica films in the early work [216] and similar results by several groups are now routinely available. This means that the thickness of very thin films of known acoustic velocity can be probed, or the elastic constants can be determined for films of known thickness. In the frequency domain, the frequency spectrum of the acoustic waves generated by the technique is the Fourier transform of the emitted pulse. This means that acoustic waves with wideband spectra centered at frequencies up to hundreds of GHz can be produced. Since the frequency-dependent attenuation can also be deduced [216], this opens the door to new ways to study physical acoustics on the interactions of high-frequency acoustic phonons.

Several studies have already been carried out on a variety of insulating, semiconducting, and metallic systems and the principal results are described briefly below.

1. Amorphous solids

   Amorphous materials (e.g., glasses) are known to display a characteristic behavior in their acoustic and thermal properties at very low temperatures [217]. The behavior below 1 K can be explained by the two-level system (TLS) model to describe interaction of the wave with localized defects. However, at higher temperatures the situation is far from clear. In this context it is important to have measurements over as much of the frequency/temperature parameter space as possible to be able to compare results with the theoretical models. Picosecond ultrasonics has been very useful in this regard.

   The first study [218] on a SiO_2 (fused silica) showed that $\alpha \sim \nu^2$ for frequencies from 75 to 450 GHz and was independent of temperature from 80 to 300 K. Additional data on amorphous polymers and metals (amorphous TiNi) [219] suggest that this behavior may be universal, as a quadratic frequency dependence was observed up to 320 GHz and the attenuation increased with temperature by a factor of two or three in the range 80 to 300 K. When data obtained using other techniques are added, the following general picture emerges. The attenuation rises rapidly with temperature from 1 to 80 K and then rises much more slowly up to 300 K. The frequency dependence is linear below 10 GHz, changing to quadratic in the range 10 to 50 GHz. This general picture was used to compare with existing theories, mainly the fracton model [219].

2. Reflectivity of high-frequency phonons at interfaces

   These studies are closely related to the well-known Kapitza boundary problem in low-temperature physics [220]. In 1941, Kapitza showed experimentally that there is a temperature jump at a copper-liquid interface in the presence of a heat flux, the so-called Kapitza resistance R_K. This thermal contact resistance exists between any two media in contact but is usually only observable at low temperatures using thermal conductivity techniques; due to its T^3 temperature variation, R_K is usually too small to measure above 4.2 K. Khalatnikov [221] showed that the thermal boundary resistance could be described theoretically by considering the partial reflection and transmission of thermal phonons (high-frequency ultrasonic waves with a frequency spectrum given by the Planck distribution), the so-called acoustic mismatch model (AMM).

Khalatnikov carried out a simple energy transmission calculation using the techniques of Chapter 7 and integrated over all incidence angles and frequencies. For the heat flux using the Planck distribution the phonon frequency at the maximum of the distribution is situated at about $3k_BT$, or 63 GHz at 1 K where k_B is Boltzmann's constant.

It turns out that the experimentally observed Kapitza resistance between solid and liquid helium above 0.1 K is much smaller than that calculated by the AMM, leading to a so-called anomalous Kapitza resistance or conductance. In contrast, results for thermal phonons below 0.1 K or for ultrasonic waves below 10 GHz at all temperatures gave consistently good agreement with AMM for all cases studied. This situation led to the belief that there was an extra heat transfer mechanism acting in parallel for sufficiently high-frequency phonons and considerable effort was put toward its discovery. The effect was also studied by thermal phonon reflectivity experiments, which also suggested an anomalously low phonon reflectivity between ordinary solids (e.g., quartz or sapphire) and condensed media exhibiting quantum effects. There is, however, convincing experimental evidence that the "anomalous" Kapitza resistance is in fact due to thin layers of imperfections at the interface, which come into play at sufficiently high phonon frequencies and are basically invisible at long wavelengths. The picosecond ultrasonic technique presents an interesting alternative approach to the problem since it can be used over a wide frequency range (10 to 700 GHz) and in principle there are no restrictions in temperature. The work thus far reported can be divided naturally into solid-solid and solid-liquid interfaces.

a. Solid-solid interfaces

In contrast to the case for liquid helium interfaces, all of the work up to now has shown good agreement between experiment and AMM [222 and 223]. This work has covered frequency variations over the full thermal phonon range up to room temperature and temperatures from 0 K up to almost room temperature. A detailed review of this work has been given in [224]. In the picosecond ultrasonic experiments, films of Al, Ti, Au, and Pb were deposited on substrates of diamond, sapphire, and BaF_2 [225]. Optically induced ultrasonic waves were excited and detected on the front surface in the usual way. Theoretical AMM curves as a function of temperature were calculated taking into account phonon dispersion and density of states in the metals. Globally, it was found that the results for Al and Ti on diamond and sapphire were in reasonable agreement with theory, but for Au and Pb films on diamond and sapphire, the measured conductances were significantly higher than the calculated values.

After consideration of various effects such as electron phonon interaction, interface quality, etc., it was concluded that the discrepancy was probably due to anharmonicity in the metal films.

b. Solid-liquid interfaces [226]

In this case, a transparent dielectric layer deposited on Al transducing films on the substrate acted as the medium forming the interface with the liquid. This served to protect the Al film from the liquid and facilitated the technical analysis of the results. A 200-fs light pulse was used, allowing frequency variations from 100 to 300 GHz at 300 K. The dielectric films were layers of Si_3N_4 or SiO_2.

In a first set of experiments, ethylene glycol was used as the liquid. It was found that the reflection coefficient was slightly lower than that predicted by the AMM model. Velocity dispersion in the liquid was discounted as a cause of the discrepancy, which was felt to be more likely due to modification of the liquid properties near the interface. Measurements were carried out on interfaces with liquid argon and nitrogen in a second series and found to be in good agreement with the AMM.

3. Other effects

Picosecond ultrasonics has also been used for a number of other studies, including electron diffusion in metals [227] and localized phonon surface modes in superlattices [228]. It would appear that picosecond ultrasonics is an emerging, powerful technique for the study of physical acoustics, particularly of microstructures. The technique has now become commercialized [229] and is being used routinely for the characterization of thin films in the microelectronics industry.

16.4 Air-Coupled Ultrasonics

In the work covered up to now, ultrasonic transducers have been coupled to the propagation medium either by direct bonding or by fluid, usually water, coupling. Ideally, one would want to excite ultrasonic waves in the sample by a noncontact method. Laser generation and electromagnetic transducers (EMATs) for metals are two ways of doing this, and they would work even *in vacuo*. An alternative approach would be to use the ambient air itself as a coupling medium and this is the subject of the present section [230].

Especially for NDE, water has many advantages as a coupling medium. Of course, it is available everywhere, it has low sound velocity and very low attenuation at low frequencies, and it is compatible with most materials. It can be used in "water-immersion" or "water-squirt" configurations. There are, however, some disadvantages; water does damage some materials, such as paper, some foams, chemicals, and some materials for electronics and aerospace. It can also fill the pores of porous materials and change their acoustic properties. Air as a coupling medium has a similar list of desirable attributes, mainly its universality and its compatibility with most industrial processes. The one big and challenging disadvantage is that air is very badly acoustically mismatched with almost all industrial and transducer materials. In this section, we look at several traditional and more recent innovative approaches to this subject to overcome this difficulty.

Some of the problems encountered in air-coupled ultrasonics can be seen by considering emission from a flat PZT transducer, which for our purposes will be operated from a few hundred kilohertz up to a few megahertz. The transducer will generally be used in resonant mode to take advantage of the high Q, hence the bandwidth will be narrow. Often a quarter wavelength matching layer with low acoustic impedance will be used on the front surface although use of such layers will be restricted by the low Q of many otherwise suitable materials. At least, on the transmitter side, the large acoustic mismatch with air can be partly compensated by using very high electrical input peak powers (up to about 10 kW). An alternative approach is to use the transducer in wide band mode, which will give rise to better pulse response. In this case, the matching layer on the front surface is deleted and a backing is used to broaden the resonance.

Focused transducers turn out to be of more interest than flat ones as they can be used for C scan imaging, and they provide excellent signal to noise in the range 100 kHz to 2 MHz. Possible configurations include using a flat transducer and a shaped plastic element or using a shaped piezoelectric ceramic or composite transducer to focus the acoustic waves directly. Considerations for matching and/or backing layers are the same as for the flat transducers. Some special applications of focused air-coupled transducers have been made in the high-frequency end and these will be described briefly.

Wickramasinghe and Petts [231] described a gas-coupled acoustic microscope with gas pressurized up to 40 kb to decrease the acoustic mismatch difference between the lens surface and the gas. This work was followed by experimental studies at 2.25 MHz in pulse echo C scan mode using argon gas at 30 atm [232]. Acoustic microscopy has also been carried out in air at standard atmospheric pressure [233], which is a significant simplification. The lens was a spherically shaped PZT-5H element with an RTV quarter wavelength matching layer operating at 2 MHz; this gives a spatial resolution equivalent to that for a 9-MHz system operating in water. The two-way insertion loss was 50 dB (excluding air losses) with a 10% bandwidth, which allowed use of tone bursts containing about 10 cycles. Phase and amplitude images were recorded. Since the reflectivity at an air-solid interface is very close to unity, this instrument finds its main application as a profilometer. This was demonstrated quantitatively by precise phase measurements of the step height of a 7.5-μm aluminum film on a quartz substrate. For the F2 lens at 2 MHz, quantitative measurements could be made with a height resolution of 0.1 μm rms and a transverse resolution of 400 μm. Qualitatively, the topographical nature of the imaging was demonstrated by images of a Lincoln penny. Thus the instrument was shown to be a noncontact surface profilometer that is basically insensitive to the material properties. Another recent air acoustic microscopy development was the air-coupled line-focused capacitive transducer, using a transducer principle to be described below [234]. The system operated in the range 200 to 900 kHz with a focal line width of about 0.67 mm. Imaging was carried out using two such transducers at right angles, which defined an image point at their intersection. As above, operation of the system was demonstrated for step height measurements and surface topography imaging.

Considerable activity took place in the 1990s on the development of microelectromechanical (MEMS) applications to ultrasonic sensors and actuators. We focus attention here on MEMS-based transducers for generation and detection of ultrasonic waves. Historically, the generic system that served as the basis for the later MEMS devices is the condenser microphone, which generally operates below 100 kHz. In this case, a steel membrane is stretched over a solid dielectric backplate with air gaps. The restoring force is provided by the tension in the membrane. The micromachined version for operation at higher frequencies was developed by Schindel et al. [235]. In this case, the silicon backplate has a series of etched holes that provide the restoring force. The etched surface is coated with gold to provide a conducting backplate. The membrane is a metallized insulating film that provides an improved impedance match to air. The emitted ultrasonic wave fields have been mapped for plane piston, annulus, and zone plate configurations, and the results are in good agreement with theory. The devices have been shown to have a wide bandwidth and to be operable above 2 MHz in air.

A different approach using surface micromachining has been adopted by Khuri-Yakub and coworkers, to obtain capacitive micromachined ultrasonic transducers (cMUTs) [236]. A thin silicon nitride membrane on silicon nitride supports is micromachined from a silicon wafer substrate. An aluminum top electrode is deposited on the membrane to form the top plate of a capacitor, the substrate acting as the bottom plate. The advantage of this geometry is that the membrane and the cavity air gap depth can be very thin (submicron) leading to very high capacitances and electric fields. This has been shown explicitly by an analysis with the Mason model, which shows that the transformer ratio n is given by

$$n = V_{DC}\frac{\varepsilon_0\varepsilon^2 S}{(\varepsilon_c l_t + \varepsilon l_a)^2} \tag{16.10}$$

where

V_{DC} is the DC bias of the electrode
ε_0 is the dielectric constant of free space

ε is the dielectric constant of the membrane material
l_t is the membrane thickness
l_a is the air gap thickness
S is the area of the transducer

Thus the transformer ratio is the product of the DC electric field and the unbiased capacitance. The capacitance can be several tens of farads and the electric field of the order of 10^8 V/m. This leads to large changes in capacitance during operation and hence very high sensitivity.

The behavior of the transducer in air and in water is quite different. In air, as DC bias is applied, the transducer has a resonance and the impedance of the transducer at resonance is comparable to that of air. The dynamic range in air is of the order of 50 dB higher than that of piezoelectric transducers, a huge increase in performance that permits much better sensitivity and operation at higher frequencies. For immersion in water, the impedance of the membrane is much smaller than that of water and can be neglected. There is no resonance and broadband operation can be assured by appropriate impedance matching. A detailed comparison with a PZT transducer for imaging applications in an underwater camera with a center frequency of 3 MHz and a bandwidth of 0.75 MHz has been carried out. It was found that the dynamic range is comparable in both cases while the cMUT system has a much wider bandwidth.

It has been shown that cMUTs have significantly improved performance compared to piezoelectric transducers for air and immersion applications. They can also be used in arrays for imaging applications and work is ongoing in this direction. Applications of air-coupled ultrasonics is a rapidly expanding field, and we can expect further developments in transducers, instrumentation, and measurement techniques. There will also undoubtedly be increased use of hybrid techniques for NDE with various transmitter/receiver combinations chosen among piezoelectric, laser, EMAT, magnetorestrictive, capacitive, air coupled, etc., where the particular advantages of each type of transducer can be exploited. Some of the principal issues that need to be addressed have been highlighted by Hayward [237]. These include transducer development, transducer noise, electronics, and the gas propagation channel. There may be a tendency to overlook the latter, but the same characteristics are required of the gas propagation as of a good buffer rod: stability, homogeneity, low attenuation, low diffraction, constant temperature, etc. It was shown in [237] that for air-coupled transducers, varying the sample temperature can have a drastic effect on the echoes in the gas column, and that for a 100°C increase they became buried in the noise. One possible solution is to use a gas jet, analogous to the water-squirt [238], to provide a homogeneous propagation channel. This solution has recently been explored by Hutchins et al. [175]. As described in [230], the type of air-coupled transducer described in this chapter can readily be fitted into a standard C scan system to provide reflection, through transmission, leaky Lamb wave measurements, etc. Some applications include examination of delaminations in fiber composites, defects in solar panels, pipeline wall thickness variations, determination of elastic moduli properties, etc. Two representative examples, NDE of gas pipelines and quality testing of paper, will be described briefly.

Since gas pipelines are buried underground, an inside-the-line automated test system for inspection of pipeline walls for defects associated with effects such as stress corrosion cracks is highly desirable. One challenge is that natural gas has a very low acoustic impedance. This is partially offset by the fact that the line is pressurized to about 70 b. A gas-coupled ultrasonic system has proven to be satisfactory for such applications [239]. A second application involved the use of capacitive transducers for testing paper [240]. Transmission experiments at 1 MHz were carried out, enabling the observation of thickness resonances, which could be correlated with paper thickness and moisture content. The possibility of imaging structure was also demonstrated.

16.5 Resonant Ultrasound Spectroscopy

Most ultrasonic phenomena described in this book would normally be carried out in the laboratory or field by the pulse echo method. This gives a direct measure of the sound velocity for the acoustic mode selected; by knowing the density one can infer the corresponding elastic constant. This information is useful for obtaining thermodynamic information on materials and to study basic properties such as phase transitions or for NDE. A given material can be studied completely and all of the elastic constants determined as outlined in Chapter 12 by studying either longitudinal or transverse waves in differently oriented samples of the material. This is a tedious business and such studies are certainly not done routinely in most ultrasonic laboratories. Recently, a new method has been developed, resonant ultrasound spectroscopy (RUS), whereby all of these elastic constants can be determined by one experiment on a single sample.

RUS is a conceptually simple technique that is nevertheless potentially very powerful. A sample of arbitrary shape is excited, usually by a piezoelectric transducer, and it exhibits a very large number of shape resonances. For most simple shapes, modern computing power is sufficient to enable the resonances to be calculated (the forward problem) and comparison with experiment then yields the elastic constants (inverse problem). Since there are now conditions on shape, size, surface condition, or orientation, the method is much more flexible than conventional ultrasonic techniques. In what follows the technique will be described. A number of applications will be covered, particularly in physical acoustics and NDE. A comprehensive review has been given by Maynard [241] and a detailed account is given in the book by Migliori and Sarrao [242].

Much of the early work in RUS was done in the earth sciences, principally in using the earth's oscillations to determine the earth's structure, as well as in the determination of the elastic moduli of the materials making up the earth's crust and believed to make up the interior. One of the more spectacular applications was by Schrieber and Anderson [243], who measured spherical lunar rocks by RUS and determined their elastic constants. In an engaging analysis, the authors showed that lunar rock moduli are surprisingly close to those of green (and other) cheese and made a parody of purely empirical, statistical studies to show that this demonstrated the moon was made out of green cheese! Following this success, other workers adopted the technique to rectangular parallelepipeds of anisotropic material. Extensions of this were soon made to other shapes and also to smaller samples, so as to apply the method to the high-temperature superconductors, which initially were only available in the form of small, irregularly shaped samples.

The forward problem involves using a set of known elastic moduli c_{ijkl}, to determine the complete set of resonant frequencies f_n, for an arbitrarily shaped specimen. While finite element analysis is potentially applicable, the most successful approach uses a Lagrangian minimization procedure, which gives a three-dimensional differential equation for the displacement and the stress-free boundary conditions. The displacement can be approximated by a linear combination of basis functions, the latter depending on the sample geometry. Legendre polynomials can be used for the uniform rectangular parallelepipeds considered in the early work; more generally, basis functions of the form $x^l y^m z^n$ can be used to describe almost all sample shapes.

The inverse problem is challenging and can be facilitated if the direct problem is simple. The most straightforward approach is to start out with a good set of initial elastic constants and carry out successive iterations until the process converges. A Levenberg-Marquadt scheme has been used successfully for inversion calculations.

The measurement technique should approximate ideal conditions as closely as possible. This mainly involves supporting the sample lightly between two piezoelectric transducers such that stress-free boundary conditions are respected. One transducer is used to set the sample into vibration and the other is used to detect the amplitude and phase of the response. The electronics can be adapted to the problem, so that CW phase-sensitive detection techniques can be used and the receiver bandwidth can be chosen to roughly match that of the resonances, which can then be recorded by sweeping the frequency. Four types of sample holder will be mentioned. The first involves the use of copolymer PVDF transducers, which have such a low Q that there is little risk in confusing transducer resonances with those of the sample. The sample is supported at its corners and no bonding agent is used. In a second approach, if a higher coupling factor transducer such as lithium niobate is used, low-frequency resonances of the transducer are suppressed by fixing the transducer on a high-velocity backing such as diamond. Metallized polymer sheets can be used to provide the electrical connections to the lithium niobate as described in [242]. A third configuration involves supporting the sample on alumina buffer rods, so high temperature measurements can be made up to almost 2000 K. Finally, RUS can also be adapted to a standard cryostat as described in [242] so that measurements can be made down to very low temperatures.

Applications of RUS, apart from in the earth sciences, have so far been carried out in two main areas: physical acoustics and NDE. The work in physical acoustics has mainly been involved in second-order phase transitions, which may be described by the Landau theory [244]. The Landau theory describes the transition in terms of an order parameter that is zero in the high-temperature symmetric phase and goes to a finite value below the transition where the system is unsymmetric. Landau wrote the thermodynamic free energy F as an expansion in the order parameter q. At a given temperature, the stable state is determined by the condition $\partial F/\partial q = 0$. One common type of phase transition that can be described in this way is the structural phase transition, in which a crystal changes symmetry when it is cooled below a transition temperature. One example is $SrTiO_3$, which undergoes a transition from cubic to tetragonal at 105 K. The Landau analysis predicts that the elastic constant involved undergoes a steplike decrease on cooling through the transition, which has been confirmed by RUS and conventional ultrasonic measurements. Another example of great interest in physical acoustics is the study of the high-temperature cuprate superconductors. When this remarkable family of high-temperature superconductors was discovered in the mid-1980s only very small samples were available, which made it difficult if not impossible to study them by conventional ultrasonic techniques. At the same time, as mentioned in Chapter 1, it was known that the conventional metallic superconductors have an interesting, characteristic response to ultrasonic waves, so it was felt that ultrasonics would be a valuable tool to study cuprate superconductors and help identify the physical mechanism involved. RUS was used successfully for a number of these superconductors, where anomalies were observed at the transition temperatures. As final examples for physical acoustics, RUS has been useful in detecting anomalies at magnetic transitions, in heavy Fermion antiferromagnetic transitions, and for characterizing quasicrystals.

Conceptually, it is easy to see how RUS could be a useful technique in NDE. In a typical sample, the RUS resonances are highly degenerate due to the symmetry. Introduction of a defect such as a crack breaks the symmetry locally and hence partially reduces the symmetry of the crystal as a whole. This will lead, for example, to the splitting of a resonant peak, and generally the size of the splitting is proportional to the size of the defect. This will only work for degenerate modes; for example, torsional modes are nondegenerate so the effect does not occur. Similarly, if a defect is in a region under strain it will reduce the stiffness constant locally, hence the resonant frequencies will be lowered. So the decrease of f_n of those modes affected by the crack provide another potential NDE tool. Of course, RUS is a laboratory NDE technique and is not adaptable to rough-and-ready field testing.

An area of some potential for RUS is that of metrology of nominally identical industrial parts. RUS can be used to determine weight, density, and size of parts; for this type of measurement it is best to use the lowest resonances, which are sensitive to the sample dimensions. This approach can also be combined with testing for flaws. Some examples include:

1. Detection of small cracks in steel roller bearings of dimensions as small as $1 \times 1 \times 300\ \mu m$
2. Length variations of nominally similar parts, with an accuracy of $\pm 5\ \mu m$ for linear pieces or the diameter of spheres
3. Mass of ceramic parts and the detection of chips and cracks

Apart from the small degree of field use for which it can be used, RUS also has a number of constraints insofar as the type of sample that can be investigated. As pointed out in [242], some limitations are listed here:

1. Samples of small size and weight. Sample size governs the f_n and if the sample is too large the resonant frequencies may be uncomfortably low. Of course, the measurement technique can always be adapted; the largest sample measured so far is a bridge across the Rio Grande River! Weight is also a constraint on the supports and in providing no-stress boundary conditions. And as with all methods in NDE, it becomes progressively more difficult to detect smaller defects in large samples.
2. The nonuniqueness of the response. A sample to be studied by RUS must be very well understood. In general, there are many possible causes of a frequency shift, for example, length change, change in elastic constants, homogeneity, or presence of a defect. Parameters must be tightly controlled to identify unambiguously the origin of the change.
3. In general simple shapes are better, as they are more easily calculable and they are highly degenerate, so that the presence of defects lifts the degeneracy.
4. High Q samples are preferable, as a low Q broadens the resonances and reduces the sensitivity.

17

Cavitation and Sonoluminescence

Cavitation, the rupture of liquids and its associated effects, is a much more general phenomenon than that caused by the propagation of an intense ultrasonic wave in a liquid. It can be engendered hydrodynamically (ship's propellers, turbines, etc.), by absorption of a laser beam, or by the passage of elementary particles in a liquid, among other possibilities. Indeed, the subject became of interest to the British Royal Navy in the late 19th century due to rapid propeller erosion of its warships. The importance of the damage ultimately led to the general study of the implosion of a liquid in an empty spherical cavity carried out by Lord Rayleigh in 1917. However, we are particularly interested in acoustic cavitation here not only because of its intrinsic interest as an acoustic phenomenon in its own right but also because of its present and potential applications. These are due, in part and principle, to the controlled erosion of nearby surfaces caused by collapsing bubbles, leading to ultrasonic cleaning, machining, etc. Other applications are in the medical area (hypothermia, lithotripsy, and the associated dosimetry concerns), sonochemistry, emulsification, etc. The actual mechanism is still incompletely understood, and in different cases almost certainly involves shock waves, imploding liquid jets, and the high temperatures and pressures associated with bubble collapse. The effect is demonstrably efficient; in some cases, one single bubble collapse is sufficient to create a deep cavitation pit.

17.1 Bubble Dynamics

17.1.1 Quasistatic Bubble Description

For a bubble of radius R_0 in a liquid of surface tension σ, the pressure inside the bubble is

$$p_i = p_0 + \frac{2\sigma}{R_0}, \tag{17.1}$$

where the hydrostatic pressure equals the pressure far from the bubble and also that in the liquid just outside the bubble. If now we take into account the vapor pressure p_v and quasistatically change the pressure in the liquid, for example, by an ultrasonic wave, we have [3]

$$p_L = \left(p_0 + \frac{2\sigma}{R_0} - p_v\right)\left(\frac{R_0}{R}\right)^{3\kappa} + p_v - \frac{2\sigma}{R} \quad (17.2)$$

where κ is the polytropic coefficient and R_0 the equilibrium radius.

This is a new condition of equilibrium. If the pressure is increased the bubble will be smaller but stable. Likewise, if it is decreased the bubble will become larger but again stable. If $P_L < 0$ and the bubble is large enough that the internal pressure can overcome surface tension, then the bubble will grow explosively; the threshold pressure at which this occurs is the Blake threshold pressure. Since the present treatment is quasistatic, it cannot describe the subsequent bubble wall evolution, which will be carried out in a later section with the Rayleigh-Plesset equation. Since the evolution at the threshold will be rapid, this justifies neglect of such effects as buoyancy and dissolution in the above discussion.

The critical radius is determined by putting $dp_L/dR = 0$ in Equation 17.2. For isothermal conditions, this gives

$$R_{crit} = \sqrt{\frac{3R_0^3}{2\sigma}\left(p_0 + \frac{2\sigma}{R_0} - p_v\right)} \quad (17.3)$$

Using this value of R_{crit} in Equation 17.2, we can find the critical value of p_L. It is customary to express this threshold by $p_L = p_0 - p_B$ where p_B is the Blake threshold. This gives [3]

$$p_L = p_0 - p_B = p_v - \frac{4\sigma}{3R_{crit}} = p_v - \frac{4\sigma}{3}\sqrt{\frac{2\sigma}{3\left(p_0 + \frac{2\sigma}{R_0} - p_v\right)}} \quad (17.4)$$

when surface tension dominates, the usual case for small bubbles,

$$p_B = p_0 + 0.77\frac{\sigma}{R_0} \quad (17.5)$$

17.1.2 Bubble Dynamics

The starting point for the calculation of the bubble dynamics is the Rayleigh-Plesset (RP) equation. This is the dynamical description of an isolated spherical bubble in an incompressible liquid with surface tension σ and viscosity η. The hydrostatic pressure is p_0 and the applied (acoustic) pressure $p_a(t) = p_{a0}\sin\omega t$. Far from the bubble, the pressure in the liquid is $p_\infty = p_0 + p_a(t)$. The derivation of the RP equation and other aspects of bubble dynamics have been described in great detail by Leighton [3] and some of the main points affecting the acoustic properties of bubbles will be summarized here.

An applied sound pressure leads to a new value of the bubble radius $R(t)$. Leighton shows that the kinetic energy acquired by the liquid in this process is $(1/2)\rho\int_R^\infty \dot{r}^2 4\pi r^2 dr$ and using $\dot{r}/\dot{R} = R^2/r^2$ and integrating, this gives for the increase in liquid kinetic energy $2\pi\rho R^3\dot{R}^2$. Equating this to the work done by p_∞ far from the bubble and the pressure p_L in the liquid near the bubble wall

$$\int_{R_0}^{R} (p_L - p_\infty) 4\pi R^2 dR = 2\pi\rho R^3 \dot{R}^2 \tag{17.6}$$

Using Equation 17.2 and adding a viscous term, we finally have the full RP equation

$$R\ddot{R} + \frac{3\dot{R}^2}{2} = \frac{1}{\rho}\left[\left(p_0 + \frac{2\sigma}{R_0} - p_v\right)\left(\frac{R_0}{R}\right)^{3\kappa} + p_v - \frac{2\sigma}{R} - \frac{4\eta\dot{R}}{R} - p_0 - p_a(t)\right] \tag{17.7}$$

This intimidating differential equation is nonlinear and can be integrated in prescribed conditions to give the time-dependent bubble radius $R(t)$. For sufficiently small amplitudes, the nonlinear terms can be dropped and the equation becomes that of a linear oscillator. In this case, writing

$$R(t) = R_0 + R_\varepsilon(t) \quad \text{with } R_\varepsilon \ll R_0$$

$$\ddot{R}_\varepsilon + \omega_0^2 R_\varepsilon = \frac{p_a}{\rho R_0} e^{j\omega t} \tag{17.8}$$

the equation for a driven harmonic oscillator with resonance frequency ω_0. The value of ω_r (Minnaert frequency) determined by the RP equation is

$$\omega_r = \frac{1}{R_0\sqrt{\rho}}\sqrt{3\kappa\left(p_0 + \frac{2\sigma}{R_0} - p_v\right) - \frac{2\sigma}{R_0} + p_v - \frac{4\eta^2}{\rho R_0^2}} \tag{17.9}$$

For large bubbles where the surface tension can be neglected

$$\omega_r = \frac{1}{R_0}\sqrt{\frac{3\kappa p_0}{\rho}} \tag{17.10}$$

For large air bubbles in water, this reduces to

$$f_0 R_0 \approx 3(Hz - m). \tag{17.11}$$

For small bubbles, where surface tension dominates

$$\omega_0^2 = \frac{2\sigma}{\rho R_0^3}(3\kappa - 1) \tag{17.12}$$

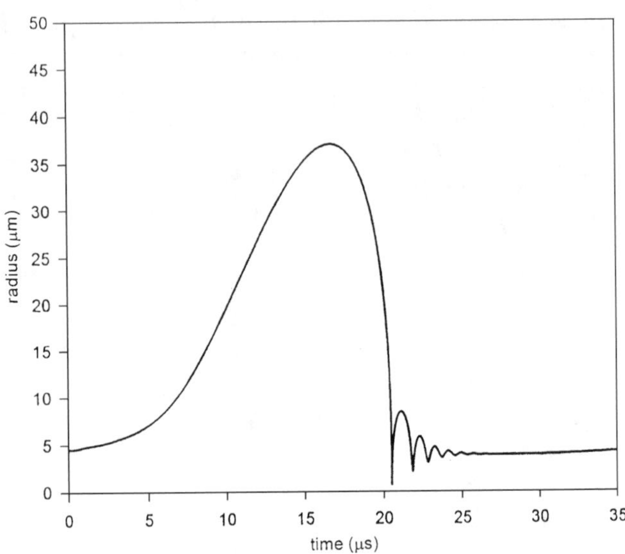

FIGURE 17.1
The radius of an air bubble in water trapped in an acoustic field with p_a = 1.275 atm, R_0 = 4.5 μm, and f_a = 26.5 kHz. The calculation is based on Equation 17.7, courtesy of Haizhong Lin. (From Cheeke, J.D.N., *Can. J. Phys.*, 75, 77, 1997. With permission.)

Solutions to the RP equation can be obtained numerically for $R(t)$ and a variety of results can be obtained depending on the precise values of ω_0, R_0, and p_a that are used. An example is shown in Figure 17.1. The response is nonlinear but in the steady state, the solutions are periodic and relatively permanent. In this case the oscillations are known as stable cavitation. In the opposite limit, the bubbles group by a factor of two or more per cycle and then collapse violently and disintegrate before or near the end of the cycle. This is known as transient cavitation. Both of these conditions can be reproduced numerically from the RP equation, which has been described in detail by Lauterborn [245], Neppiras [247], Walton and Reynolds [246], and Leighton [3]. The model has been extended to take into account damping by sound radiation [248] and further refinements have been added by Prosperetti et al. [249].

17.1.2.1 Bjerknes Forces

The RP equation shows that the bubble carries out damped-driven nonlinear motion under the influence of an applied acoustic pressure. At low amplitudes, it will act like a linear oscillator and we will apply the results of the variation of the phase of such an oscillator in this section.

We know that a bubble of volume V will be subjected to the time-averaged radiation pressure $\langle -V\vec{\nabla}_p \rangle$. We suppose that the bubble is at position x in a standing wave field of the form

$$p(x, t) = p_0 + 2p_{a0}\sin kx \cos \omega_a t \tag{17.13}$$

For $2p_{a0} \ll p_0$, the bubble will undergo harmonic oscillations of the form

$$R(t) = R_0 + \varepsilon_0 \cos(\omega_a t + \alpha)$$

where α is the phase angle with the applied pressure. The volume of the bubble is

$$V(t) = V_0\left(1 + \frac{3\varepsilon_0}{R_0}\cos(\omega_a t + \alpha)\right) \tag{17.14}$$

where $V_0 = (4/3)\pi R_0^3$ is the equilibrium volume. For the driven oscillator, bubbles larger than resonant size are π out of phase with the driving pressure ($\alpha = \pi$) and smaller bubbles are in phase ($\alpha = 0$). Using these results for α with Equations 17.13 and 17.14, we obtain

$$F_x = -3p_a k\varepsilon_0 V_0 |\cos kx| R_0 \tag{17.15}$$

for large bubbles, and

$$F_x = \frac{3p_a k\varepsilon_0 V_0 |\cos kx|}{R_0} \tag{17.16}$$

for small bubbles.

Comparing these results to the form of the pressure field leads to the conclusion that bubbles larger than the resonant value are subjected to a force from a pressure antinode to a pressure node. Conversely, small bubbles are pushed toward a pressure antinode. The force responsible is called the primary Bjerknes force. According to this picture, small bubbles experience a force pushing them to the pressure antinode where they grow under the influence of the ultrasonic field to resonant size, at which point they will eventually be pushed back to a pressure node and ultimately dissolve or disintegrate. Of course, Bjerknes forces are only one piece of the puzzle and bubble growth by rectified diffusion (to be discussed in the next section) is also important.

Bjerknes forces exist not only in standing waves but in any sound field where there is a pressure gradient. Thus, for example, small bubbles will migrate to the pressure antinode in a focused ultrasonic field. In this and other geometries, the bubbles can agglomerate to form "streamers." Finally, there are also secondary Bjerknes forces between two bubbles or a bubble and its image at a boundary. Detailed calculations show that if two bubble oscillations are in phase the bubbles attract while the force is repulsive for bubbles in antiphase. A very full and complete treatment has been given by Leighton [3].

17.1.2.2 Rectified Diffusion

We suppose a state of stable cavitation. Then the gas content in the bubble can change due to the variation in diffusion conditions over a cycle, leading to a net pumping of gas dissolved in the liquid into the bubble. As described by Leighton [3], there are two components:

1. Area effect. The basic idea is very simple. If $R < R_0$, then the gas inside the bubble is compressed to concentrations greater than the equilibrium value so there is a net flow of gas out of the bubble. By the same token, when the bubble radius is larger than the equilibrium value, the gas concentration in the bubble is decreased and there is a flow of gas into the bubble. These flows do not cancel; although the relevant fractions of a cycle for the two processes are roughly comparable, the area is much bigger in the latter case, leading to a net inflow of gas atoms into the bubble over a cycle.
2. Shell effect. The diffusion rate of gas in a liquid varies as the concentration gradient. We consider two thin liquid shells around the bubble. When the bubble expands the distance between shells contracts, so the concentration gradient increases, leading to a high rate of diffusion in the liquid. Conversely, when the bubble contracts the diffusion rate is diminished due to a lower concentration gradient.

The area and the shell effect work together to increase the net inflow of gas into the bubble from the liquid. This one-way pumping effect is called rectified diffusion. For bubbles above a certain minimum size there is a threshold for rectified diffusion in the acoustic field, so that the bubble grows progressively up to the resonant size. If the acoustic wave amplitude is further increased, the bubble will encounter another threshold for transient collapse. Rectified diffusion plays a central role in bubble growth in an acoustic field and is particularly important in the behavior of trapped single bubbles, to be discussed later.

17.1.3 Acoustic Emission

Oscillating bubbles in a liquid act as a point source of sound. At low drive levels, the bubble motion is linear and acoustic emission occurs at the fundamental frequency. As the acoustic power is increased, harmonics appear in the stable cavitation regime. In the early part of the transient cavitation regime, subharmonics and ultraharmonics (odd harmonics of subharmonics) appear superposed on a background continuum. This white noise gives rise to a very audible, characteristic sound. A better understanding of cavitation noise is important for the extending applications of the phenomenon as well as for a better understanding of the mechanisms of cavitation itself.

Esche [250] was the first to observe subharmonics and their existence has since created considerable interest. The situation up to 1980 has been summarized by Neppiras [247] and more recently by Leighton [3]. Some of the suggested origins of subharmonics are parametric surface vibrations of the bubble, oscillations associated with larger bubbles, parametric amplification in the liquid, two- or three-period transient cavitation, etc. From a practical point of view, it has been established that generation of the subharmonic is not an acceptable indicator of the onset of transient cavitation. On the fundamental side, there has been much recent interest in the subharmonic $f_0/2$ as being a step on the way to chaos in the system.

Lauterborn and Cramer [251] have observed a progressive sequence from period doubling to broadband noise in water with increase of the acoustic power. A characteristic period doubling bifurcation behavior has been simulated by Ilychev et al. [252] and also by Kamath et al. [253], who presented the results on a bifurcation diagram. Lauterborn and Cramer [251] have quantitatively analyzed the experimental power spectrum and found a period doubling route to chaos. After a cascade of period, doubling bifurcations occurred a chaotic noise attractor was found. The calculated Lyapunov exponent for the spectrum was positive, confirming the existence of a chaotic system with a fractal dimension of three. This suggests a highly correlated bubble structure in agreement with observations.

17.1.4 Acoustic Response of Bubbly Liquids

Since bubbles are highly nonlinear acoustically, it is not surprising that a bubbly liquid is a highly dispersive and attenuating medium. The response of the medium is very important in areas such as oceanography and high-power ultrasonics in liquids. Since it involves a large number of bubbles of different sizes in various states of interaction, it is a highly complex subject. In fact, the theory for even a very simplified system is complicated. In the following, we give a brief summary of some of the main features. Recent surveys of the subject have been given by Leighton [3] and Medwin and Clay [254].

If we consider the attenuation of a plane wave incident on a bubble, the total cross-section can be written as the sum of scattering and absorption cross-sections. At resonance these scattering cross-sections are hundreds or thousands of times larger than the geometrical cross-section. Bubbles at resonance are described as presenting a "hole" to the incident wave. This region of low acoustic impedance distorts the incident sound field and sets up a power flow toward the bubble. This huge distortion of the incoming wave corresponds to the extremely high scattering cross-section. Moreover, bubbles larger than the resonant size scatter sound relatively even more, due to a geometrical shadow effect [3].

There is a similar, large effect on the sound velocity of bubbly liquids. In this case, where the velocity is given by $V_0 = \sqrt{K/\rho}$, the origin of the modification of V_0 is not due to the effect on the density but rather that on the compressibility $\chi = 1/K$. The analysis shows a strong variation of V_0 near the resonant frequency of the form

$$\mathrm{Re}(V_P) = V_0\left(1 - \frac{Y^2 - 1}{(Y^2 - 1) + \delta^2} \cdot \frac{3UY^2}{2a^2k_R^2}\right), \quad ka < 1 \tag{17.17}$$

where

$Y = \omega_r/\omega$
$U = N(\frac{4}{3}\pi a^3)$ = void fraction
δ = damping constant
$k_R = \omega_r/V_0$

for N bubbles per unit volume of radius a.

For $f \gg f_r$, $V_p \to V_0$ while for low frequencies

$$V_P = V_0\left(1 - \frac{3U}{2a^2k_R^2}\right), \quad f \ll f_R \tag{17.18}$$

In oceanographic applications, ak_R is a constant, so the low-frequency velocity depends only on the void fraction. This fact has been used to develop an effective medium theory [255] for oceanography. The result is

$$\rho_A = U\rho_b + (1 - U)\rho_\omega \tag{17.19}$$

$$\frac{1}{K_A} = \frac{U}{K_b} + \frac{1 - U}{K_\omega} \tag{17.20}$$

$$V_P = \sqrt{\frac{K_a}{\rho_A}} = \sqrt{\frac{K_b K_\omega}{UK_\omega + (1 - U)K_b[U\rho_b + (1 - U)\rho_\omega]}} \tag{17.21}$$

where ρ_b and ρ_ω are the densities of air in bubbles and of water and K_b and K_ω are the respective bulk moduli.

These results agree with the previous formulation for $U \leq 10^{-5}$.

17.2 Multibubble Sonoluminescence (MBSL)

MBSL was discovered in 1934 by Frenzel and Schultes shortly after the introduction of high-power ultrasonic generator technology. Basically, it is the emission of light by bubbles in a liquid undergoing cavitation. Despite nearly 70 years of investigation, the phenomenon is still not completely understood. One of the difficulties is that most if not all of the experimental parameters are badly defined experimentally. We are dealing with millions of bubbles of varying sizes and phases of collapse in an inhomogeneous, unspecified acoustic field, for a phenomenon where the collapse is very sensitive to bubble size, acoustic pressure, and frequency. Cavitation is also notoriously sensitive to the initial state and history of the liquid. When the parameters have been controlled, it has been a case of either physicists using a well-defined sound field on a poorly characterized liquid, or chemists investigating a superbly specified liquid with an unknown sound field. Nevertheless, the subject is important industrially and it has led to a new area of application (sonochemistry) and a new, controlled experimental configuration, single bubble sonoluminescence (SBSL). Before studying the latter, it behooves us then to become familiar with the principles of bubble dynamics, cavitation, and some of the main results of MBSL. The subject has been reviewed in [246] and more recently in [256], some of which is included in the following.

Following the discovery of SL by Frenzel and Schultes, Zimakov proposed the first explanation for the phenomenon, namely that it was caused by an electrical discharge between the vapor cavities and the glass wall of the container. The first formal theory was put forward by Chambers. At the time liquids were thought to have a quasi-crystalline structure similar to solids, and in this triboluminescent model, it was proposed that SL was similar to the emission of light by many crystals when they are crushed. Levsin and Rzevkin suggested that SL was due to an electric discharge associated with liquid rupture. This idea was extended by Harvey in 1939 with the balloelectric theory, which was based on the collection of electric charge at the liquid-vapor interface, leading to an electrical discharge upon compression of the bubble. An alternative electrical model, the electrical microdischarge theory, was presented by Frenkel in 1940. The model involved statistical fluctuations of charge on the surface of nonspherical cavities, leading to electrical discharge, this time during the expansion phase of the bubble.

Other models followed in quick succession. In the mechanochemical theory in 1939, Weyl and Marboe proposed that molecules were fractured during expansion of the bubble, their radiative recombination giving rise to SL. Griffing ascribed the effect to chemiluminescence; the high temperatures caused by the cavity collapse were supposed to give rise to oxidizing agents such as H_2O_2, which would dissolve in the surrounding liquid, causing chemiluminescent reactions. In 1950, Neppiras and Noltink advanced the hot-spot theory in which the adiabatically compressed gas gave rise to blackbody radiation. Jarman proposed a variant of the hot-spot model by postulating that shock waves were formed inside the bubble and that these lead to SL. Hickling incorporated the thermal conductivity of the gas in the hot-spot model and was able to explain SL results for several gases. Finally, further variants of the electrical discharge model were proposed by Degrois and Balso and by Margulis, but these efforts proved unfruitful.

Most of these early theories were forcefully qualitative and speculative in nature, mainly due to the lack of systematic experimental results obtained in controlled conditions. Most of them were rejected over time as more experiments were carried out; in this context, the resurrection of Jarman's shock wave model is somewhat ironic as it was rejected by the opinion of the day more than 30 years before its now increasing acceptance. Nevertheless, several of these models directly sowed the seeds for further experimental and theoretical work, particularly the hot-spot and chemiluminescent models. The hot-spot theory pointed to the need to attempt a fit to a black-body spectrum and generated serious interest in the measurement of experimental spectra. Gunther et al. showed that the spectra for water-xenon mixtures could be fitted to a black-body spectrum at 600 K for the spectral range 300 to 700 nm. Furthermore, in a related study, Gunther et al. also showed that SL was emitted as sharp flashes lasting about 1/50 of a period and with the same frequency as the acoustic wave, setting the stage for later studies of SBSL.

17.2.1 Summary of Experimental Results

In a typical MBSL experiment, acoustic waves are excited in the liquid of interest by a high-power ultrasonic source. A piezoelectric sonicator (Mason horn), spark coil, or laser are some of the most common sources used. Standing waves, typically near 20 kHz, are set up in the basin and the emitted light is collected either by photographic images or by a spectrometer to measure the visible spectrum. The spectrum obtained depends, of course, on the gas-liquid combination studied. For example, Gunther et al. [257] found that the MBSL spectrum of water saturated with noble gases was a continuum from 300 to 700 nm that could be fitted to a black body for a temperature of about 600 K. Flint and Suslick [258] have made extensive studies of aqueous, hydrocarbon, and halocarbon liquids and have been able to correlate the observed line spectra unambiguously to excited state molecules created during cavitation. Line spectra were also observed by Gunther et al. for gas-saturated salt solutions and these authors identified the metal ion transitions giving rise to the line spectra. In what follows, the total intensity of the sonoluminescence is often used to determine the dependence of MBSL on various parameters of the measurement. Given the uncertain reliability of the data, these will be summarized very briefly so as to give an overview of the phenomenon and a basis for comparison with SBSL.

1. Dissolved gas

   Dissolved gases are present in all liquids. If the liquid is degassed, higher power levels must be used to obtain MBSL. The gas in the bubble is a mixture of dissolved gas and liquid vapor. The physical properties of the gas influence the bubble dynamics; high thermal conductivity reduces the bubble maximum temperature and high compressibility increases it. Chemically, the gas composition strongly affects the free radicals present, hence the sonoluminescence.

2. Liquid properties

   Again, the physical properties of the liquid directly affect bubble dynamics: surface tension, density, and vapor pressure. However, no convincing dependencies of sonoluminescence on these parameters have been confirmed, other than semi-empirical correlations.

3. Liquid history

   It has been known for a long time that the presence of nuclei strongly reduces the liquid tensile strength, and hence lowers the cavitation threshold. The driving intensity and pulse length also affect subsequent cavitation properties.

4. Temperature

   A general tendency of decreasing MBSL with increasing temperature has been observed. However, a critical discussion by Leighton [3] suggests that the evidence provided by the data is unclear in this regard.

5. Hydrostatic pressure

   An increase of hydrostatic pressure has two opposing effects, namely limiting the maximum radius but increasing the compressive forces on collapse. For the total population of bubbles, an increase in p_0 increases the Blake threshold and hence reduces the number of bubbles that can undergo transient cavitation. Experimentally the MBSL intensity typically goes through a maximum with p_0.

6. Acoustic pressure amplitude

   Most results show a linear increase of MBSL intensity with acoustic power, which is expected qualitatively.

7. Acoustic frequency

   There are competing factors as the frequency is increased. The number of antinodes increases although their size decreases, but attenuation reduces the effective power at the antinodes. Moreover, the acoustic field distribution changes with frequency.

The early studies on SL spectra and ideas on chemiluminescence led to a series of systematic studies by Flint and Suslick during the 1980s, which culminated in the first quantitative and systematic studies of SL and an experimental determination of the temperature at the center of the bubble for MBSL systems. They carried out a series of methodical experiments on MBSL for the various molecular species, demonstrating that the observed spectra are consistent with the excitation of various vapor species inside the bubble. Coupled with parallel work described below, this work showed that MBSL is fundamentally a thermal chemiluminescent effect and is not due to electrical discharge or other competing models described earlier.

A quantitative determination of the MBSL emission temperature was made for the first time by Flint and Suslick [258] in 1991. The ro-vibronic spectra of excited states of C_2, from silicone oil were identified and compared with synthetic spectra generated by the Speir method, which is the standard approach for adding a set of overlapping, dense, spectral lines, each of the form

$$I \sim \mathrm{v}^4 AS \exp\left(\frac{-hc}{k}\right)\left[\left(\frac{G}{T_\mathrm{v}}\right)+\left(\frac{F}{T_\mathrm{r}}\right)\right] \tag{17.22}$$

where

- v = energy of the transition in cm^{-1}
- A = the Franck-Condon factor for the vibrational transition
- S = line strength
- G = energy of the vibrational state
- F = energy of the rotational state above the vibrational state
- T_v = vibrational temperature and
- T_r = rotational temperature

The systematic spectra were fitted to experiment using three adjustable parameters: T_v, T_r, and the spectrometer aperture. For thermal equilibrium ($T_\mathrm{v} = T_\mathrm{r}$), this procedure gave a best fit for $T_{SL} = 5075 \pm 156$ K. The identification of well-known spectral lines and the ability to fit them using standard theoretical procedures gives strong credibility to these experimentally deduced temperatures.

17.3 Single Bubble Sonoluminescence (SBSL)

17.3.1 Introduction

Felipe Gaitan joins the select ranks of graduate students who have made major discoveries in the course of their Ph.D. research studies. The precursor to his work was the study by Crum and Reynolds [259] using an ultrasonic horn in a water bath. They set up acoustic standing waves and obtained clear evidence of light emission from bubbles trapped at the pressure antinodes that they associated with stable cavitation. Using a levitation cell, Gaitan et al. [260] were able to trap a single bubble at the pressure antinode, and they observed that it emitted a tiny burst of light once per cycle. They carried out a detailed study of the SBSL associated with the bubble including pulse width and synchronicity.

This work was continued by Barber and Putterman [261], who found that the light pulses were surprisingly sharp and synchronous. The three hallmarks of SBSL emerged from this early work by Gaitan and Barber:

1. Extremely sharp pulses of light, of the order of 50 ps wide.
2. Amazing synchronicity, on a much finer time scale than the phase noise of the ultrasonic source

3. Enormous concentration of energy, of the order of 12 orders of magnitude. It is amazing that very low frequency acoustic waves ($\hbar\omega \sim 10^{-12}$ eV) combine on their own in such a way as to emit photons with an energy of the order of 1 eV.
4. Since that time there has been a flurry of experimental and theoretical work to elucidate the nature of SBSL and, more specifically, identify the mechanism of light emission. It is a happy coincidence of nature that although the parameter space available to the phenomenon is small and very restrictive, this space includes water near room temperature and pressure. An intriguing result is that noble gas atoms appear to be an essential ingredient of the phenomenon. With increasing information on these points and the hallmarks mentioned above, together with new insights, experiment and theory have danced a tantalizing tango to the tune of SBSL throughout the 1990s. Experiment has thrown at least one curve ball to the theorists, obliging them to explain more than they in fact had to. At present, experiment and theory find themselves on a plateau of convergence, so this is a suitable time to review the main developments in the field.

There are a number of reasons why it is interesting and possibly profitable to study SBSL. Some of these, not necessarily in order of importance, include the following:

1. Understand cavitation better by being able to do controlled experiments on a stable system.
2. Create a microchemical laboratory with temperature, pressure, and other parameters controlled by bubble dynamics, i.e., a refined and controlled version of sonochemistry.
3. Explain universal bubble phenomena by comparing the results for those for bubbles created by other techniques.
4. Enrich the study of thermodynamics by the addition of a unique system.
5. Set up one of the best performance/cost physics undergraduate laboratory experiments.
6. Work toward the goal of controlled fusion.

The plan of our discussion of SBSL is as follows. In the next section, we describe the basic experimental setup and the simple physics of the phenomenon. In Section 17.3, we apply the RP equations to the bubble dynamics and describe the parameter determination and fitting. This is followed by a brief summary of the main experimental results, then a more detailed critique of some of the more important features including spectrum, line width, shock waves, pressure effects, and the dissociation hypothesis. Finally, two of the more successful models are described; they share some common assumptions and conclusions and are both able to account for the available experimental facts. At this point we will be well placed to decide if SBSL can be accounted for by known mechanisms, or it is an exotic phenomenon involving radically new physics.

17.3.2 Experimental Setup

A typical experimental setup for SBSL is shown in Figure 17.2. The liquid is contained in a 500-ml flask and the piezoelectric transducer is glued to the sidewall. It is driven by the amplified sinusoidal signal from a function generator. To avoid excessive power dissipation due to standing waves, it is advisable to tune out the transducer capacitance with a series inductance (the author has seen smoke coming out of the BNC connectors of an untuned system). The liquid must be degassed, which can be done by heating and pumping. For an open system the desired gas can be injected at the center by use of a syringe. Once the bubble is stabilized and emitting as described below, the light can be detected by a photomultiplier tube (PMT). In good conditions, the experienced eye can easily see the bubble in dim light conditions. A good account of a simple experimental system is given by Hiller and Barber [262].

Bubble trapping follows the general principles of acoustic levitation. In the case indicated, the transducer frequency is chosen to set up a fundamental standing wave resonance, which has a pressure antinode (velocity node) at the center. Radial nodes are set up along all diameters so that the bubble is trapped at the center. In the vertical direction, the buoyancy force is balanced by the Bjerknes force, so the bubble is actually trapped slightly above center. The same sort of reasoning applies to cylindrical resonators except that the symmetry is cylindrical instead of spherical.

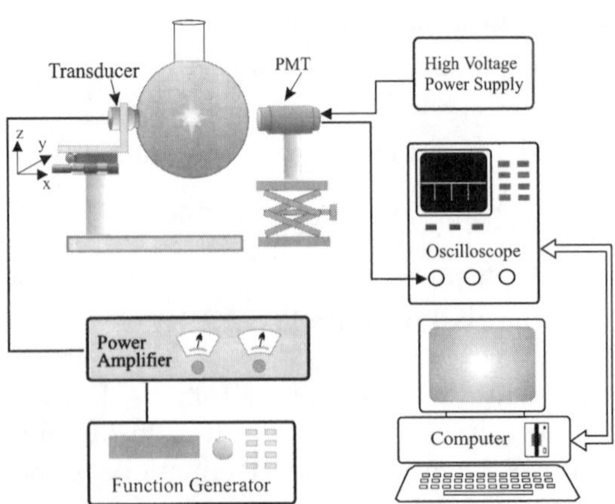

FIGURE 17.2
Experimental setup for single bubble sonoluminescence. The laser and photomultiplier tube (PMT) are used for MIE scattering to determine the bubble radius as a function of time. (From Cheeke, J.D.N., *Can. J. Phys.*, 75, 77, 1997. With permission.)

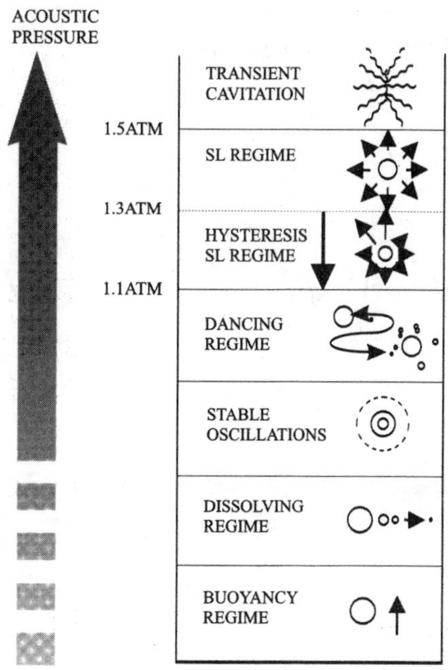

FIGURE 17.3
Various bubble regimes in a stationary acoustic field in a water-glycerin mixture. With increasing pressure for this mixture the SL regime occurs for an acoustic pressure from 1.3 to 1.5 atm. With decreasing pressure there is a hysteresis regime as shown. (From Cheeke, J.D.N., *Can. J. Phys.*, 75, 77, 1997. With permission.)

As the acoustic pressure amplitude p_a is increased, the bubble passes through several well-defined regimes, as shown in Figure 17.3. At very low amplitude, the buoyant force dominates and the bubble floats to the surface. p_a must also be above the dissolving threshold, otherwise the bubble will dissolve in the liquid. Above the trapping threshold, the bubble undergoes stable oscillations near the flask center. As p_a is increased, the bubble goes into the dancing regime, which has been described as representing a shuttlecock motion. At the upper end of this regime the bubble becomes very small and almost disappears. Once it crosses the SBSL threshold, however, it becomes very stable, apparently locked in place, and it glows with a faintly bluish hue. For pure water, this region is roughly $1.1 < p_a < 1.3$ bars. At the upper bound, the bubble drifts off and disappears. If the driving pressure p_a is now lowered, there is hysteresis and SL persists below the original SBSL threshold.

Assuming that the bubble is stable and emitting, we now follow it through one cycle as shown in Figure 17.4. When the acoustic pressure is negative, the bubble expands to its maximum radius (R_{max} ~ 40 μm for f ~ 20 kHz); at this point, there is almost a vacuum inside the bubble. As p_a turns positive, the bubble undergoes a violent collapse since there is no force opposing it. There is a deep overshoot past the equilibrium radius (5 μm) to a minimum

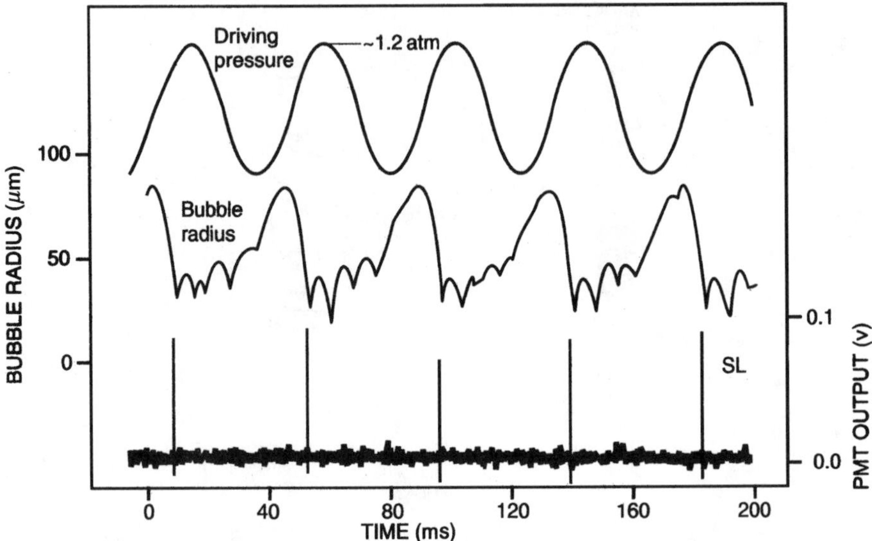

FIGURE 17.4
Simultaneous measurements of the sound field (top), bubble radius (middle), and SBSL (bottom) in a water-glycerin sample at $p_a = 1.2$ atm and $f = 22.3$ kHz. (From Gaitan et al., *J. Acoust. Soc. Am.*, 91, 3166, 1992. With permission.)

radius $R_{min} \sim 0.1$ μm, which is close to the van der Waals limit. This is followed by a series of afterbounces, which occur very approximately at the Minnaert frequency, the free bubble resonance frequency.

The moment of emission of the light pulse occurs almost exactly at the first, deep minimum. This point was a source of some confusion in the early work on MBSL; for SBSL, determination of the precise moment in the cycle of luminous emission is straightforward if correct triggering techniques are used. It has been established unambiguously that the light is emitted just before and almost exactly at the deep minimum, as will be clarified in the later discussion. From cycle to cycle, the emission occurs precisely at this point, leading to the precise synchronicity of the phenomenon. The flashes are rather weak, with 10^7 photons per flash. The narrow width and the continuous spectrum will be discussed later.

In addition to optical emission, there is also a well-defined acoustic emission (ae). The ae pulses can be detected by placing a needle hydrophone a millimeter or two from the bubble. There is a strong ae pulse corresponding to emission at the principal minimum as can be confirmed by time of flight from the bubble to the hydrophone using the velocity of sound in water. There are also subsequent weaker ae pulses that originate from the weaker collapses of the afterbounces. It is difficult to get useful data from the hydrophone measurements [263] due to the finite frequency and temporal response of the hydrophone and the fact that it cannot be placed too close to the bubble to avoid perturbing its motion. Recently, important results on ae have been

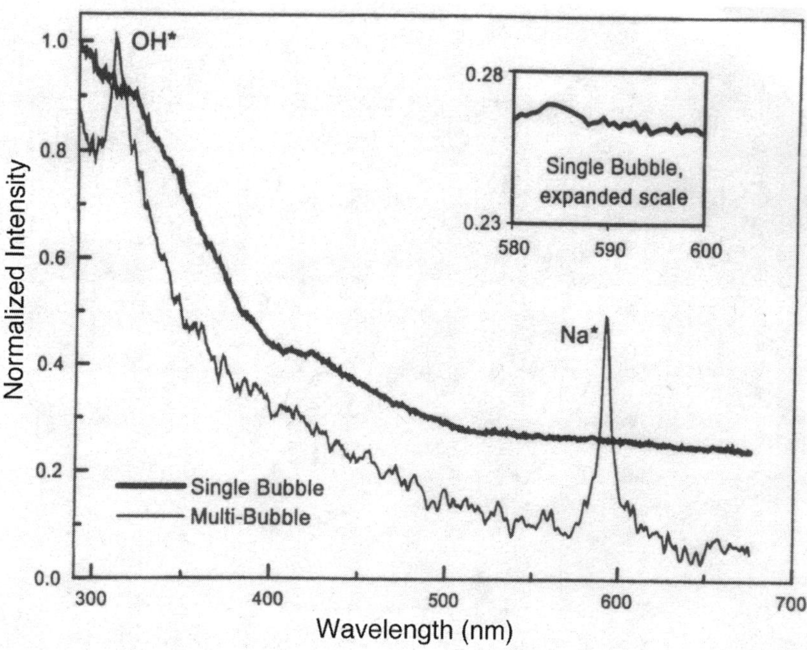

FIGURE 17.5
Comparison of the background-subtracted spectra of MBSL and SBSL in a 0.1 *M* NaCl solution. Each spectrum was normalized to its highest intensity. In MBSL there is a peak for the sodium emission line at 589 nm, which is absent in MBSL. (From Matula, T.J. et al., *Phys. Rev. Lett.*, 75, 2602, 1995. With permission.)

obtained with a streak camera and these will be reported later. It is estimated [264] that at least 99% of the energy radiated from the bubble occurs by acoustic emission, which in fact starts in the dancing regime. The change in the bubble dynamics in the dancing and SBSL regimes is shown in Figure 17.6.

One of the most important pieces of experimental information is the quantitative determination of $R(t)$ by Mie scattering. Mie scattering of a wave by a spherical object provides an exact solution for the scattered intensity is measured over an angular range of 30 to 60° to average out diffraction effects. Fits to the theory at different times in the bubble cycle give the variation $R(t)$. The absolute determination of bubble radius parameters will be discussed next.

17.3.3 Bubble Dynamics

The numerical solutions of the RP equation give the bubble motion $R(t)$, providing that we have at our disposition a thermodynamic model to describe the gas. Since the bubble motion is relatively slow over most of the cycle, the overall shape of the $R(t)$ curve is not sensitive to the details of the

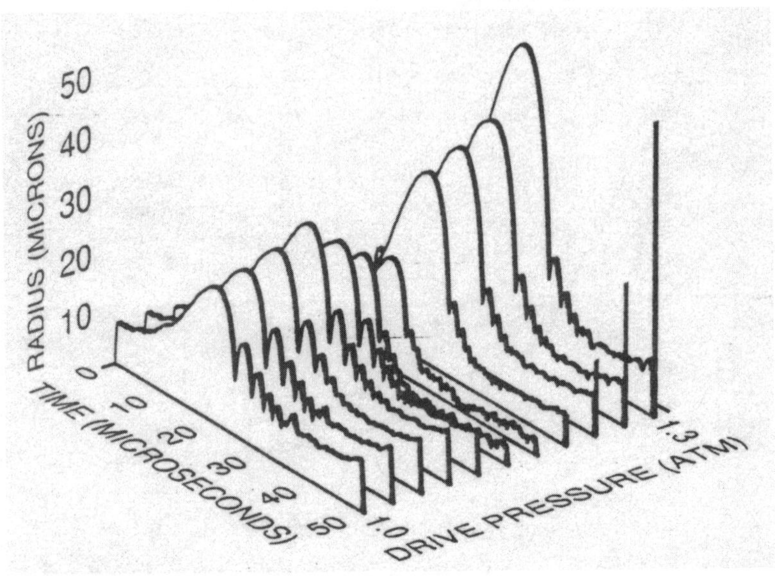

FIGURE 17.6
Bubble radius vs. time for about one cycle of the acoustic field as a function of increasing drive level for an air bubble in water. The relative intensity of the emitted light as a function of drive level is indicated by the vertical lines. (From Barber, B.P. et al., *Phys. Rep.*, 281, 65, 1997. With permission.)

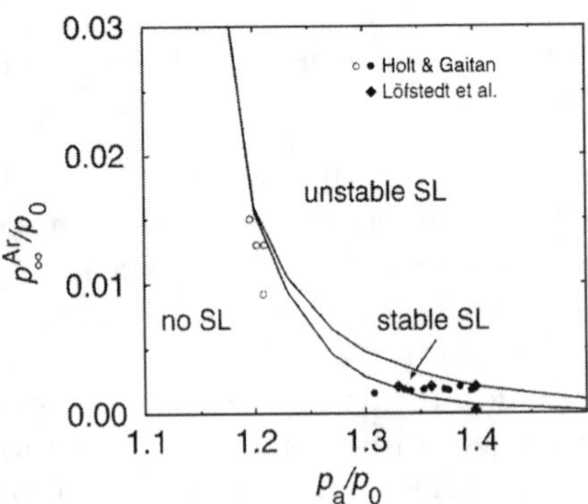

FIGURE 17.7
Phase diagram for pure argon bubbles in the plane p_∞^{Ar}/p_0 vs. p_a/p_0, together with experimental data. The filled symbols are for stable SBSL bubbles and the open symbols are for stable non-SBSL bubbles. (From Lohse, D. et al., *Phys. Rev. Lett.*, 78, 1359, 1997. With permission.)

particular model chosen. However, the choice of model is absolutely critical in the region of collapse, and in the last 100 ps of collapse the hydrodynamic model of the RP equation is no longer valid.

Regarding first the overall shape of the $R(t)$ curve, one of the simplest choices is the adiabatic van der Waals equation, with the assumption that the gas pressure inside the bubble, p_g, is spatially uniform. Then

$$\frac{p_g}{p_0} = \left(\frac{R_0^3 - H^3}{R(t)^3 - H^3}\right)^{\gamma} \tag{17.23}$$

where

$H = R_0(b/v_m)^{1/3}$
b = van der Waals excluded volume
v_m = specific molar volume at STP
R_0 = equilibrium radius of the bubble
$\gamma = C_p/C_v$

An $R(t)$ curve based on such a model is shown in Figure 17.1; it exhibits all the usual features that have already been discussed. A number of variants on the simple adiabatic model have been studied by different authors and each variant shares the common feature of compressional heating and pressure rise during collapse. One treatment that is qualitatively different specifically allows $p_g = p_g(r,t)$ and $T_g = T_g(r,t)$, which allows spatially nonuniform solutions, in particular the excitation of shock waves [265]. It has still not been clearly established whether shock waves exist inside the bubble during collapse, and this point will be discussed separately.

The determination of the bubble parameters in SBSL is a nontrivial question. The important parameters of the liquid (σ, η, p_v) and the chemical composition of the gas depend, evidently, on the particular choice made. The operating frequency is determined by the fundamental resonance of the acoustic trapping cell. In the actual experiment, parameters p_0, T, p_a, and gas concentration c_∞/c_o are chosen by the experimenter. This leaves R_0, which cannot, in fact, be independently controlled but is determined by all of the other parameters and self-consistently by the dynamical equations.

In order to establish the absolute $R(t)$ curve, the parameters R_0 and p_a must be determined. After much study, it is now generally accepted that the most accurate approach is to determine them by a direct fit to the RP equation solutions as outlined by Barber et al. [266]. Since the relative $R(t)$ curve is known by Mie scattering, the absolute values of the other parameters, e.g., R_{max} and R_{min}, follow directly. A key parameter is the expansion ratio R_{max}/R_{min}. It has been established empirically that expansion ratios of the order of ten or greater are needed to produce SBSL. It should also be noted that R_0 and p_a are the two main parameters used to describe the stability conditions for SBSL given in the next section.

17.3.3.1 Bubble Stability

SBSL can only be established if the following four conditions are satisfied [267]:

1. Energy processing

   There must be sufficient energy transfer between the acoustic wave and the bubble, that is, between the moving bubble wall and the enclosed gas. The criterion that appears to satisfy all models is that the bubble wall Mach number be greater than one, i.e.,

$$|M_g| = \left|\frac{\dot{R}}{V_g}\right| \geq 1$$

2. Shape stability

   A sufficiently violent collapse can only be attained by the collapse of spherical bubbles. If the bubble becomes too large (≥10 μm) then various shape oscillations and instabilities are set up that destroy the sphericity.
3. Diffusive stability

   The diffusive time scale is much longer than that for the RP dynamics, but the solutions of the latter control the boundary conditions for the diffusion problem. In $p_a - R_0$ space the stability condition is that $dR_0/dp_a > 0$ for stability. Following the earlier discussion on bubble dynamics, if R_0 is below the equilibrium value the bubble shrinks and dissolves. Above it, the bubble grows by rectified diffusion. If it becomes too large, as in the second condition, then shape instabilities occur and the bubble becomes unstable. While the R_0 vs. p_a curves are useful for theoretical considerations, they are not applicable to the laboratory results as R_0 cannot be controlled directly experimentally. It will be shown shortly that for air bubbles only the argon concentration p_∞^{Ar}/p_0 is relevant. Hence, the stable regions can be identified in the plane p_∞^{Ar}/p_0 vs. p_a as shown in Figure 17.6, which does not involve any fitted parameters. From an experimental point of view SBSL can only occur in a tiny region of parameter space.
4. Chemical stability

   Chemical stability comes into play because of the high temperatures known to exist inside the bubble at collapse, at least of the order of 10000 to 20000 K. Such temperatures will cause the dissociation of the molecular constituents of the gas; for example, for an air bubble, N_2 and O_2 are dissociated at these high temperatures and the radicals will recombine to form such products as NO, NH, NO_2, and HNO_3, which are water soluble. In fact, for an air bubble only argon (about 1% of normal air) is stable. The above process will be repeated cycle after cycle, and it is easily seen that this corresponds to a process of argon rectification. When the bubble is big, dissolved gases diffuse into the bubble and after collapse the reaction products dissolve in water as the argon steadily accumulates.

In this picture, SBSL of an air bubble is, in fact, that of an argon bubble. By a happy accident of nature, it is seen from Figure 17.7 that the normal concentration of argon in natural air corresponds to the small available stable area in phase space. The model in this section is known as the dissociation hypothesis (DH).

17.3.4 Key Experimental Results

Experimental results for SBSL have been summarized briefly elsewhere [256, 266, 268]. Here we focus attention on very recent results that are relevant to a critical understanding of the models for bubble dynamics and light emission.

17.3.4.1 SBSL Spectrum

It has been known from early on [266] that the spectrum was continuous in the visible and that there are no indications of the presence of line spectra. This result has recently been confirmed with nm resolution [266]. An interesting set of controlled spectral measurements of MBSL and SBSL on identical fluids and gases with the same calibrated spectrometer was carried out by Matula et al. [269]. The spectra of dilute NaCl solutions show sharp emission lines for OH$*$ and Na$*$ for MBSL but a very continuous spectrum for SBSL. These results tend to confirm the generally accepted picture that MBSL emission lines involve dissociation of both gas and liquid molecules, and SBSL involves only the spectrum of gases dissolved in the liquid.

Typical SBSL spectra for rare gases dissolved in water are shown in Figures 17.8 and 17.9. Figure 17.8 demonstrates a strong increase of the SBSL radiance with decreasing temperature. It has been shown by Hilgenfeldt et al. [270] that this increase in SBSL at lower temperatures is due to the temperature dependence of the water viscosity and vapor pressure and the argon solubility.

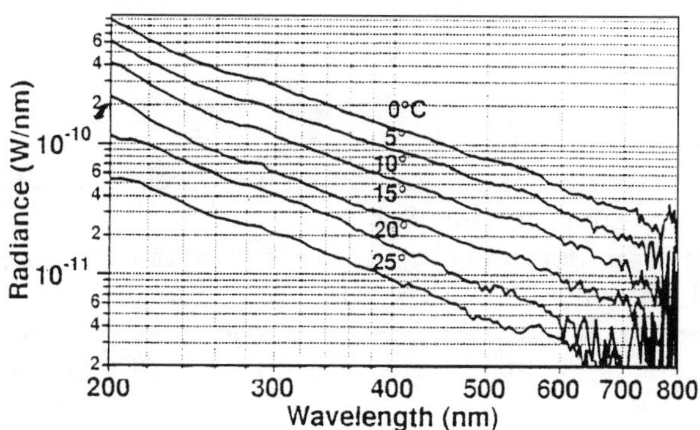

FIGURE 17.8
Corrected spectra for a 150-mm partial pressure bubble of helium in water at various temperatures. (From Barber, B.P. et al., *Phys. Rep.*, 281, 65, 1997. With permission.)

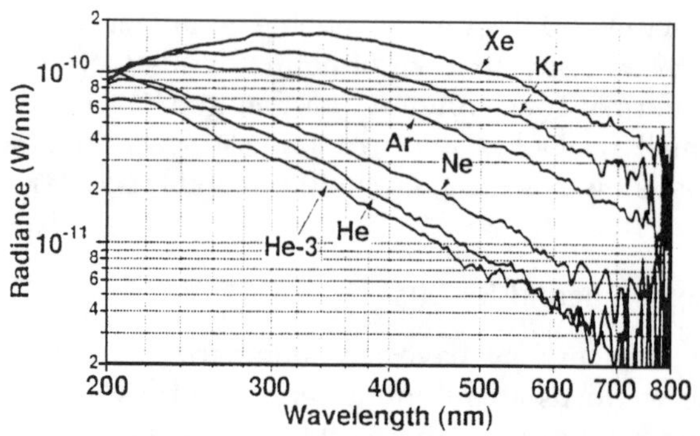

FIGURE 17.9
Room temperature spectra of various noble gases in a cylindrical resonator. No transmission corrections have been made. The gases were dissolved at 3 mm pressure. (From Barber, B.P. et al., *Phys. Rep.*, 281, 65, 1997. With permission.)

Lower temperatures also allow larger stable bubbles and larger driving pressures. Thus it is proposed that the temperature effect is mainly a bubble dynamics effect. The variation with wavelength shows several characteristic features. Above 800 nm, no spectra can be observed due to absorption by the water from 800 nm down to 300 nm. The spectrum shows a monotonic increase with a broad maximum for the case of xenon from 300 to 200 nm. Important corrections must be made due to absorption in the glass and water, yet in the UV below 200 nm the water absorbs all of the emitted light.

According to the DH hypothesis, the SBSL spectrum for an air bubble should be the same as that for a stable argon bubble. The latter has been calculated for two successful models to be described later, and good agreement is found with Figure 17.9. Similar agreement has been found by Hammer and Frommhold [268]. In fact, the DH hypothesis has been verified experimentally by several direct tests that will now be described.

17.3.4.2 Direct Test of the DH Hypothesis

The DH hypothesis has been verified directly by several experimental studies and many others have given indirect supporting evidence. We describe briefly the first reported direct verification by Matula and Crum [271] and then list the other supporting evidence.

Matula and Crum developed a technique for monitoring $R(t)$ and SBSL emission cycle by cycle. In this way, they were able to show that a bubble that had already been above the SBSL threshold sonoluminesces easily and that such a bubble resembles an argon bubble in its SL properties. Two sets of experiments were carried out:

1. A virgin air bubble was compared to one that had been stabilized for 30 s in the SL state. p_a was then lowered below the threshold and after several thousand cycles it was then raised above the

threshold. SL occurred almost immediately. If, however, the bubble is kept too long below the threshold it reverts to the virgin state. These observations strongly support the hypothesis of accumulation of argon above the SBSL threshold and depletion by diffusion below it.

2. In a second set of experiments, a pure N_2 bubble was compared to a pure argon bubble under the same conditions. The pure N_2 bubble behaved as the virgin air bubble in the first experiment while the pure argon bubble behaved as the "recycled" air bubble in the second part of the above experiment. This strongly supports the conclusion that the latter had transformed into an argon bubble by a progressive rectification process. The authors draw an additional conclusion from these experiments. Since argon rectification requires several thousand cycles of SBSL and MBSL bubbles only exist for several cycles, this is a fundamental difference between the two processes. Further confirmatory experimental studies of the DH hypothesis using a second harmonic added to the drive signal were carried out by Holzfuss et al. [272] and Ketterling and Apfel [273].

The following studies also support the DH hypothesis:

1. Experimental confirmation of the theoretical phase diagram for argon bubbles by Barber et al. [266]
2. Direct measurement of the phase diagram by Holt and Gaitan [274]
3. Ambient pressure variation of SBSL by Dan et al. [275]

17.3.4.3 SBSL Pulse Width

The early work indicated that the pulse width was too narrow to be measured: in one case less than 50 ps [261] and in another inferior to 12 ps [276]. However, recent elegant experiments by the group of W. Eisenmenger indicate that the actual pulse width is in the range of 60 to 300 ps depending on the experimental conditions [277]. This work has had a considerable effect on the evaluation and evolution of the theoretical models.

The principle used is that of time-correlated single photon counting (TC-SPC). A time-to-amplitude converter (TAC) is started by the first SBSL photon, stopped, and reset by the second, and the process continues. Thus a statistical count is made of the arrival times of independent single photons, leading to a measure of the auto correlation function of the pulse shape. Since this depends on the experimental conditions, it is important to control the main parameters such as driving amplitude, gas concentration, etc. It was found that the full width half maximum (FWHM) increases with p_a and the gas concentration. Most importantly, it was found that the FWHM was independent of wavelength over the visible spectrum. This drives a nail into the coffin of the black-body model, which predicts a much larger pulse width at the red end than in the UV. The results are, however, compatible with a Bremsstrahlung model as the mechanism for SBSL emission.

The above results were confirmed by other workers [278] and also supported by streak camera measurements by the same group [264]. It had been observed that the pulse shape was asymmetrical, but TC-SPC measurements could not distinguish if the slower part was on the rising or the trailing edge. The steak camera results showed directly that the trailing edge had a slower decay, and the other results were compatible with those obtained by TC-SPC. It was observed that the slower decay and its increase with p_a were consistent with the conclusion that the energy emission is almost entirely due to emission of acoustic waves.

17.3.4.4 Shock Waves

The violent nature of the bubble collapse understandably led to much speculation on the possible role of shock waves. This was particularly true in the early period, when it was thought the the SBSL flashes were much narrower than 50 ps. A detailed shock wave model by Wu and Roberts [265] was able to predict pulse widths compatible with this feature. However, with the more recent work showing pulse widths of the order of 50 to 300 ps, shock waves are no longer seen as an essential component of successful theories. The situation has also been complicated by the success of the DH hypothesis. Finally, there are two quite separate stories to discuss, namely the existence of shock waves inside the bubble as opposed to their existence outside the bubble. These will be considered separately, as the experimental and theoretical implications are quite different in the two cases.

Wu and Roberts assumed a spatially and temporally varying pressure and temperature inside the bubble. They solved the RP equation together with the hydrodynamic conservation equations for a van der Waals air bubble. The system was solved with a fine grid of points with a temporal resolution of about 4.10^{-4} ps near the principal minimum of the bubble radius. They obtained detailed solutions for all of the relevant thermodynamic parameters in a region 400 ps around the minimum. Solutions were found for a shock wave spherically converging on the bubble center, giving rise to an extremely sharp temperature spike. Using a Bremsstrahlung model, they calculated a very sharp SBSL spike emitted near the principal minimum.

The role of shock waves was considered later by Cheng et al. [279] in the light of the DH hypothesis. They considered a much more complete description of the physical processes than that of Wu and Roberts by including diffusion effects, variable gas content, surface tension, and compressibility. They used a range of equations of state for nitrogen and argon bubbles. The inclusion of a hard core affects the compressibility; a higher compressibility favors shock formation. In fact, they found that whether shocks were excited or not depends in a sensitive fashion on the choice of parameters. Globally, it was found relatively feasible to excite shocks in nitrogen (air) but not in argon. In the context of the DH model, they suggest that shocks may well occur during the argon rectification stage during the period that the bubble is being cleansed of air, and smooth compressional waves dominate the situation during the argon-rich part of the process. There is little experimental input on the question; it has been observed that the bubble wall collapse speed well exceeds Mach 1 (/M/ ~ 3) but no shock waves that could exist inside the bubble could be detected by the technique used [266].

The status of shock formation in the liquid is equally fascinating. A number of studies [263] using needle hydrophones placed close to the bubble report observation of ae, but any shock wave that was present at emission would have transformed into any ordinary sound pulse at distance much less than the 1 to 2 mm hydrophone distance. Also, extrapolation of measured sound pressures back to the bubble center is too uncertain a process to allow any conclusions to be drawn. The question was settled by the elegant experiments of Pecha et al. [264], who used a streak camera to image the emission of a shock wave from the bubble. They found a variation of the velocity from 4000 m/s at emission to 1430 m/s, the velocity of sound in water at 60°C, the ambient temperature, after a propagation distance of 50 μm. The imaging mechanism was provided by the refractive index gradient at the shock front. The Cole formula was used to estimate the acoustic pressure gradient $p(z)$ from the measured velocity gradient extrapolating back to the bubble to obtain a pressure of about 60 kb at emission.

17.3.4.5 Ambient Pressure Variation

Ambient pressure was seen to have an effect on MBSL and this is also true for SBSL. The theoretical situation was studied by Kondic et al. [280] in the framework of the RP equation. They focused attention on the relation between p_0 and R_0. At first, they found a decrease in the expansion ratio when p_0 is increased, predicting a decrease in R_0 at constant p_a. They also studied the dependence of R_0 on p_a and p_0 for different gas concentrations c_i/c_0. In the stable regions of the solution, for air bubbles this implies an increase of R_0 with p_0. For concentrations corresponding to argon bubbles (c_i/c_0 ~ 0.002), the theory predicts a decrease in R_0 with increase of p_0. Thus, measurement of the variation of R_0 with p_0 provides a direct test of the DH hypothesis. An experimental study by Dan et al. [275] for air bubbles in the accessible range of p_0: 0.8 to 1.0 b confirmed both of these predictions: an increase of SBSL by a factor of about five when decreasing p_0 to 0.8 b at constant p_a, and an increase of R_0 from 7 to 9 μm, in support of the DH hypothesis. The bubble disappeared below 0.8 b, presumably due to shape instabilities.

There is a second way to study the influence of p_0 indirectly. Young et al. [281] applied magnetic fields B up to 20 T to air bubbles in water in the range 10 to 20°C. The objective was to see if the field had an effect on the plasma at the bubble center that was predicted by many models. Experimentally, they found that the thresholds for SBSL increased with B, leading them to propose that B acted as a kind of negative acoustic pressure. Yasui [282] provided a detailed theoretical foundation. He showed that for a polar liquid,

B gives rise to a Lorentz force on the molecular dipole moment. Incorporating this in the RP equation, he found that it adds a term that is formally equivalent to increasing p_0. Comparison with the experimental results of Young indicates that the application of a field of 6 T corresponds to a 10% change in p_0. The model predicts that the effect should increase with the magnetic flux density and the amount of water in the cell. It also predicts that there should be no effect on nonpolar liquids.

17.3.5 Successful Models

One of the most difficult challenges in SBSL remains the determination of the origin of the light-emitting mechanism. In part, this is due to the difficulty in probing inside the bubble. As a consequence, most of the main evidence is provided by the details of the optical emission spectrum. The latter being continuous, *a priori* there is no unique matching of a model spectrum to experiment. However, enough important parameters have emerged so critical comparisons can be made. We briefly describe two successful models that have so far passed all of the tests provided by experiment.

Hilgenfeldt et al. [283] put forward a simple model that correctly predicts the parameter dependences of the temporal and frequency properties of the light emission. The approach is based on the use of simple bubble dynamics, assuming a spatially uniform temperature, isothermal during most of the collapse, and adiabatic just near the minimum. Applying the DH hypothesis they assume only noble gases are in the bubble in the steady state. Using typical parameters for the experiment, they calculate maximum temperatures of the order of 20000 to 30000 K, leading to a small degree of ionization of the noble gases (~3% for argon and 10% for xenon). The absorption and emission processes are assumed to be the following:

1. Bremsstrahlung due to electrons near ions
2. Bremsstrahlung due to electrons near neutrals
3. Ionization/recombination

The calculated spectrum is in good agreement with experiment regarding FWHM, FWHM wavelength independence, spectral variation of intensity, relative behavior of argon and xenon, and their partial pressure dependence. The model is simple, does not rely on extraordinarily high temperatures or pressures in the bubble, and does not need to invoke a new and exotic mechanism for light emission.

An alternative approach that has also successfully met comparison with experiment is by Moss et al. [284] in which the bubble is modeled as a thermally conducting partially ionized plasma. The model incorporates shock wave generation on collapse, leading to excess heating at the bubble center, and local ionization and creation of a two-component plasma of ions and electrons. An energy cascade occurs from ions to electrons to photons via a

Bremsstrahlung emission mechanism. Comparison is made with "a star in a jar," with a hot, optically opaque center, and a cooler, optically thin outer region. As in most of the models, the action predominantly occurs in the final 100 ps. The calculated spectrum is in good agreement with experiment. The main difference between the two models is the assumption of uniform heating in the first case and shock waves in the second. Both models are firmly based on the DH hypothesis. It may be that in practice both mechanisms may be operative during different phases of the compression cycle, as proposed by Cheng et al. [279].

References

1. Graff, K.F., A history of ultrasonics, in *Physical Acoustics,* XV, Mason, W.P. and Thurston, R.N., Eds., Academic Press, New York, 1981, chap. 1.
2. Minnaert, M., *The Nature of Light and Color in the Open Air,* Dover Publications, New York, 1954.
3. Leighton, T.G., *The Acoustic Bubble,* Academic Press, San Diego, 1994.
4. Rayleigh, J.W.S., *The Theory of Sound,* Vols. 1 and 2, Dover Publications, New York, 1945.
5. Pain, H.J., *The Physics of Vibrations and Waves,* John Wiley & Sons, New York, 1968.
6. Kinsler, L.E., Frey, A.R., Coppens, A.B., and Sanders, J.V., *Fundamentals of Acoustics,* John Wiley & Sons, New York, 2000.
7. Beyer, R.T. and Letcher, S.V., *Physical Ultrasonics,* Academic Press, New York, 1969.
8. Schaaffs, W., Zur Bestimmung von Molekulradien organischer Flüssigkeiten aus Schallgeschwindigkeit und Dichte, *Z. Physik,* 114, 110, 1939.
9. Herzfeld, K.F. and Litovitz, T.A., *Absorption and Dispersion of Ultrasonic Waves,* Academic Press, New York, 1959.
10. Hall, L., The origin of ultrasonic absorption in water, *Phys. Rev.,* 73, A775, 1948.
11. Nye, J.F., *Physical Properties of Crystals,* Clarendon Press, Oxford, 1957.
12. Landau, L.D. and Lifshitz, E.M., *Theory of Elasticity,* Pergamon Press, London, 1959.
13. Truell, R., Elbaum, C., and Chick, B.B., *Ultrasonic Methods in Solid State Physics,* Academic Press, New York, 1969.
14. Bommel, H. and Dransfeld, K., Attenuation of hypersonic waves in quartz, *Phys. Rev. Lett.,* 2, 1959.
15. Woodruff, T.O. and Ehrenreich, H., Absorption of sound in insulators, *Phys. Rev.,* 123, 1553, 1961.
16. Mason, W.P. and Bateman, T.B., Ultrasonic wave propagation in pure silicon and germanium, *J. Acoust. Soc. Am.,* 36, 644, 1964.
17. Zener, C., Internal friction in solids. Pt. II: General theory of thermoelastic internal friction, *Phys. Rev.,* 53, 90, 1938.
18. Papadakis, E.M., Scattering in polycrystalline media, in *Methods of Experimental Physics: Ultrasonics,* Edmonds, P.D., Ed., Academic Press, New York, 1981, chap. 5.
19. Morse, P.M., *Vibration and Sound,* McGraw-Hill, New York, 1948.
20. Kino, G.S., *Acoustic Waves,* Prentice-Hall, Englewood Cliffs, NJ, 1987.
21. Williams, Jr., A.O., Acoustic intensity distribution from a "piston" source, *J. Acoust. Soc. Am.,* 219, 1946.
22. O'Neil, H.T., Theory of focusing radiators, *J. Acoust. Soc. Am.,* 21, 516, 1949.
23. Lucas, B.G. and Muir, T.G., The field of a focusing source, *J. Acoust. Soc. Am.,* 1289, 1982.
24. Chen, X., Schwartz, K.Q., and Parker, K.J., Radiation pattern of a focused transducer: a numerically convergent solution, *J. Acoust. Soc. Am.,* 94, 2979, 1993.

25. Born, M. and Wolf, E., *Principles of Optics*, Pergamon Press, Oxford, 1970.
26. Royer, D. and Dieulesaint, E., *Elastic Waves in Solids I*, Springer-Verlag, Berlin, 1999.
27. Torr, G. R., The acoustic radiation force, *Am. J. Phys.*, 52, 402, 1984.
28. Pierce, A. D., *Acoustics*, McGraw-Hill, New York, 1981.
29. Brekhovskikh, L.M. and Godin, O.A., *Acoustics of Layered Media I*, Springer-Verlag, Berlin, 1998.
30. Brekhovskikh, L.M., *Waves in Layered Media*, Academic Press, New York, 1980.
31. Ristic, V.M., *Principles of Acoustic Devices*, John Wiley & Sons, New York, 1983.
32. Auld, B.A., *Acoustic Fields and Waves in Solids*, Vol. II, Krieger Publishing Company, Malabar, 1990.
33. Rayleigh, J.W.S., On waves propagating along the plane surface of an elastic solid, *Proc. Lond. Math. Soc.*, 17, 4, 1885.
34. Farnell, G.W., Properties of elastic surface waves, in *Physical Acoustics*, IX, Mason, W.P., and Thurston, R.N., Eds., Academic Press, New York, 1972 chap. 3.
35. Viktorov, I.A., *Rayleigh and Lamb Waves*, Plenum Press, New York, 1967.
36. Dransfeld, K. and Salzmann, E., Excitation, detection, and attenuation of high-frequency elastic surface waves, in *Physical Acoustics*, VII, Mason, W.P. and Thurston, R.N., Eds., Academic Press, New York, 1970, chap. 4.
37. Goos, F. and Haenchen, H., *Ann. Physik*, 1, 333, 1947.
38. Lotsch, H.K.V., *Optik*, 32, 116, 1970.
39. Schoch, A., *Acoustics*, 2, 18, 1952.
40. Mott, G., Reflection and refraction coefficients at a fluid-solid interface, *J. Acoust. Soc. Am.*, 50, 819, 1970.
41. Neubauer, W.G., Ultrasonic reflection of a bounded beam at Rayleigh and critical angles for a plane liquid-solid interface, *J. Appl. Phys.*, 44, 48, 1973.
42. Breazeale, M.A., Adler, L., and Scott, G.W., Interaction of ultrasonic waves incident at the Rayleigh angle onto a liquid–solid interface, *J. Acoust. Soc. Am.*, 48, 530, 1977.
43. Becker, F.L. and Richardson, R.L., Influence of material properties on Rayleigh critical-angle reflectivity, *J. Acoust. Soc. Am.*, 51, 1609, 1971.
44. Bertoni, H.L. and Tamir, T., Unified theory of Rayleigh angle phenomena for acoustic beams at liquid–solid interfaces, *Appl. Phys.*, 2, 157, 1973.
45. Uberall, H., Surface waves in acoustics, in *Physical Acoustics*, X, Mason, W.P. and Thurston, R.N., Eds., Academic Press, New York, 1973, chap. 1.
46. Von Schmidt, O., *Z. Phys.* 39, 868, 1938.
47. Ewing, W.M., Jardetzky, W.S., and Press, F., *Elastic Waves in Layered Media*, McGraw-Hill, New York, 1957.
48. Lamb, H., On waves in an elastic plate, *Proc. Roy. Soc.*, Ser. A, 93, 114, 1917.
49. Rose, J.L., *Ultrasonic Waves in Solid Media*, Cambridge University Press, Cambridge, 1999.
50. Uberall, H., Hosten, B., Deschamps, M., and Gerard, A., Repulsion of phase velocity dispersion curves and the nature of plate vibrations, *J. Acoust. Soc. Am.*, 96, 908, 1994.
51. Osborne, M.F.N. and Hart, S.D., Transmission, reflection, and guiding of an exponential pulse by a steel plate in water, I. Theory, *J. Acoust. Soc. Am.*, 17, 1, 1945.
52. Bao, X.L., Franklin, H., Raju, P.K., and Uberall, H., The splitting of dispersion curves for plates fluid-loaded on both sides, *J. Acoust. Soc. Am.*, 102, 1246, 1997.
53. Neubauer, W.G., Observation of acoustic radiation from plane and curved surfaces, in *Physical Acoustics*, X, Mason, W.P. and Thurston, R.N., Eds., Academic Press, New York, 1973, chap. 2.

54. Maze, G., Leon, F., Ripoche, J., and Uberall, H., Repulsion phenomena in the phase velocity dispersion curves of circumferential waves on elastic cylindrical shells, *J. Acoust. Soc. Am.*, 105, 1695, 1999.
55. Sessarego, J.P., Sageloli, J., Gazanhes, C., and Uberall, H., Two Scholte-Stoneley waves on doubly fluid-loaded plates and shells, *J. Acoust. Soc. Am.*, 101, 135, 1997.
56. Bao, X.L., Raju, P.K., and Uberall, H., Circumferential waves on an immersed, fluid-filled elastic cylindrical shell, *J. Acoust. Soc. Am.*, 105, 2704, 1999.
57. Farnell, G.W. and Adler, E.L., Elastic wave propagation in thin layers, in *Physical Acoustics*, IX, Mason, W.P. and Thurston, R.N., Eds., Academic Press, New York, 1972, chap. 2.
58. Love, A.E.H., *Some Problems in Geodynamics*, Cambridge University Press, London, 1911, 1926.
59. Tournois, P. and Lardat, C., Love wave dispersive delay lines for wide-band pulse compression, *IEEE Trans.*, SU-16, 107, 1969.
60. Tiersten, H.F., Elastic surface waves guided by thin films, *J. Appl. Phys.*, 40, 770, 1969.
61. Sezawa, K. and Kanai, K., *Bull. Earth. Res. Inst. Tokyo*, 13, 237, 1935.
62. Stoneley, R., Elastic surface waves at the surface of separation of two solids, *Proc. Roy. Soc.*, Ser. A, 106, 416, 1924.
63. Lowe, M.J.S., Matrix techniques for modeling ultrasonic waves in multilayered media, *IEEE Trans. UFFC*, 42, 525, 1995.
64. Achenbach, J.D., Kim, J.O., and Lee, Y.-C., Measuring thin-film elastic constants by line-focus acoustic microscopy, in *Advances in Acoustic Microscopy*, 1, Briggs, A., Ed., Plenum Press, New York, 1995, chap. 5.
65. Knopoff, L., A matrix method for elastic wave problems, *Bull. Seism. Soc. Am.*, 54, 431, 1964.
66. Thomson, W.T., Transmission of elastic waves through a stratified medium, *J. Appl. Phys.*, 21, 89, 1950.
67. Haskell, N.A., The dispersion of surface waves on multilayered media, *Bull. Seism. Soc. Am.*, 43, 17, 1953.
68. Meeker, T.R. and Meitzler. A.H., Guided wave propagation in elongated cylinders and plates, in *Physical Acoustics*, IA, Mason, W.P. and Thurston, R.N., Eds., Academic Press, New York, 1964, chap. 2.
69. Rosenberg, R.L., Schmidt, R.V., and Coldren, L.A., Interior-surface acoustic waveguides in capillaries, *Appl. Phys. Letts.*, 25, 324, 1974.
70. Oliner, A. A., Waveguides for surface waves, in *Acoustic Surface Waves*, Oliner, A.A., Ed., Springer-Verlag, Berlin, 1978, chap. 5.
71. Jen, C.K., Acoustic fibers, in *Proc. 1987 IEEE Ultrasonics Symp.*, McAvoy, B.R., Ed., IEEE, New York, 1987, 443.
72. Boyd, G.D., Coldren, L.A., and Thurston, R.N., Acoustic clad fiber delay lines, *IEEE Trans.*, SU-24, 246, 1977.
73. Borgnis F.E., Specific directions of longitudinal wave propagation in anisotropic media, *Phys. Rev.*, 98, 1000, 1955.
74. Federov, F.I., *Theory of Elastic Waves in Crystals*, Plenum Press, New York, 1968.
75. Gooberman, G.L., *Ultrasonics*, The English Universities Press Ltd., London, 1968.
76. ANSI/IEEE Standard on Piezoelectricity, *IEEE Trans. UFFC*, 43, 717, 1996.
77. Eveleth, J.H., A survey of ultrasonic delay lines operating below 100 Mc/s, *Proc. IEEE*, 53, 1406, 1965.
78. Smith, W.R., Circuit-model analysis and design of interdigital transducers for surface acoustic wave devics, in *Physical Acoustics*, XV, Mason, W.P. and Thurston, R.N., Eds., Academic Press, New York, 1981, chap. 2.

79. Datta, S., *Surface Acoustic Wave Devices,* Prentice-Hall, Englewood Cliffs, NJ, 1986.
80. Ash, E.A., Fundamentals of signal processing, in *Acoustic Surface Waves,* Oliner, A.A., Ed., Springer-Verlag, Berlin, 1978, chap. 4.
81. Ballantine, D.S. et al., *Acoustic Wave Sensors,* Academic Press, San Diego, 1997.
82. Sauerbrey, G., Verwendung von Schwingquarzen zur Wagung dünner Schichten und zur Mikrowagung, *Z. Phys.*, 155, 206, 1959.
83. Miller, J.G. and Bolef, D.I., Acoustic wave analysis of the operation of quartz-crystal film-thickness monitors, *J. Appl. Phys.*, 5815, 1968.
84. Kanazaw, K.K. and Gordon, II, J.G., Frequency of a quartz microbalance in contact with a liquid, *Anal. Chem.*, 57, 1770, 1985.
85. Ricco, A.J. Martin, S.J., and Zipperian, T.E., Surface acoustic wave gas sensor based on film conductivity changes, *Sensors Actuators*, 8, 319, 1985.
86. Adler, A., Simple theory of acoustic amplification, *IEEE Trans.*, SU-18, 115, 1971.
87. Shiokawa, S. and Kondoh, J., Surface acoustic wave sensor for liquid-phase application, in *Proc. IEEE Ultrasonics Symp.*, Schneider, S.C., Levy, M., and McAvoy, B.R., Eds., IEEE, New York, 1999, 445.
88. Tiersten, H.F. and Sinha, B.K., A perturbation analysis of the attenuation and dispersion of surface waves, *J. Appl. Phys.*, 49, 87, 1978.
89. Martin, S.J., Frye, G.C., and Senturia, S.D., Dynamics and response of polymer-coated surface acoustic wave devices: Effect of viscoelastic properties and film resonance, *Anal. Chem.*, 66, 2201, 1994.
90. Hou, J. and van de Vaart, H., Mass sensitivity of plate modes in surface acoustic wave devices and their potential as chemical sensors, in *Proc. IEEE Ultrasonics Symp.*, McAvoy, B.R., Ed., IEEE, New York, 1987, 573.
91. Martin, S.J., Ricco, A.J., Niemzyk, T.M., and Frye, G.C., Characterization of SH plate mode liquid sensors, *Sensors Actuators*, 20, 253, 1989.
92. Campbell, J.J. and Jones, W.R., A method for estimating optimal crystal cuts and propagation directions for excitation of piezoelectric surface waves, *IEEE Trans. Sonics Ultrason.*, SU-15, 209,1968.
93. Jacoby, B. and Vellekoop, M.J., Viscous losses of shear waves in layered structures used for biosensing, in *Proc. IEEE Ultrasonics Symp.*, Schneider, S.C., Levy, M., and McAvoy, B.R., Eds., IEEE, New York, 1998, 493.
94. Baer, R.L. et al., STW chemical sensors, in *Proc. IEEE Ultrasonics Symp.*, McAvoy, B.R., Ed., IEEE, New York, 1992, 293.
95. Wang, Z., Cheeke, J.D.N., and Jen, C.K., Sensitivity analysis for Love mode acoustic gravimetric sensors, *Appl. Phys. Letts.*, 64, 2940, 1994.
96. Wenzel, S.W. and White, R.M., Flexural plate-wave gravimetric chemical sensor, *IEEE Trans. Electron Devices*, Ed-35, 735, 1988.
97. Wenzel, S.W., Applications of Ultrasonic Lamb Waves, Ph.D. thesis, University of California, Berkeley, 1992.
98. Viens, M. et al., Mass sensitivity of thin rod acoustic wave sensors, *IEEE Trans. UFFC*, 43, 852, 1996.
99. Wang, Z., Jen, C.K., and Cheeke, J.D.N., Unified approach to analyze mass sensitivities of acoustic gravimetric sensors, *Electron. Letts.*, 26, 1511, 1990.
100. Wang, Z., Jen, C.K., and Cheeke, J.D.N., Mass sensitivities of two-layer sagittal plane plate wave acoustic sensors, *Ultrasonics*, 32, 201, 1994.
101. Cheeke, J.D.N. and Wang, Z., Acoustic wave gas sensors, *Sensors Actuators*, 59, 146, 1999.
102. Lynnworth, L.C., *Ultrasonic Measurements for Process Control*, Academic Press, New York, 1989.

103. Puttmer, A. and Hauptman, P., Ultrasonic density sensor for liquids, in *Proc. IEEE Ultrasonics Symposium*, Schneider, S.C., Levy, M., and McAvoy, B.R., Eds., IEEE, 1998, New York, 497.
104. Costello, B.J., Wenzel, S.W., and White, R.M., Density and viscosity sensing with ultrasonic flexural plate waves, in *Proc. 7th Intl. Conf. Solid State Sensors Actuators, Transducers 93,* IEEE of Japan, 1993, 704.
105. Ricco, A.J. and Martin, S.J., Acoustic wave viscosity sensor, *Appl. Phys. Letts.*, 50, 1474, 1987.
106. Martin, S.J., Frye, G.C., Ricco, A.J., and Senturia, S.D., Effect of surface roughness on the response of thickness-mode resonators in liquids, *Anal. Chem.*, 65, 2910, 1993.
107. Herrmann, F., Hahn, D., and Buttenbach, S., Separate determination of liquid density and viscosity with sagitally corrugated Love mode sensors, *Sensors Actuators*, 78, 99, 1999.
108. Bell, J.F.W., The velocity of sound in metals at high temperatures, *Phil. Mag.*, 2, 1113, 1957.
109. Hauden, D., Miniaturized bulk and surface acoustic wave quartz oscillators used as sensors, *IEEE Trans.*, UFFC-34, 253, 1987.
110. Viens, M. and Cheeke, J.D.N., Highly sensitive temperature sensor using SAW oscillator, *Sensors Actuators,* A, 24, 209, 1990.
111. Magori, V., Ultrasonic sensors in air, in *Proc. IEEE Ultrasonics Symp.*, Levy, M., Schneider, S.H., and McAvoy, B.R., Eds., IEEE, New York, 1994, 471.
112. Ahmad, N., Surface acoustic wave flow sensor, *Proc. 1985 IEEE Ultrasonics Symp.*, McAvoy, B.R., Ed., IEEE, New York, 1985, 483.
113. Joshi, S.G., Use of a SAW device to measure gas flow, *IEEE Trans. Instrum. Meas.*, 38, 824, 1989.
114. Brace, J.G., Sanfelippo, T.S., and Joshi, S.G., Mass-flow sensing using surface acoustic waves, in *Proc. 1989 IEEE Ultrasonics Symp.*, McAvoy, B.R., Ed., IEEE, New York, 1989, 573.
115. Dieulesaint, E., Royer, D., Legras, O., and Boubenider, F., A guided acoustic wave level sensor, in *Proc. 1987 IEEE Ultrasonics Symp.*, McAvoy, B.R., Ed., IEEE, New York, 1987, 569.
116. Motegi, R. et al., Remote measurement method of acoustic velocity I liquids by a pair of leaky acoustic waveguides, in *Proc. 1987 IEEE Ultrasonics Symp.*, McAvoy, B.R., Ed., IEEE, New York, 1987, 625.
117. Gillespie, A.B. et al., A new ultrasonic technique for the measurement of liquid level, *Ultrasonics*, 20, 13, 1982.
118. Liu, Y. and Lynnworth, L.C., Flexural wave sidewall sensor for noninvasive measurement of discrete liquid levels in large storage tanks, in *Proc. 1993 IEEE Ultrasonics Symp.*, Levy, M. and Mcavoy, B.R., Eds., IEEE, New York, 1993, 385.
119. Lynnworth, L.C. et al., Improved shear wave hockey stick transducer measures liquid flow and liquid level, in *Proc. 1997 IEEE Ultrasonics Symp.*, Levy, M., Schneider, S.C., and McAvoy, B.R., Eds., IEEE, New York, 1997, 865.
120. Cheeke, J.D.N., Shannon, K., and Wang, Z., Loading effects on A_0 Lamb-like waves in full and partially filled thin-walled tubes, *Sensors Actuators*, 59, 180, 1999.
121. Shannon, K., Li, X., Wang, Z., and Cheeke, J.D.N., Mode conversion and the path of acoustic energy in a partially water-filled aluminum tube, *Ultrasonics*, 37, 303, 1999.
122. Martin, S.J. et al. Gas sensing with acoustic devices, in *Proc. 1996 IEEE Ultrasonics Symp.*, Levy, M., Schneider, S.C., and McAvoy, B.R., Eds., IEEE, New York, 1996, 423.

123. Yan, Y. and Bein, T., Zeolite thin films with tunable molecular sieve function, *J. Am. Chem. Soc.*, 117, 9990, 1995.
124. Persaud, K. and Dodd, G., Analysis of discrimination mechanisms in the mammalian olfactory system using a model nose, *Nature*, 299, 352, 1982.
125. King, Jr., W.H., *Anal. Chem.*, 36, 1735, 1964.
126. Karasek, F.W. and Gibbins, K.R., *J. Chromatography*, 9, 535, 1975.
127. Staples, E.J., Electronic nose simulation of olfactory response containing 500 orthogonal sensors in 10 seconds, in *Proc. 1999 IEEE Ultrasonics Symp.*, Schneider, S.C., Levy, M., and McAvoy, B.R., Eds., IEEE, New York, 1999, 417.
128. Black, J.P. et al., Comparison of the performance of flexural plate wave and surface acoustic wave devices as the detector in a gas chromatograph, in *Proc. 2000 IEEE Ultrasonics Symp.*, Schneider, S.C., Levy, M., and McAvoy, B.R., Eds., IEEE, New York, 1999, 435.
129. Andle, J.C. and Vetelino, J.F., Acoustic wave biosensors, in *Proc. 1995 IEEE Ultrasonics Symp.*, Levy, M., Schneider, S.C., and McAvoy, B.R., Eds., IEEE, New York, 1995, 451.
130. Sokolov, S.Y., *Dokl. Akad. Nauk. SSSR*, 64, 333, 1949.
131. Lemons, R.A. and Quate, C.F., Acoustic microscopy, in *Physical Acoustics*, XIV, Mason, W.P. and Thurston, R.N., Eds., Academic Press, London, 1979, chap. 1.
132. Weglein, R.D. and Wilson, R.G., Characteristic material signatures by acoustic microscopy, *Elect. Letts.*, 14, 352, 1978.
133. Kushibiki, J. and Chubachi, N., Material characterization by line-focus beam acoustic microscope, *IEEE Trans. Sonics Ultrason.*, SU-33, 189, 1985.
134. Germain, L. and Cheeke, J.D.N., Generation and detection of high order harmonics in liquids using a scanning acoustic microscope, *J. Acoust. Soc. Am.*, 83, 942, 1988.
135. Kessler, L.W., Acoustic microscopy—an industrial view, in *Proc. 1988 IEEE Ultrasonics Symp.*, McAvoy, B.R., Ed., IEEE, New York, 1988, 725.
136. Lemons, R.A., Acoustic Microscopy by Mechanical Scanning, Ph.D. thesis, E.L. Ginzton Laboratory, Stanford University, Stanford, CA, 1975.
137. Attal, J. and Quate, C.F., Investigation of some low ultrasonic absorption liquids, *J. Acoust. Soc. Am.*, 59, 69, 1976.
138. Briggs, G.A.D., *Acoustic Microscopy*, Clarendon Press, Oxford, 1992.
139. Heiserman, J., Rugar, D., and Quate, C.F., Cryogenic acoustic microscopy, *J. Acoust. Soc. Am.*, 67, 1629, 1980.
140. Rugar, D., Resolution beyond the diffraction limit in the acoustic microscope, *J. Appl. Phys.*, 56, 1338, 1984.
141. Germain, L. and Cheeke, J.D.N., Acoustic microscopy applied to nonlinear characterization of biological media, *J. Acoust. Soc.*, A, 86, 1560, 1989.
142. Chou, C.H., Khuri-Yakub, B.T., and Kino, G.S., Lens design for acoustic microscopy, *IEEE Trans. UFFC*, 35, 464, 1988.
143. Rugar, D., Cryogenic Acoustic Microscopy, Ph.D. thesis, E.L. Ginzton Laboratory, Stanford University, Stanford, CA, 1981.
144. Liang, K.K., Kino, G.S., and Khuri-Yakub, B.T., Material characterization by the inversion of V(z), *IEEE Trans. Sonics Ultrasonics*, SU-33, 213, 1985.
145. Chubachi, N., ZnO film concave transducers for focusing radiation of microwave ultrasound, *Elec. Letts.*, 12, 595, 1976.
146. Labreche, A. et al., Scanning acoustic microscopy using PVDF concave lenses, *Elec. Letts.*, 21, 990, 1985.
147. Atalar, A., Degertekin, F.L., and Koymen, H., Recent advances in acoustic microscopy, in *Proc. 1991 IEEE Ultrasonics Symp.*, McAvoy, B.R., Ed., IEEE, New York, 1991, 719.

148. Davids, D.A. and Bertoni, H.L., Bow-tie transducers for measurement of anisotropic materials in acoustic microscopy, in *Proc. 1986 IEEE Ultrasonics Symp.*, McAvoy, B.R., Ed., IEEE, New York, 1986, 735.
149. Khuri-Yakub, B.T. and Chou, C.-H., Acoustic microscope lenses with shear wave transducers, in *Proc. 1986 IEEE Ultrasonics Symp.*, McAvoy, B.R., Ed., IEEE, New York, 1986, 741.
150. Yamada, K., Sugiyama, T., and Shimuzu, H., Planar-structure microscope lens for simultaneous acoustic and optical imaging, in *Proc. 1988 IEEE Ultrasonics Symp.*, McAvoy, B.R., Ed., IEEE 1988, 779.
151. Atalar, A., An angular spectrum approach to contrast in the reflection acoustic microscope, *J. Appl. Phys.*, 49, 5130, 1978.
152. Bertoni, H.L., Ray-optical evaluation of V(z) in the reflection acoustic microscope, *IEEE Trans. Sonics Ultrasonics*, SU-31, 105, 1984.
153. Pino, F., Sinclair, D.A., and Ash, E.A., New technique for subsurface imaging using scanning acoustic microscopy, in *Ultrasonics Int.*, 81, 93, 1981.
154. Achenbach, J.D., Kim, J.O., and Lee, Y.C., Measuring thin-film constants by line-focus acoustic microscopy, in *Advances in Acoustic Microscopy,* Vol. 1, Briggs, A., Ed., Plenum Press, 1995, chap. 5.
155. Tsukahara, Y. et al., An acoustic micrometer and its application to layer thickness measurements, *IEEE Trans. UFFC,* 36, 326, 1989.
156. Richard, P., Gremaud, G., and Kulik, A., Thin film adhesion investigations with the acoustic microscope, in *Proc. 1994 IEEE Ultrasonics Symp.*, Levy, M., Schneider, S.C., and McAvoy, B.R., Eds., IEEE, New York, 1994, 1425.
157. Khuri-Yakub, B.T. and Rheinholdsten, Nondestructive evaluation of composite materials using acoustic microscopy, in *Review of Progress in Quantitative Nondestructive Evaluation*, Vol. 5B, Thompson, D.O. and Chimenti, D.E., Eds., Plenum Press, New York, 1986, 1093.
158. Lee, Y.C., Kim, J.O., and Achenbach, J.D., Measurement of stresses by line-focus acoustic microscopy, *Ultrasonics*, 32, 359, 1994.
159. Liang, K.K. et al., Precise phase measurements with the acoustic microscope, *IEEE Trans. Sonics Ultrasonics*, SU-32, 266, 1985.
160. Drescher-Krasicka, E., Scanning acoustic imaging of stress in the interior of solid materials, *J. Acoust. Soc. Am.*, 94, 435, 1993.
161. Gilmore, R.S. et al., Acoustic microscopy from 10 to 100 MHz for industrial applications, *Phil. Trans. R. Soc. Lond.*, A, 320, 215, 1986.
162. Crean, G.M., Flannery, C.M., and Mathuna, S.C.O., Acoustic microscopy analysis of microelectronic interconnection and packaging technologies, in *Advances in Acoustic Microscopy,* Vol. 1, Briggs, A., Ed., Plenum Press, 1995, chap. 1.
163. Pfannschmidt, G., Characterization of electronic components by acoustic microscopy, in *Advances in Acoustic Microscopy,* Vol. 2, Briggs, A. and Arnold, W., Eds., Plenum Press, 1996, chap. 1.
164. Kushibiki, J. et al., Characterization of 36° YX-$LiTaO_3$ wafers by line-focus-beam acoustic microscopy, *IEEE Trans. UFFC*, 42, 83, 1995.
165. Jen, C.K. et al., Characterization of cladded glass fibers using acoustic microscopy, *App. Phys. Lett.*, 55, 2485, 1989.
166. Ash, E.A. and Nicholls, N., Super resolution aperture scanning microscope, *Nature*, 237, 1972.
167. Zieniuk, J.K. and Latuszek, L., Nonconventional pin scanning ultrasonic microscope, in *Acoustical Imaging 17*, Schimuzu, H., Chubachi, N., and Kushibiki, J., Eds., Plenum Press, New York, 1989, 219.

168. Takata, K. et al., Tunneling acoustic microscope, *Appl. Phys. Lett.*, 55, 1718, 1989.
169. Cretin, B. and Stahl, F., Scanning microdeformation microscopy, *Appl. Phys. Lett.*, 62, 829, 1993.
170. Yamanaka, K., New approaches in acoustic microscopy for noncontact measurement and ultra high resolution, in *Advances in Acoustic Microscopy*, Vol. 1, Briggs, A., Ed., Plenum Press, New York, 1995, chap. 8.
171. Rabe, U., Janser, K., and Arnold, W., Acoustic microscopy with resolution in the nm range, in *Acoustical Imaging*, 22, Tortolli, P. and Masotti, L., Eds., Plenum Press, New York, 1996, 669.
172. Burnham, N.A., Kulik, A.J., and Gremaud, G., Scanning local acceleration microscopy, *J. Vac. Sci. Technol.*, B, 14, 794, 1996.
173. Gordon, G.A. and Tittmann, B.R., Forward models for surface wave prediction of material property profiles, in *Proc. 1993 IEEE Ultrasonics Symp.*, Levy, M. and McAvoy, B.R., Eds., IEEE, New York, 1993, 301.
174. Chaloner, C.A. and Bond, L.J., Ultrasonic inversion: A direct and an indirect method, in *Review of Progress in Quantitative NDE*, 6A, Thompson, D.O. and Chimenti, D.E., Eds., Plenum Press, New York, 1987, 563.
175. Hutchins, D.A. et al., The propagation of ultrasound within a gas jet, *J. Acoust. Soc. Am.*, 110, 2964, 2001.
176. Mayer, W.G., Determination of ultrasonic velocities by measurement of angles of total reflection, *J. Acoust. Soc. Am.*, 32, 1213, 1960.
177. Rollins, Jr., F.R., Ultrasonic reflectivity at a liquid-solid inreface near the angle of incidence for total reflection, *Appl. Phys. Lett.*, 7, 212, 1965.
178. Rollins, Jr., F.R., Critical angle reflectivity—a neglected tool for material evaluation, *Mater. Evaluation*, 24, 683, 1966.
179. Rokhlin, S.I. and Wang, W., Critical angle measurement of elastic constants in composite material, *J. Acoust. Soc. Am.*, 86, 1876, 1989.
180. Chimenti, D.E., Ultrasonic plate wave evaluation of composite plates, in *Proc. 1993 IEEE Ultrasonics Symp.*, Levy, M. and McAvoy, B.R., Eds., IEEE, New York, 1993, 863.
181. Lobkis, O.I. and Chimenti, D.E., Elastic guided waves in plates with surface roughness. II. Experiments, *J. Acoust. Soc. Am.*, 102, 150, 1997.
182. Xu, P.-C., Lindenschmidt, K.E., and Meguid, S.A., A new high-frequency analysis of coatings using leaky Lamb waves, *J. Acoust. Soc. Am.*, 94, 2954, 1993.
183. Karim, M.R., Mal, A.K., and Bar-Cohen, Y., Inversion of leaky Lamb wave data by Simplex algorithm, *J. Acoust. Soc. Am.*, 88, 482, 1990.
184. Wang, Z. et al., Material characterization using leaky Lamb waves, in *Proc. 1994 IEEE Ultrasonics Symp.*, Levy, M., Schneider, S.C., and McAvoy, B.R., Eds., IEEE, New York, 1994, 1227.
185. Cheeke, J.D.N., Wang, Z., and Viens, M., Thermal sprayed coatings characterization by acoustic waves, in *Proc. 1996 IEEE Ultrasonics Symp.*, Levy, M., Schneider, S.C., and McAvoy, B.R., Eds., IEEE, New York, 1996, 749.
186. Thompson, R.B. and Thompson, D.O., *J. Adhesion Sci. Technol.*, 5, 583, 1991.
187. Cawley, P., Ultrasonic measurements for the qualitative NDE of adhesive joints—potential and challenges, in *Proc. 1992 IEEE Ultrasonics Symp.*, McAvoy, B.R., Ed., IEEE, New York, 1992, 767.
188. Cawley, P. and Pialucha, T., The prediction and measurements of the ultrasonic reflection coefficient from interlayers in adhesive joints, in *Proc. 1993 IEEE Ultrasonics Symp.*, Levy, M. and McAvoy, B.R., Eds., IEEE, 1993, 729.

189. Wang, Y.-J. et al., Lamb wave modes in a two-layered solid medium with a weak interface, *Acta Physica Sinica,* 3, 561, 1994.
190. Breazeale, M.A., Cantrell, J.H., and Heyman, J.S., Ultrasonic wave velocity and attenuation measurements, in *Methods of Experimental Physics: Ultrasonics,* 19, Edmonds, P.D., Ed., Academic Press, New York, 1981, chap. 2.
191. Lee, C.C. and Tsai, C.S., Complete characterization of thin and thick film matrerials using wideband reflection acoustic microscopy, *IEEE Trans. Sonics Ultrasonics,* SU-32, 248, 1985.
192. Huang, W. and Rokhlin, S.I., Low-frequency normal incidence ultrasonic method for thin layer characterization, *Mater. Evaluation,* 1279, 1993, 1993.
193. Kushibiki, J., Ishikawa, T., and Chubachi, N., Cut-off characteristics of leaky Sezawa and pseudo-Sezawa wave modes for thin-film characterization, *Appl. Phys. Lett.,* 57, 1967, 1990.
194. Kushibiki, J. and Chubachi, N., Application of LFB acoustic microscope to film thickness measurements, *Elect. Letts.,* 23, 652, 1987.
195. Rose, J.L. and Barshinger, J., Using ultrasonic guided wave mode cutoff for corrosion detection and classification, in *Proc. 1998 IEEE Ultrasonics Symp.,* Schneider, S.C., Levy, M., and McAvoy, B.R., Eds., IEEE, New York, 1998, 851.
196. Jen, C.K. et al., Long acoustic imaging probes, in *Proc. 1990 IEEE Ultrasonics Symp.,* McAvoy, B.R., Ed., IEEE, New York, 1990, 875.
197. Jen, C. K., Legoux, J.-G., and Parent, L., Experimental evaluation of clad metallic buffer rods for high-temperature ultrasonic measurements, *NDT & E Intl.,* 33, 145, 2000.
198. Biot, M.A., *J. Acoust. Soc. Am.,* 28, 168, 1956.
199. Berryman, J.G., Confirmation of Biot's theory, *Appl. Phys. Lett.,* 37, 382, 1980.
200. Johnson, D.L., Equivalence between fourth sound in liquid helium II at low temperatures and the Biot sound wave in consolidated porous media, *Appl. Phys. Lett.,* 37, 1065, 1980.
201. Plona, T.J., Observation of a second bulk compressional wave in a porous medium at ultrasonic frequencies, *Appl. Phys. Lett.,* 36, 259, 1980.
202. Chandler, R., Transient streaming potential measurements on fluid saturated porous structures; an experimental verification of Biot's slow wave in the quasi-static limit, *J. Acoust. Soc. Am.,* 70, 116, 1981.
203. Stoll, R.D. and Bryan, G.M., Wave attenuation in saturated sediments, *J. Acoust. Soc. Am.,* 47, 1440, 1970.
204. Stoll, R.D., Experimental studies of attenuation in sediments, *J. Acoust. Soc. Am.,* 66, 1152, 1979.
205. Sheng, P., *Introduction to Wave Scattering, Localization, and Mesoscopic Phenomena,* Academic Press, New York, 1995.
206. Page, J.H. et al., Group velocity in strongly scattering media, *Science,* 271, 634, 1996.
207. Lebowitz, J.L., Boltzmann's entropy and time's arrow, *Physics Today,* 32, 1993.
208. Fink, M., Time reversal of ultrasonic fields—Part I: basic principles, *IEEE Trans. UFFC,* 39, 555, 1992.
209. Fink, M., Time reversed acoustics, *Physics Today,* 34, 1997.
210. Derode, A. et al., Robust acoustic time reversal with high-order multiple scattering, *Phys. Rev. Lett.,* 75, 4206, 1995.
211. Hutchins, D. A., Mechanisms of pulsed photoacoustic generation, *Can. J. Phys.,* 64, 1247, 1986.

212. Thomsen, C. et al., Surface generation and detection of phonons by picosecond light pulses, *Phys. Rev.*, B, 34, 4219, 1986.
213. Wright, O.B., Local acoustic probing using mechanical and ultrafast optical technques, in *Proc. 1995 IEEE Ultrasonics Symp.*, Levy, M., Schneider, S.C., and McAvoy, B.R., Eds., IEEE, New York, 1995, 567.
214. Smith, S., Orr, B.G., Kopelman, R., and Norris, T., 100 femtosecond/100 nanometer near-field probe, *Ultramicroscopy*, 57, 173, 1995.
215. Hurley, D.H. et al., Laser picosecond acoustics in isotropic and anisotropic materials, *Ultrasonics*, 38, 470, 2000.
216. Lin, H.-N., Stoner, R.J., and Maris, H.J., Ultrasonic experiments at ultra-high frequency with picosecond time-resolution, in *Proc. 1990 IEEE Ultrasonics Symp.*, McAvoy, B.R., Ed., IEEE, New York, 1990, 1301.
217. Zeller, R.C. and Pohl, R.O., Thermal conductivity and specific heat of non-crystalline solids, *Phys. Rev.*, B, 4, 2029, 1971.
218. Zhu T.C., Maris, H.J., and Tauc, J., Attenuation of longitudinal acoustic phonons in amorphous SiO_2 at frequencies up to 440 Ghz, *Phys. Rev.*, B, 44, 4281, 1991.
219. Morath, C. J. and Maris, H. J., Phonon attenuation in amorphous solids studied by picosecond ultrasonics, *Phys. Rev.*, B, 54, 203, 1996.
220. Pollack, G.L., Kapitza resistance, *Rev. Mod. Phys.*, 41, 48, 1969.
221. Khalatnikov, I. M., *Zh. Eksp. Teor. Fiz.*, 22, 687, 1952.
222. Cheeke, J.D.N., Hebral, B., and Martinon, C., Transfert de chaleur entre deux solides en dessous de 100 K, *J. Phys.* (*Paris*), 34, 257, 1973.
223. Martinon, C. and Weis, O., α-Quartz as a substrate in thermal phonon radiation, *Z. Phys.*, B, 48, 258, 1979.
224. Swartz, E.T. and Pohl, R.O., Thermal boundary resistance, *Rev. Mod. Phys.*, 61, 605, 1989.
225. Stoner, R.J. and Maris, H.J., Kapitza, conductance and heat flow between solids at temperatures from 50 to 300 K, *Phys. Rev.*, B, 48, 16 373, 1993.
226. Tas, G. and Maris, H.J., Picosecond ultrasonic study of phonon reflection from solid-liquid interfaces, *Phys. Rev.*, B, 55, 1852, 1997.
227. Tas, G. and Maris, H.J., Electron diffusion in metals studied by picosecond ultrasonics, *Phys. Rev.*, B, 49, 15046, 1994.
228. Chen, W., Lu, Y., Maris, H.J., and Xiao, G., Picosecond ultrasonic study of localized phonon surface modes in Al/ag superlattices, *Phys. Rev.*, B, 50, 14506, 1994.
229. Rudolph Technologies, Inc., Flanders, New Jersey.
230. Grandia, W.A. and Fortunko, C.M., NDE applications of air-coupled ultrasonic transducers, in *Proc. 1995 IEEE Ultrasonics Symp.*, Levy, M., Schneider, S.C., and McAvoy, B.R., Eds., IEEE, New York, 1995, 697.
231. Wickramasinghe, H.K. and Petts, C.R., Acoustic microscopy in high-pressure gases, in *Proc. 1980 IEEE Ultrasonics Symp.*, McAvoy, B.R., Ed., IEEE, 1980, 668.
232. Chimenti, D.E. and Fortunko, C.M., Characterization of composite prepreg with gas-coupled ultrasonics, *Ultrasonics*, 32, 261, 1994.
233. Fox, J.D., Kino, G.S., and Khuri-Yakub, B.T., Acoustic microscopy in air at 2 MHz, *Appl. Phys. Lett.*, 47, 465, 1985.
234. Robertson, T.J., Hutchins, D.A., and Billson, D.R., An air-coupled line-focused capacitive ultrasonic transducer, in *Proc. 2000 IEEE Ultrasonics Symp.*, Schneider, S.C., Levy, M., and McAvoy, B.R., Eds., IEEE, New York, 1993, 667.
235. Schindel, D.W., Hutchins, D.A., Zou, L., and Sayer, M., The design and characterization of micro-machined air-coupled capacitance transducers, *IEEE Trans. UFFC*, 42, 42, 1995.

236. Ladabaum, I., Jin, X., Soh, H., Atalar, A., and Khuri-Yakub, B.T., Surface micromachined capacitive ultrasonic transducers, *IEEE Trans. UFFC*, 45, 678, 1998.
237. Hayward, G., Air-coupled NDE-constraints and solutions for industrial implementation, in *Proc. 1997 IEEE Ultrasonics Symp.*, Schneider, S.C., Levy, M., and McAvoy, B.R., Eds., IEEE, New York, 1997, 665.
238. Krautkramer, J. and Krautkramer, H., *Ultrasonic Testing of Materials*, Springer-Verlag, Berlin, 1990.
239. Fortunko, C.M., Dube, W.P., and McColskey, J.D., Gas-coupled acoustic microscopy in the pulse-echo mode, in *Proc. 1993 IEEE Ultrasonics Symp.*, Levy, M. and McAvoy, B.R., Eds., IEEE, New York, 1993, 667.
240. McIntyre, C.S. et al., The use of air-coupled ultrasound to test paper, *IEEE Trans. UFFC*, 48, 717, 2001.
241. Maynard, J., Resonant ultrasound spectroscopy, *Phys. Today*, 26, 1996.
242. Migliori, A. and Sarrao, J.L., *Resonant Ultrasound Spectroscopy*, John Wiley & Sons, New York, 1997.
243. Schreiber, E. and Anderson, O.L., *Science*, 168, 1579, 1970.
244. Landau, L.D. and Lifshitz, E.M., *Statistical Physics*, Pergamon Press, London, 1958.
245. Lauterborn, W.J., Numerical investigation of nonlinear oscillations of gas bubbles in liquids, *J. Acoust. Soc. Am.*, 59, 283, 1976.
246. Walton, A.J. and Reynolds, G.T., Sonoluminescence, *Adv. Phys.*, 33, 595, 1984.
247. Neppiras, E.A., Acoustic cavitation, *Phys. Rep.*, 61, 159, 1980.
248. Keller, J.B. and Miksis, M., Bubble oscillations of large amplitude, *J. Acoust. Soc. Am.*, 68, 628, 1980.
249. Prosperetti, A., Crum, L.A., and Commander, K.W., Nonlinear bubble dynamics, *J. Acoust. Soc. Am.*, 83, 502, 1986.
250. Esche, R., *Acustica*, 2, 208, 1952.
251. Lauterborn, W. and Cramer, E., Subharmonic route to chaos observed in acoustics, *Phys. Rev. Lett.*, 47, 1445, 1981.
252. Ilychev, V.I., Koretz, V.L., and Melnikov, N.P., Spectral characteristics of acoustic cavitation, *Ultrasonics*, 27, 357, 1989.
253. Kamath, V., Prosperetti, A., and Egolopoulos, F.N., A theoretical study of sonoluminescence, *J. Acoust. Soc. Am.*, 94, 248, 1993.
254. Medwin, H. and Clay, C.S., *Fundamentals of Oceanography*, Academic Press, San Diego, 1998.
255. Wood, A.B., *A Textbook of Sound*, MacMillan, New York, 1955.
256. Cheeke, J.D.N., Single-bubble sonoluminescence: bubble, bubble, toil and trouble, *Can. J. Phys.*, 75, 77, 1997.
257. Gunther, P., Heim, E., and Borgstedt, H.U., *Z. Electrochem.*, 63, 43, 1959.
258. Flint, E.B. and Suslick, K.S., The temperature of cavitation, *Science*, 253, 1397, 1991.
259. Crum, L.A. and Reynolds, G.A., Sonoluminescence produced by stable cavitation, *J. Acoust. Soc. Am.*, 78, 137, 1985.
260. Gaitan, D.F., Crum, L.A., Church, C.C., and Roy, R.A., Sonoluminescence and bubble dynamics for a single, stable cavitation bubble, *J. Acoust. Soc. Am.*, 91, 3166, 1992.
261. Barber, B.P. and Putterman, S.J., Observation of synchronous picosecond sonoluminescence, *Nature*, 352, 1991.
262. Hiller, R.A. and Barber, B.P., Producing light from a bubble of air, *Sci. Am.*, 96, 1995.
263. Matula, T.J. et al., The acoustic emissions from single-bubble sonoluminescence, *J. Acoust. Soc. Am.*, 103, 1377, 1998.

264. Pecha, R. and Gompf, B., Microimplosions: cavitation collapse and shock wave emission on a nanosecond scale, *Phys. Rev. Lett.*, 84, 1328, 2000.
265. Wu, C.C. and Roberts, P.H., A model of sonoluminescence, *Proc. Roy. Soc. Lond.*, 445, 323, 1994.
266. Barber, B.P. et al., Defining the unknowns of sonoluminescence, *Phys. Rep.*, 281, 65, 1997.
267. Lohse, D. et al., Sonoluminescing air bubbles rectify argon, *Phys. Rev. Lett.*, 78, 1359, 1997.
268. Hammer, D. and Frommhold, L., Sonoluminescence: how bubbles glow, *J. Mod. Opt.*, 48, 239, 2001.
269. Matula, T.J. et al., Comparison of multibubble and single-bubble sonoluminescence spectra, *Phys. Rev. Lett.*, 75, 2602, 1995.
270. Hilgenfeldt, S. et al., Analysis of Rayleigh-Plesset dynamics for sonoluminescing bubbles, *J. Fluid Mech.*, 365, 171, 1998.
271. Matula, T.J. and Crum, L.A., Evidence for gas exchange in single bubble sonoluminescence, *Phys. Rev. Lett.*, 809, 865, 1998.
272. Holzfuss, J., Ruggberg, M., and Mettin, R., Boosting sonoluminescence, *Phys. Rev. Lett.*, 81, 1961, 1998.
273. Ketterling, J.A. and Apfel, R.E., Using phase space diagrams to interpret multiple frequency drive sonoluminescence, *J. Acoust. Soc. Am.*, 107, 819, 2000.
274. Holt, R.G. and Gaitan, D.F., Observation of stability boundaries in the parameter space of single bubble sonoluminescence, *Phys. Rev. Lett.*, 77, 3791, 1996.
275. Dan, M., Cheeke, J.D.N., and Kondic, L., Ambient pressure effect on single-bubble sonoluminescence, *Phys. Rev. Lett.*, 83, 1870, 1999.
276. Moran, M.J. et al., *Nucl. Instrum. Methods Phys. Res. Sect.*, B, 96, 651, 1995.
277. Gompf, B. et al., Resolving sonoluminescing pulse width with time-correlated single photo counting, *Phys. Rev. Lett.*, 79, 1405, 1997.
278. Moran, M.J. and Sweider, D., Measurements of sonoluminescence temporal pulse shape, *Phys. Rev. Lett.*, 80, 4987, 1998.
279. Cheng, H.Y. et al., How important are shock waves to single-bubble sonoluminescence, *Phys. Rev.*, B, 58, R2705, 1998.
280. Kondic, L., Yuan, C., and Chan, C.K., Ambient pressure and single-bubble sonoluminescence, *Phys. Rev.*, E, 57, R32, 1998.
281. Young, J.B., Schmiedel, T., and Kang, W., Sonoluminescence in high magnetic fields, *Phys. Rev. Lett.*, 77, 4816, 1996.
282. Yasui, K., Effect of a magnetic field on sonoluminescence, *Phys. Rev.*, E, 60, 1759, 1999.
283. Hilgenfeldt, S., Grossman, S., and Lohse, D., A simple explanation of light emission in sonoluminescence, *Nature*, 398, 402, 1999.
284. Moss, W.C., Clarke, D.B., and Young, D.A, Calculated pulse widths and spectra of a single sonoluminescing bubble, *Science*, 276, 1398, 1997.
285. Foster, F.S. et al., Principles and applications of ultrasound backscatter microscopy, *IEEE Trans. UFFC*, 40, 608, 1993.
286. Lethiecq, M. et al., Principles and applications of high frequency medical imaging, in *Advances in Acoustic Microscopy*, Vol. 2, Briggs, A. and Arnold, W., Eds., Plenum Press, New York, 1996, chap. 2.

Appendix A

TABLE A.1

Bessel Functions of the First Kind of Order 0 and 1, Together with the Directivity Function for a Piston

x	$J_0(x)$	$J_1(x)$	Pressure $\frac{2J_1(x)}{x}$	Intensity $\left(\frac{2J_1(x)}{x}\right)^2$
0.0	1.0000	0.0000	1.0000	1.0000
0.1	0.9975	0.0499	0.9988	0.9975
0.2	0.9900	0.0995	0.9950	0.9900
0.3	0.9776	0.1483	0.9888	0.9777
0.4	0.9604	0.1960	0.9801	0.9607
0.5	0.9385	0.2423	0.9691	0.9391
0.6	0.9120	0.2867	0.9557	0.9133
0.7	0.8812	0.3290	0.9400	0.8836
0.8	0.8463	0.3688	0.9221	0.8503
0.9	0.8075	0.4059	0.9021	0.8138
1.0	0.7652	0.4401	0.8801	0.7746
1.1	0.7196	0.4709	0.8562	0.7331
1.2	0.6711	0.4983	0.8305	0.6897
1.3	0.6201	0.5220	0.8031	0.6450
1.4	0.5669	0.5419	0.7742	0.5994
1.5	0.5118	0.5579	0.7439	0.5534
1.6	0.4554	0.5699	0.7124	0.5075
1.7	0.3980	0.5778	0.6797	0.4620
1.8	0.3400	0.5815	0.6461	0.4175
1.9	0.2818	0.5812	0.6117	0.3742
2.0	0.2239	0.5767	0.5767	0.3326
2.1	0.1666	0.5683	0.5412	0.2929
2.2	0.1104	0.5560	0.5054	0.2555
2.3	0.0555	0.5399	0.4695	0.2204
2.4	0.0025	0.5202	0.4335	0.1879
2.5	−0.0484	0.4971	0.3977	0.1581
2.6	−0.0968	0.4708	0.3622	0.1312
2.7	−0.1424	0.4416	0.3271	0.1070
2.8	−0.1850	0.4097	0.2926	0.0856
2.9	−0.2243	0.3754	0.2589	0.0670
3.0	−0.2601	0.3391	0.2260	0.0511
3.1	−0.2921	0.3009	0.1941	0.0377
3.2	−0.3202	0.2613	0.1633	0.0267
3.3	−0.3443	0.2207	0.1337	0.0179
3.4	−0.3643	0.1792	0.1054	0.0111

(continued)

TABLE A.1 (*continued*)

Bessel Functions of the First Kind of Order 0 and 1, Together with the Directivity Function for a Piston

x	$J_0(x)$	$J_1(x)$	Pressure $\frac{2J_1(x)}{x}$	Intensity $\left(\frac{2J_1(x)}{x}\right)^2$
3.5	−0.3801	0.1374	0.0785	0.0062
3.6	−0.3918	0.0955	0.0530	0.0028
3.7	−0.3992	0.0538	0.0291	0.0008
3.8	−0.4026	0.0128	0.0067	0.0000
3.9	−0.4018	−0.0272	−0.0140	0.0002
4.0	−0.3971	−0.0660	−0.0330	0.0011
4.1	−0.3887	−0.1033	−0.0504	0.0025
4.2	−0.3766	−0.1386	−0.0660	0.0044
4.3	−0.3610	−0.1719	−0.0800	0.0064
4.4	−0.3423	−0.2028	−0.0922	0.0085
4.5	−0.3205	−0.2311	−0.1027	0.0105
4.6	−0.2961	−0.2566	−0.1115	0.0124
4.7	−0.2693	−0.2791	−0.1188	0.0141
4.8	−0.2404	−0.2985	−0.1244	0.0155
4.9	−0.2097	−0.3147	−0.1284	0.0165
5.0	−0.1776	−0.3276	−0.1310	0.0172
5.1	−0.1443	−0.3371	−0.1322	0.0175
5.2	−0.1103	−0.3432	−0.1320	0.0174
5.3	−0.0758	−0.3460	−0.1306	0.0170
5.4	−0.0412	−0.3453	−0.1279	0.0164
5.5	−0.0068	−0.3414	−0.1242	0.0154
5.6	0.0270	−0.3343	−0.1194	0.0143
5.7	0.0599	−0.3241	−0.1137	0.0129
5.8	0.0917	−0.3110	−0.1073	0.0115
5.9	0.1220	−0.2951	−0.1000	0.0100
6.0	0.1506	−0.2767	−0.0922	0.0085
6.1	0.1773	−0.2559	−0.0839	0.0070
6.2	0.2017	−0.2329	−0.0751	0.0056
6.3	0.2238	−0.2081	−0.0661	0.0044
6.4	0.2433	−0.1816	−0.0568	0.0032
6.5	0.2601	−0.1538	−0.0473	0.0022
6.6	0.2740	−0.1250	−0.0379	0.0014
6.7	0.2851	−0.0953	−0.0285	0.0008
6.8	0.2931	−0.0652	−0.0192	0.0004
6.9	0.2981	−0.0349	−0.0101	0.0001
7.0	0.3001	−0.0047	−0.0013	0.0000
7.1	0.2991	0.0252	0.0071	0.0001
7.2	0.2951	0.0543	0.0151	0.0002
7.3	0.2882	0.0826	0.0226	0.0005
7.4	0.2786	0.1096	0.0296	0.0009
7.5	0.2663	0.1352	0.0361	0.0013
7.6	0.2516	0.1592	0.0419	0.0018
7.7	0.2346	0.1813	0.0471	0.0022
7.8	0.2154	0.2014	0.0516	0.0027
7.9	0.1944	0.2192	0.0555	0.0031
8.0	0.1717	0.2346	0.0587	0.0034

Appendix B

Acoustic Properties of Materials

The following tables are reprinted from the Specialty Engineering Associates (SEA) Web site (www.ultrasonic.com) with the permission of Johnson-Selfridge, P., and Selfridge, R. A., Approximate materials properties in isotropic materials, *IEEE Trans.*, UFFC SU-32, 381, 1985 (© IEEE, with permission). Notes and references on the abbreviations used are given at the end of the tables. Except where noted, the notation is the same as has been used throughout this book. For a list of vendors consult the SEA Web site.

Note that the units as originally expressed by the author have been modified to respect the convention used in this book.

TABLE B.1

Acoustic Properties of Solids and Epoxies

	Solid/Epoxy	V_L (10^3 m/s)	V_S (10^3 m/s)	ρ (10^3 kg/m^3)	Z_L (MRayl)	Poisson Ratio (σ)	Loss (dB/cm)
AS	Alumina	10.52		3.86	40.6		
CRC	Aluminum - rolled	6.42	3.04	2.7	17.33	0.355	
	AMD Res-in-all - 502/118, 5:1	2.67		1.35	3.61		
	AMD Res-in-all - 502/118, 9:1	2.73		1.35	3.68		
JA	Araldite - 502/956	2.62		1.16	3.04		
JA	Araldite - 502/956, 10phe C5W	2.6		1.23	3.19		
JA	Araldite - 502/956, 20phe C5W	2.54		1.39	3.52		
JA	Araldite - 502/956, 30phe C5W	2.41		1.5	3.62		
JA	Araldite - 502/956, 40phe C5W	2.31		1.67	3.86		
JA	Araldite - 502/956, 50phe C5W	2.13		1.95	4.14		
JA	Araldite - 502/956, 60phe C5W	2.1		2.24	4.7		
JA	Araldite - 502/956, 70phe C5W	1.88		3.17	5.95		
JA	Araldite - 502/956, 80phe C5W	1.72		4.71	8.11		
JA	Araldite - 502/956, 50phe 325mesh W	2.16		2.86	6.17		
JA	Araldite - 502/956, 60phe 325mesh W	1.91		2.78	5.33		
JA	Araldite - 502/956, 70phe 325mesh W	1.82		3.21	5.84		
JA	Araldite - 502/956, 80phe 325mesh W	1.64		4.55	7.45		
JA	Araldite - 502/956, 90phe 325mesh W	1.52		8.4	12.81		
AS	Arsenic tri sulphide As_2S_3	2.58	1.4	3.2	8.25	0.29	
	Bacon P38	4	2.17	1.9	7.6	0.29	13.5 @ 5
M	Bearing babbit	2.3		10.1	23.2		
CRC	Beryllium	12.89	8.88	1.87	24.1	0.046	
	Bismuth	2.2	1.1	9.8	21.5	0.33	
	Boron carbide	11		2.4	26.4		
PK	Boron nitride	5.03	3.86	1.965	9.88		
	Brass - yellow, 70% Cu, 30% Zn	4.7	2.1	8.64	40.6	0.38	
	Brick	4.3		1.7	7.4		

	Cadmium	2.8	1.5	8.6	24	0.3	
AS	Carbon aerogel	3.5		1.15	4.02		1.69 @ 5
AS	Carbon aerogel	3.14		0.85	2.67		5.68 @ 5
AS	Carbon - pyrolytic, soft, variable properties	3.31		2.21	7.31		
AS	Carbon - vitreous, very hard material	4.26	2.68	1.47	6.26	0.17	
AS	Carbon - vitreous, Sigradur K	4.63		1.59	7.38		
	Columbium (same as Niobium) m.p. 2468°C	4.92	2.1	8.57	42.4	0.39	
KF	Concrete	3.1		2.6	8		
CRC	Copper, rolled	5.01	2.27	8.93	44.6	0.37	
AS	DER317 - 9phr DEH20, 110phr W, r3	2.18	0.96	2.04	4.45	0.38	6.6 @ 2
AS	DER317 - 9phr DEH20, 115phr W, r3	1.93		2.37	4.58		
AS	DER317 - 9phr DEH20, 910phr T1167, r3	1.5		7.27	10.91		13.2 @ 2
AS	DER317 - 10.5phr DEH20 rt, outgass	2.75		1.18	3.25		
AS	DER317 - 10.5phr DEH20, 110phr W, r3	2.07		2.23	4.61		8.3 @ 2
AS	DER317 - 13.5phr mpda, 50phr W, r1	2.4		1.6	3.84		
AS	DER317 - 13.5phr mpda, 100phr W, r1	2.19		2.03	4.44		
AS	DER317 - 13.5phr mpda, 250phr W, r1	1.86	0.93	3.4	6.4	0.33	
AS	DER332 - 10phr DEH20, rt cure 48 hours	2.6		1.2	3.11		
AS	DER332 - 10.5phr DEH20, 10phr alumina, r2	2.61		1.26	3.29		
AS	DER332 - 10.5phr DEH20, 30phr alumina, r2	2.75		1.37	3.78		
AS	DER332 - 11phr DEH20, 150phr alumina, r2	3.25		1.83	5.95		
AS	DER332 - 14phr mpda, 30phr LP3, 70°C cure	2.59		1.25	3.24		8.3 @ 2
AS	DER332 - 15phr mpda, 25phr LP3, 76°C cure	2.55	1.18	1.24	3.16	0.36	7.4 @ 1.3
AS	DER332 - 15phr mpda, 30phr LP3, 80°C cure	2.66		1.24	3.3		8.8 @ 2
AS	DER332 - 15phr mpda, 50phr alumina, 60°C cure	2.8	1.43	1.49	4.18	0.32	
AS	DER332 - 15phr mpda, 60phr alumina, 80°C cure	2.78	1.45	1.54	4.27	0.31	
AS	DER332 - 15phr mpda, SiC, r5	3.9		2.24	8.74		
AS	DER332 - 15phr mpda, SiC, 25phr LP3, r5	3.75		2.15	8.06		
AS	DER332 - 15phr mpda, 6 micron W, r5	1.75		6.45	11.3		
AS	DER332 - 50phr V140, rt cure	2.34	0.97	1.13	2.64	0.4	
AS	DER332 - 64phr V140, rt cure	2.36		1.13	2.65		
AS	DER332 - 75phr V140, rt cure	2.35		1.12	2.62		

(continued)

TABLE B.1 ***(continued)***

Acoustic Properties of Solids and Epoxies

	Solid/Epoxy	V_L (10^3 m/s)	V_S (10^3 m/s)	ρ (10^3 kg/m^3)	Z_L (MRayl)	Poisson Ratio (σ)	Loss (dB/cm)
AS	DER332 - 100phr V140, rt cure	2.32		1.1	2.55		
AS	DER332 - 100phr V140, 30phr LP3, r8	2.27	1.13	2.55			7.5 @ 2, 11.2 @ 2.5
AS	DER332 - 100phr V140, 30phr LP3, r9	2.36		1.16	2.74		9.6 @ 2
AS	DER332 - 100phr V140, 50phr LP3, r8	2.32		1.13	2.63		12.0 @ 2
AS	DER332 - 50phr V140, 50phr St. Helens Ash, 60°C	2.43		1.94	6.24		
CRC	Duraluminin 17S	6.32	3.13	2.79	17.63	0.34	
AS	Duxseal	1.49		1.68	2.5		13.3 @ 0.5
AS	E.pox.e glue, EPX-1 or EPX-2, 100phA of B	2.44		1.1	2.68		8.4 @ 5
AS	Eccosorb - CR 124 - 2PHX of Y	2.62		4.59	12.01		9.4 @ 5
AH	Ecosorb - MF 110	2.61		1.6	4.2		
AH	Ecosorb - MF 112	2.4		2.19	5.25		
AH	Ecosorb - MF 114	2.29		2.9	6.65		
AH	Ecosorb - MF 116	2.45		3.69	9.02		
AH19	Ecosorb - MF 124	2.6		4.5	12		
AS	Eccosorb - MF 190	2.67		4.45	11.88		15.9 @ 4
	Epon - 828, mpda	2.829	1.23	1.21	3.4	0.45	
	Epotek - 301	2.64		1.08	2.85		
	Epotek - 330	2.57		1.14	2.94		
	Epotek - H70S	2.91		1.68	4.88		
AS	Epotek - V6, 10phA of B, r6	2.61		1.23	3.21		4.5 @ 2
AS	Epotek - V6, 10phA of B, r7	2.55		1.23	3.14		8 @ 2
AS	Epotek - V6, 10phA of B, 20phA LP3, r6	2.6		1.25	3.25		6 @ 2
AS	Epotek - V6, 10phA of B, 20phA LP3, r7	2.55		1.26	3.22		6 @ 2
DYNA	Fused silica	5.7	3.75	2.2	12.55	0.17	6.2e-5 @ 2
M	Germanium, mp = 937.4°C, transparent to infared	5.41		5.47	29.6		
	Glass - corning 0215 sheet	5.66		2.49	14.09		

	Glass - crown	5.1	2.8	2.24	11.4	0.28	
	Glass - FK3	4.91	2.85	2.26	11.1	0.245	
	Glass - FK6 (large minimum order)	4.43	2.54	2.28	10.1	0.25	
AE	Glass - flint	4.5		3.6	16		
	Glass - macor machinable code 9658	5.51		2.54	14		
	Glass - pyrex	5.64	3.28	2.24	13.1	0.24	
AE	Glass - quartz	5.5		2.2	12.1		
AE	Glass - silica	5.9		2.2	13		
	Glass - soda lime	6		2.24	13.4		
	Glass - TIK	4.38		2.38	10.5		
RB	Glucose	3.2		1.56	5		
CRC	Gold - hard drawn	3.24	1.2	19.7	63.8	0.42	
EM	Granite	6.5	2.7		17.6		
M	Hafnium, mp = 2150°C, used in reactor control rods C	3.84		13.29	51		
	Hydrogen, solid at 4.2 K	2.19		0.089	0.19		
AS	Hysol - CAW795/25 phr HW796 50°C	2.7		1.18	3.19		17.0 @ 5
BB	Hysol - C8-4143/3404	2.85		1.58	4.52		
BB	Hysol - C9-4183/3561	2.92		1.48	4.3		
BB	Hysol - C9-4183/3561, 15phe C5W	2.62		1.8	4.7		
BB	Hysol - C9-4183/3561, 30phe C5W	2.49		2.14	5.33		
BB	Hysol - C9-4183/3561, 45phe C5W	2.3		2.66	6.1		
BB	Hysol - C9-4183/3561, 57.5phe C5W	2.16		3.27	7.04		
AS	Hysol - EE0067/H3719 76°C, formerly C9-H905	2.53		1.93	4.88		22.4 @ 5
AS	Hysol - EE4183/HD3469 90°C	2.99		1.57	4.7		15.1 @ 5
AS	Hysol - EE4183/HD3469, 20phr 3μ Alumina	3.07		1.76	5.4		14.9 @ 5
	Hysol - ES 4212, 1:1	2.32		1.5	3.49		
	Hysol - ES 4412, 1:1	2.02		1.68	3.39		
BB	Hysol R8-2038/3404	2.59		1.18	3.05		
BB	Hysol R9-2039/3404	2.59		1.13	2.92		
BB	Hysol R9-2039/3469	2.61		1.17	3.07		
BB	Hysol R9-2039/3561	2.53		1.18	3		

(*continued*)

TABLE B.1 (*continued*)

Acoustic Properties of Solids and Epoxies

	Solid/Epoxy	V_L (10^3 m/s)	V_S (10^3 m/s)	ρ (10^3 kg/m^3)	Z_L (MRayl)	Poisson Ratio (σ)	Loss (dB/cm)
RLB	Hysol R9-2039/3561, 427phr WO3	2.15		3.51	7.54		33.5 @ 5
	Ice	3.99	1.98	0.917	3.66	0.34	
	Inconel	5.7	3	8.28	47.2	0.31	
	Indium	2.56		7.3	18.7		
	Iron	5.9	3.2	7.69	46.4	0.29	
	Iron - cast	4.6	2.6	7.22	33.2	0.27	
	Lead	2.2	0.7	11.2	24.6	0.44	
	Lead metaniobate	3.3		6.2	20.5		Q = 15
	Lithium niobate - 36° rotated Y-cut	7.08		4.7	33		
	Magnesium - various types listed in ref 'M'	5.8	3	1.738	10	0.32	
AE	Marble	3.8		2.8	10.5		
	Molybdenum	6.3	3.4	10	63.1	0.29	
	Monel	5.4	2.7	8.82	47.6	0.33	
	Nickel	5.6	3	8.84	49.5	0.3	
M	Niobium, m.p. = 2468°C	4.92	2.1	8.57	42.2	0.39	
AS	Paraffin	1.94		0.91	1.76		10.5 @ 1
RLB	Phillips 66 "Crystallor"	2.17	1.03	0.83	1.79	0.36	5.3 @ 5
CRC	Platinum	3.26	1.73	21.4	69.8	0.32	
RLB	Poco - DFP-1	3.09	1.73	1.81	5.61	0.27	1.2 @ 5
RLB	Poco - DFP-1C	3.2	1.81	3.2	11	0.31	2.0 @ 5
	Polyester casting resin	2.29		1.07	2.86		

AE	Porcelain	5.9		2.3	13.5		
	PSN, potassium sodium niobate	6.94		4.46	31		
	Pressed graphite	2.4		1.8	4.1		
	PZT 5H - Vernitron	4.44		7.43	33		
AS	PZT - Murata	4.72		7.95	37.5		
	PVDF	2.3		1.79	4.2		Q = 10
KF	Quartz - X-cut	5.75	2.2	2.65	15.3	0.42	
AS	Resin Formulators - RF 5407	3.06		2.16	6.61		14.9 @ 5
AS	Resin Formulators - RF 5407, 30 PHR LP3	2.56		1.92	4.92		54.7 @ 5
M	Rubidium, mp = 38.9, a 'getter' in vacuum tubes	1.26		1.53	1.93		
M	Salt - NaCl, crystalline, X-direction	4.78		2.17	10.37		
PK	Sapphire (aluminum oxide) Z-axis	11.1	6.04	3.99	44.3		
DP	Scotch tape - 0.0025" thick	1.9		1.16	2.08		
AS	Scotchcast XR5235, 38 pha B, rt cure	2.48		1.49	3.7		3.8 @ 1.3
	Scotchply SP1002 (a laminate with fibers)	3.25		1.94	6.24		
	Scotchply XP 241	2.84		0.65	1.84		
AS	Silicon - very anisotropic, values are approximate	8.43	5.84	2.34	19.7		
PK	Silicon carbide	13.06	7.27	3.217	42		

TABLE B.2

Longitudinal Wave Transducer Materials

Material	Z_3^D (MRayl)	V_3^D (10^3 m/s)	V_3^E (10^3 m/s)	Q_m	ε_{33}^s	k_t^2	k_p^2	ρ (10^3 kg/m^3)	tan δ	T_c (°C)	Ref.
Lithium niobate - 36°Y-cut	34.2	7.36		100	39	0.24	0.188	4.64	0.001	1150	[5]
K83 - modified lead metaniobate, after poling	25.6	5.95		110	150	0.169	finite	4.3		570	
K350 - lead zirconate titanate	33.7	4.381		75	790	0.249	0.307	7.7	0.024	360	[8]
PCM	35.7	4.82		150	270	0.291	strong	7.4	0.006		[5]
P3 - an inexpensive barium titanate	31.3	5.75		200	885	0.179	0.083	5.45	0.003	110	[5]
P5 - lead zirconate titanate	31.6	4.33		80	847	0.127	0.125	7.3	0.011	260	[5]
P6 - lead zirconate titanate	35.1	4.78		70	883	0.24	0.216	7.34	0.014	290	[5]
P7 - lead zirconate titanate	36	4.68		65	1000	0.259	0.315	7.69	0.019	320	[5]
"surface wave material"	37.4	4.709		1000	240	0.23	0.062	7.95	0.0014	280	[5]
"surface wave material"	37.2	4.683		1000	230	0.231	0.063	7.95	0.0016	280	[8]
"surface wave material" after repoling @200°C, 50V/0.001" for 5 min	37.4	4.706		1000	200	0.251	0.062	7.95	0.0016	280	[8]
LTZ1 - with plain electrode	35.6	4.682		500	640	0.254	0.294	7.6	0.007	350	[8]
LTZ1 - with wrap-around electrode	35.6	4.679		200	600	0.254	0.287	7.6	0.007	350	[8]
LTZ2 - with plain electrode	35.4	4.717		75	920	0.262	0.301	7.5	0.02	360	[8]
LTZ2 - with wrap-around electrode	34.4	4.583		100	830	0.259	0.29	7.5	0.019	360	[8]
LTZ5 - lead zirconate titanate	36.8	4.84		186	450	0.154	0.135	7.6	0.008	350	[5]
LTZ5 - lead zirconate titanate	36.5	4.803		200	370	0.157	0.13	7.6	0.01	350	[8]
PZT4 - lead zirconate titanate	36.1	4.82		500	635	0.233	0.219	7.5	0.008	328	[5]
PZT4 - lead zirconate titanate	34.5	4.6		500	635	0.263	0.336	7.5	0.004	328	[6]

PZT5A - lead zirconate titanate	34.5	4.445	3.97	75	870	0.24	0.285	7.75	0.023	365	[8]
PZT5A - lead zirconate titanate	33.7	4.35		75	830	0.236	0.36	7.75	0.02	365	[6]
PZT5H - lead zirconate titanate	32.6	4.35		50	1260	0.292	0.36	7.5	0.025	193	[5]
PZT5H - lead zirconate titanate	34.2	4.6		65	1470	0.255	0.423	7.5	0.02	193	[6]
PZT5H - lead zirconate titanate, pillar mode	27.4	3.66	2.59	65	1450	0.549	n.a.	7.5	0.02	193	[5]
PZT5H - lead zirconate titanate, array element mode	28.5	3.8	2.81	65	1365	0.502	n.a.	7.5	0.02	193	[5]
PZT8 - lead zirconate titanate, not as uniform as other Vernitron ceramics, brittle	35	4.6		1000	600	0.23	0.26	7.6	0.004	300	[6]
Pz 11 - lead barium titanate	30.5	5.5		500	1150	0.158	0.099	5.55	0.007	125	[5]
Pz 23 - lead zirconate titanate	34.4	4.56		100	900	0.241	0.259	7.55	0.02	350	[5]
Pz 24 - lead zirconate titanate	35.9	4.72		2000	310	0.246	0.243	7.6	0.014	330	[5]
Pz 25 - lead zirconate titanate	34	4.56		80	975	0.282	0.3	7.45	0.039	280	[5]
Pz 26 - lead zirconate titanate	35.2	4.62		1000	790	0.256	0.276	7.6	0.002	320	[5]
Pz 27 - lead zirconate titanate	34.8	4.51		60	930	0.257	0.298	7.7	0.024	350	[5]
Pz 29 - lead zirconate titanate	33.2	4.49		60	1300	0.296	0.332	7.4	0.03	235	[5]
Pz 32 - modified lead titanate	37.1	4.82		1000	250	0:181	0.02	7.7	0.0024	400	[5]
Pz 45 - bismuth titanate	34.8	4.83		1000	205	0.016	0!	7.2	0.004	500	[5]
Nova 7A - lead titanate	35.2	4.61		800	140	0.196	~0	7.63	0.009		[5]
PC11 - lead titanate	37.2	4.89		800	140	0.223	0!	7.6	0.0035	355	[5]
PC23 - lead zirconate titanate	35.9	4.68		60	2700	0.224	0.352	7.67	0.035	140	[5]
PC24 - lead zirconate titanate	36.7	4.76		100	1150	0.277	0.446	7.71	0.017	210	[5]
PC25 - lead zirconate titanate	37	4.64		120	530	0.23	0.455	7.97	0.016	360	[5]
PC26 - lead zirconate titanate	37	4.63		100	700	0.229	0.426	7.98	0.016	315	[5]

(continued)

TABLE B.2 (*continued*)

Longitudinal Wave Transducer Materials

Material	Z_3^D (MRayl)	V_3^D (10^3 m/s)	V_3^E (10^3 m/s)	Q_m	ε_{33}^s	k_t^2	k_p^2	ρ (10^3 kg/m^3)	tan δ	T_c (°C)	Ref.
EC64 - lead zirconate titanate, pillar mode	29.4	3.924	3.046	1800	668	0.447	n.a.	7.5	0.0016	320	[5]
EC64 - lead zirconate titanate, array element mode, PZT4D equivalent	30.5	4.065	3.155	400	650	0.447	n.a.	7.5	0.004	320	[5]
EC97 - lead titanate	34	5.08	4.38	950	188	0.295	0	6.7	0.009	240	[11]
EC98 - lead magnesium niobate	33.4	4.26	3.82	70	3230	0.231	0.263	7.85	0.02	170	[11]
EC69 - lead zirconate titanate, plate mode	41.5	5.53		80	619	0.265		7.5	0.038	300	[11]
Quartz - X-cut	15.21	5.74		106	4.5	0.0087	0.01	2.65	0.0001	575	[9]
ZnO, single crystal, hexagonal 6 mm Z-cut thin film	36	6.33			8.8	0.078		5.68	small		[10]
LT01 - lead titanate, plate mode	37.37	4.854		250	144.5	0.277	small	7.7	0.0033	300	[11]
SEA3 - lead zirconate titanate	34.94	4.48	4.02	35	1100	0.26	0.52	7.8	0.007	260	RLB
C5800 pillar mode	30.05	3.981	3.111	7740	500	0.4389		7.55	0.0103	300	AS

TABLE B.3

Shear Wave Transducer Materials

Material/Comments	Z_s (MRayl)	V_s (10^3 m/s)	Q_m	ε_r	k_t	ρ (10^3 kg/m^3)	tan δ	T_c (°C)	Ref.
Lithium niobate 163° Y-cut	20.6	4.44	100	58.1	0.305	4.64	0.001	1150	[5]
"surface wave material"	22.1	2.78	1000	360	0.25	7.95	0.0024	280	[7]
C5500	16.55	2.18	35	800	0.436	7.6	0.03	350	[5]
PZT-4	19.72	2.63	500	730	0.504	7.5	0.004	328	[6]
PZT-5A	17.52	2.26	75	916	0.469	7.75	0.02	365	[6]
PZT-5H	17.85	2.38	65	1700	0.456	7.5	0.02	193	[6]
PZT-8 not as uniform as other Vernitron ceramics, brittle	18.32	2.41	1000	900	0.303	7.6	0.004	300	[6]

TABLE B.4

Acoustic Properties of Plastics

	Plastic	V_L (10^3 m/s)	V_S (10^3 m/s)	ρ (10^3 kg/m^3)	Z_L (MRayl)	Poisson Ratio (σ)	Loss (dB/cm)
AS	ABS, Beige	2.23		1.03	2.31		11.1 @ 5
AS	ABS, Black, Injection molded (Grade T, Color #4500, "Cycolac")	2.25		1.05	2.36		10.9 @ 5
AS	ABS, Grey, Injection molded (Grade T, Color #GSM 32627)	2.17		1.07	2.32		11.3 @ 5
AS	Acrylic, Clear, Plexiglas G Safety Glazing	2.75		1.19	3.26	0.4	6.4 @ 5
AS	Acrylic, Plexiglas MI-7	2.61		1.18	3.08	0.4	12.4 @ 5
M	Bakelite	1.59		1.4	3.63		
AS	Cellulose Butyrate	2.14		1.19	2.56		21.9 @ 5
AS	Delrin, Black	2.43		1.42	3.45		30.3 @ 5
JA	Ethyl vinyl acetate, VE-630 (18% Acetate)	1.8		0.94	1.69		
JA	Ethyl vinyl acetate, VE-634 (28% Acetate)	1.68		0.95	1.6		
JA	Kydex, PVC Acrylic Alloy Sheet	2.218		1.35	2.99		
AS	Lexan, Polycarbonate	2.3		1.2	2.75		23.2 @ 5
AS	Lustran, SAN	2.51		1.06	2.68		5.1 @ 5
	Mylar	2.54		1.18	3		
AS	Kodar PETG, 6763, Copolyester	2.34		1.27	2.97		20.0 @ 5
	Melopas	2.9		1.7	4.93		7.2 @ 2.5
	Nylon, 6/6	2.6	1.1	1.12	2.9	0.39	2.9 @ 5
AS	Nylon, Black, 6/6	2.77		1.14	3.15		16.0 @ 5
PKY	Parylene C	2.15		1.4	3		0.1 @ 1
PKY	Parylene C	2.2		1.18	2.6		
PKY	Parylene D	2.1		1.36	2.85		
AS	Polycarbonate, Black, Injection molded (Grade 141R, Color No. 701, "Lexan")	2.27		1.22	2.77		22.1 @ 5
AS	Polycarbonate, Blue, Injection molded (Grade M-40, Color No. 8087, "Merlon")	2.26		1.2	2.72		23.5 @ 5

AS	Polycarbonate, Clear, Sheet Material	2.27		1.18	2.69		24.9 @ 5
CRC	Polyethylene	1.95	0.54	0.9	1.76		
	Polyethylene, high density, LB-861	2.43		0.96	2.33		
	Polyethylene, low density, NA-117	1.95	0.54	0.92	1.79	0.46	2.4 @ 5
AH	Polyethylene DFDA 1137 NT7	1.9		0.9	1.7		
	Polyethylene oxide, WSR 301	2.25		1.21	2.72		
	Polypropylene, Profax 6432, Hercules	2.74		0.88	2.4		5.1 @ 5
AS	Polypropylene, White, Sheet Material	2.66		0.89	2.36		18.2 @ 5
	Polystyrene, "Fostarene 50"	2.45		1.04	2.55		
AS	Polystyrene, "Lustrex," Injection molded (Resin #HF55-2020-347)	2.32	1.15	1.04	2.42		3.6 @ 5
	Polystyrene, Styron 666	2.4	1.15	1.05	2.52	0.35	1.8 @ 5
	Polyvinyl butyral, Butacite (used to laminate safety glass together)	2.35		1.11	2.6	0.37	
	PSO, Polysulfone	2.24		1.24	2.78		4.25 @ 2
AS	PVC, Grey, Rod Stock (normal impact grade)	2.38		1.38	3.27		11.2 @ 5
AS	Styrene Butadiene, KR 05 NW	1.92		1.02	1.95		24.3 @ 5
AS	TPX-DX845, Dimethyl pentene polymer	2.22		0.83	1.84		3.8 @ 1.3, 4.4 @ 4
AS	Valox, Black (glass filled polybutalene teraphlate "PBT")	2.53		1.52	3.83		15.7 @ 5
AS	Vinyl, Rigid	2.23		1.33	2.96		12.8 @ 5

TABLE B.5

Acoustic Properties of Rubbers

	Rubber	V_L (10^3 m/s)	ρ (10^3 kg/m^3)	Z_L (MRayl)	Loss (dB/cm)
	Adiprene LW-520	1.68	1.16	1.94	
	Butyl rubber	1.80	1.11	2.0	
AS	Dow Silastic Rubber GP45, 45 Durometer	1.02	1.14	1.16	23.4 @ 4
AS	Dow Silastic Rubber GP70, 70 Durometer	1.04	1.25	1.3	33.7 @ 4
AS	Ecogel 1265, 100PHA OF B, outgass, 80C	1.96	1.1	2.16	33.4 @ 2
AS	Ecogel 1265, 100PHA OF B, 100PHA Alumina, R4	1.7	1.4	2.38	>24.0 @ 1.3
AS	Ecogel 1265, 100PHA OF B, 1940PHA T1167, R4	1.32	9.19	12.16	14 @ 0.4
	Ecothane CPC-39	1.53	1.06	1.63	
	Ecothane CPC-41	1.52	1.01	1.54	
	Neoprene	1.6	1.31	2.1	
AS	Pellathane, Thermoplastic Urethane Rubber (55D durometer)	2.18	1.2	2.62	32.0 @ 5
AS	Polyurethane, GC1090	1.76	1.11	1.96	46.1 @ 4
BB	Polyurethane, RP-6400	1.5	1.04	1.56	
BB	Polyurethane, RP-6401	1.63	1.07	1.74	
LP	Polyurethane, RP-6401	1.71	1.07	1.83	100 @ 5
BB	Polyurethane, RP-6402	1.77	1.08	1.91	
BB	Polyurethane, RP-6403	1.87	1.1	2.05	
BB	Polyurethane, RP-6405	2.09	1.3	2.36	
BB	Polyurethane, RP-6410	1.33	1.04	1.38	
LP	Polyurethane, RP-6410	1.49	1.04	1.55	73.0 @ 5
BB	Polyurethane, RP-6413	1.65	1.04	1.71	
LP	Polyurethane, RP-6413	1.71	1.04	1.78	35.2 @ 5
BB	Polyurethane, RP-6414	1.78	1.05	1.86	
LP	Polyurethane, RP-6414	1.85	1.04	1.92	35.2 @ 5
BB	Polyurethane, RP-6422	1.6	1.04	1.66	
LP	Polyurethane, RP-6422	1.62	1.04	1.68	27.6 @ 5

AS	PR-1201-Q (MEDIUM), PHR 10, RT Cure	1.45	1.79	2.59	12.2 @ 2
	RTV-11	1.05	1.18	1.24	2.5 @ 0.8
	RTV-21	1.01	1.31	1.32	2.8 @ 0.8
	RTV-30	0.97	1.45	1.41	2.8 @ 0.8
	RTV-41	1.01	1.31	1.32	3.2 @ 0.8
	RTV-60	0.96	1.47	1.41	2.8 @ 0.8
AS	RTV-60/0.5% DBT @ 5.00 MHz	0.92	1.49	1.37	34.0 @ 5.00
AS	RTV-60/0.5% DBT @ 2.25 MHz	0.92	1.49	1.37	11.25 @ 2.25
AS	RTV-60/0.5% DBT @ 1.00 MHz	0.92	1.49	1.37	3.69 @ 1.00
AS	RTV-60/0.5% DBT @ 5.00 MHz/10 PHR Toluene	0.92	1.48	1.36	43.2 @ 5.00
AS	RTV-60/0.5% DBT @ 2.25 MHz/10 PHR Toluene	0.91	1.48	1.35	10.8 @ 2.25
AS	RTV-60/0.5% DBT @ 1.00 MHz/10 PHR Toluene	0.91	1.48	1.35	3.76 @ 1.00
AS	RTV-60/0.5% DBT @ 2.25 MHz/5 PHR Vitreous C	0.94	1.49	1.41	22.2 @ 2.25
AS	RTV-60/0.5% DBT @ 2.25 MHZ/10 PHR Vitreous C	0.96	1.51	1.45	13.1 @ 2.25
AS	RTV-60/0.5% DBT @ 1.00 MHz/13.6 PHR W, R11	0.86	1.68	1.44	
AS	RTV-60/0.5% DBT @ 1.00 MHz/21.3 PHR W, R11	0.83	1.87	1.55	
AS	RTV-60/0.5% DBT @ 1.00 MHz/40.8 PHR W, R11	0.8	2.04	1.64	
AS	RTV-60/0.5% DBT @ 1.00 MHz/69.5 PHR W, R11	0.73	2.39	1.73	
AS	RTV-60/0.5% DBT @ 1.00 MHz/85.2 PHR W, R11	0.71	2.52	1.78	
AS	RTV-60/0.5% DBT @ 1.00 MHz/100.0 PHR W, R11	0.69	2.75	1.89	

(*continued*)

TABLE B.5

Acoustic Properties of Rubbers

	Rubber	V_L (10^3 m/s)	ρ (10^3 kg/m^3)	Z_L (MRayl)	Loss (dB/cm)
AS	RTV-60/0.5% DBT @ 1.00 MHz/117.4 PHR W, R11	0.67	2.83	1.88	
	RTV-77	1.02	1.33	1.36	3.2 @ 0.8
	RTV-90	0.96	1.5	1.44	4.2 @ 0.8
	RTV-112	0.94	1.05	0.99	
	RTV-116	1.02	1.1	1.12	
	RTV-118	1.03	1.04	1.07	
	RTV-511	1.11	1.18	1.31	2.5 @ 0.8
AS	RTV-560, 0.6% DBT	0.99	1.41	1.4	2.2 @ 0.8, 8.4 @ 2
	RTV-577	1.08	1.35	1.46	3.8 @ 0.8
	RTV-602	1.16	1.02	1.18	4.35 @ 0.8
	RTV-615, use with 4155 primer	1.08	1.02	1.1	1 @ 0.8
	RTV-616	1.06	1.22	1.29	2.2 @ 0.8
	RTV-630	1.05	1.24	1.3	
AS	SOAB	1.6	1.09	1.74	15.5 @ 1
JF	Silly Putty, very lossy, hard to measure	1	1	1	
JA	Sylgard 170, a silicon rubber	0.974	1.38	1.34	
JA	Sylgard 182	1.027	1.05	1.07	
JA	Sylgard 184	1.027	1.05	1.04	
JA	Sylgard 186	1.027	1.12	1.15	

TABLE B.6

Acoustic Properties of Liquids

	Liquid	V_L (10^3 m/s)	$\Delta V/\Delta T$ (m/s°C)	ρ (10^3 kg/m^3)	Z_L (MRayl)	Loss, α (Np/cm)
	Acetate, butyl	1.27		0.871	1.02	
	Acetate, ethyl, $C_4H_8O_2$	1.19		0.9	1.069	
	Acetate, methyl, $C_3H_6O_2$	1.21		0.934	1.131	
M	Acetate, propyl	1.18		0.891	1.05	
LB	Acetone, $(CH_3)_2CO$ at 25°C	1.174	−4.5	0.791	1.07	54
M	Acetonitrile, C_2H_3N	1.29		0.783	1.01	
M	Acetonyl acetone, $C_6H_{10}O_2$	1.4		0.729	1.359	
M	Acetylendichloride, $C_2H_2Cl_2$	1.02		1.26	1.28	
M	Alcohol, butyl, C_4H_9OH at 30°C	1.24		0.81	1.003	74.3
CRC	Alcohol, ethanol, C_2H_5OH, at 25°C	1.207	−4	0.79	0.95	48.5
M	Alcohol, furfuryl, $C_5H_4O_2$	1.45		1.135	1.645	
LB	Alcohol, isopropyl, 2-Propanol, at 20°C	1.17		0.786	0.92	92
CRC	Alcohol, methanol, CH_3OH, at 25°C	1.103	−3.2	0.791	0.872	30.2
M	Alcohol, propyl (n) C_3H_7OH at 30°C	1.22		0.804	0.983	64.5
M	Alcohol, t-amyl, C_5H_9OH	1.2		0.81	0.976	
M	Alkazene 13, $C_{15}H_{24}$	1.32		0.86	1.132	
M	Aniline, $C_6H_5NH_2$	1.69		1.022	1.675	
DR	Argon, liquid at 87K	0.84		1.43	1.2	15.2
CRC	Benzene, C_6H_6, at 25°C	1.295	−4.65	0.87	1.12	873
M	Benzol	1.33		0.878	1.16	
M	Benzol, ethyl	1.34		0.868	1.16	
	Bromobenzene C_6H_5Br at 22°C	1.167		1.522	1.776	1.63
M	Bromoform, $CHBr_3$	0.92		2.89	2.67	
M	t-Butyl chloride, C_4H_9Cl	0.98		0.84	0.827	
M	Butyrate, ethyl	1.17		0.877	1.03	
M	CARBITOL, $C_6H_{14}O_3$	1.46		0.988	1.431	

(continued)

TABLE B.6 (*continued*)

Acoustic Properties of Liquids

	Liquid	V_L (10^3 m/s)	$\Delta V/\Delta T$ (m/s°C)	ρ (10^3 kg/m^3)	Z_L (MRayl)	Loss, α (Np/cm)
CRC, M	Carbon disulphide, CS_2 at 25°C	1.149		1.26	1.448	
DR	Carbon disulphide, CS_2, 25°C, 3 GHz	1.31		1.221	1.65	10.1
CRC, M	Carbon tetrachloride, CCl_4, at 25°C	0.926	−2.7	1.594	1.48	538
M	Cesium at 28.5°C the melting point	0.967		1.88	1.82	
LB	Chloro-benzene, C_6H_5Cl, at 22°C	1.304		1.106	1.442	167
M	Chloro-benzene, C_6H_5Cl	1.3		1.1	1.432	
CRC,M	Chloroform, $CHCl_3$, at 25°C	0.987	−3.4	1.49	1.47	
M	Cyclohexanol, $C_6H_{12}O$	1.45		0.962	1.4	
M	Cyclohexanone, $C_6H_{10}O$	1.42		0.948	1.391	
M	Diacetyl, $C_4H_6O_2$	1.24		0.99	1.222	
M	1, 3 Dichloroisobutane $C_4H_{18}Cl_2$	1.22		1.14	1.39	
M	Diethyl ketone	1.31		0.813	1.07	
M	Dimethyl phthalate, $C_8H_{10}O_4$	1.46		1.2	1.758	
M	Dioxane	1.38		1.033	1.425	
CRC, M	Ethanol amide, C_2H_7NO, at 25°C	1.724	−3.4	1.018	1.755	
CRC, M	Ethyl ether, $C_4H_{10}O$, at 25°C	0.985	−4.87	0.713	0.7023	
M	d-Fenchone	1.32		0.94	1.241	
M	Florosilicone oil, Dow FS-1265	0.76				
M	Formamide, CH_3NO	1.62		1.134	1.842	
M	Furfural, $C_5H_4O_2$	1.45		1.157	1.67	
3m	Fluorinert FC-40	0.64		1.19	1.86	
3m	Fluorinert FC-70	0.687		1.94	1.33	
3m	Fluorinert FC-72	0.512		1.68	0.86	
3m	Fluorinert FC-75	0.585		1.76	1.02	
3m	Fluorinert FC-77	0.595		1.78	V	
3m	Fluorinert FC-104	0.575		1.76	1.01	
3m	Fluorinert FG-43	0.655		1.85	1.21	
LB	Fluoro-benzene, C_6H_5F, at 22°C	1.18		1.024	1.205	317

AS	Freon, TF	0.716		1.57	1.12	
DR	Gallium at 30°C mp = 28.8°C (expands 3% when it freezes)	2.87		6.09	17.5	1.58
M	Gasoline	1.25		0.803	1	
CRC	Glycerin - $CH_2OHCHOHCH_2OH$, at 25°C	1.904	−2.2	1.26	2.34	
M	Glycol - 2,3 butylene	1.48		1.019	1.511	
M	Glycol - diethylene $C_4H_{10}O_3$	1.58		1.116	1.77	
CRC	Glycol - ethylene 1,2-ethanediol @ 25°C	1.658	−2.1	1.113	1.845	120
JA	Glycol - ethylene Preston II	1.59		1.108	1.76	
JA	Glycol - polyethylene 200	1.62		1.087	1.75	
JA	Glycol - polyethylene 400	1.62		1.06	1.71	
M	Glycol - polypropylene (Polyglycol P-400) at 38°C	1.3				
M	Glycol - polypropylene (Polyglycol P-1200) at 38°C	1.3				
M	Glycol - polypropylene (Polyglycol E-200) at 29°C	1.57				
M	Glycol - tetraethylene $C_8H_{18}O_6$	1.58		1.12	1.784	
M	Glycol, triethylene, $C_6H_{14}O_4$	1.61		1.123	1.81	
DR	Helium-4, liquid at 0.4 K	0.238		0.147	0.035	1.73
DR	Helium-4, liquid at 2 K	0.227		0.145	0.033	70
DR	Helium-4, liquid at 4.2 K	0.183		0.126	0.023	226
	n-Hexane, C_6H_{14}, liquid at 30°C	1.103		0.659	0.727	87
M	n-Hexanol, $C_6H_{14}O$	1.3		0.819	1.065	
AS	Honey, Sue Bee Orange	2.03		1.42	2.89	
	Hydrogen, liquid at 20 K	1.19		0.07	0.08	5.6
LB	Iodo-benzene, C_6H_5I, at 22°C	1.104		1.183	2.012	242
M	Isopentane, C_5H_{12}	0.992		0.62	0.615	
CRC, M	Kerosene	1.324	−3.6	0.81	1.072	
M	Linalool	1.4		0.884	1.23	
CRC	Mercury at 25.0°C	1.45		13.5	19.58	5.8
M	Mesityloxide, $C_6H_{16}O$	1.31		0.85	1.115	
M	Methylethylketone	1.21		0.805	0.972	
M	Methylene iodide	0.98				
M	Methyl napthalene, $C_{11}H_{10}$	1.51		1.09	1.645	
M	Monochlorobenzene, C_6H_5Cl	1.27		1.107	1.411	

(continued)

TABLE B.6 (*continued*)

Acoustic Properties of Liquids

	Liquid	V_L (10^3 m/s)	$\Delta V/\Delta T$ (m/s°C)	ρ (10^3 kg/m^3)	Z_L (MRayl)	Loss, α (Np/cm)
M	Morpholine, C_4H_9NO	1.44		1	1.442	
DR	Neon, liquid at 27 K	1.2		1.2	0.72	23.1
LB	Nicotin, $C_{10}H_{14}N_2$, at 20°C	1.49		1.01	1.505	
CRC,M	Nitrobenzene, $C_6H_6NO_2$, at 25°C	1.463	−3.6	1.2	1.756	
DR	Nitrogen, N_2, liquid at 77 K	0.86		0.8	0.68	13.8
M	Nitromethane CH_3NC_2	1.33		1.13	1.504	
JA	Oil - baby	1.43		0.821	1.17	
CRC,M	Oil - castor, $C_{11}H_{10}O_{10}$ @ 25°C	1.477	−3.6	0.969	1.431	
GD	Oil - castor, @ 20.2°C @ 4.224 MHz	1.507		0.942	1.42	10100
JA	Oil - corn	1.46		0.922	1.34	
M	Oil - diesel	1.25				
M	Oil - gravity fuel AA	1.49		0.99	1.472	
MH	Oil - jojoba	1.455		1.17	1.24	
JA	Oil - linseed	1.46		0.94	1.37	
M	Oil - linseed	1.77		0.922	1.63	
JA	Oil - mineral, light	1.44		0.825	1.19	
JA	Oil - mineral, heavy	1.46		0.843	1.23	
JA	Oil - olive	1.445		0.918	1.32	
M	Oil - paraffin	1.42		0.835	1.86	
JA	Oil - peanut	1.436		0.914	1.31	
M	Oil - SAE 20	1.74		0.87	1.51	
	Oil - SAE 30	1.7		0.88	1.5	
JA	Oil - silicon Dow 200, 1 centistoke	0.96		0.818	0.74	
JA	Oil - silicon Dow 200, 10 centistoke	0.968		0.94	0.91	
JA	Oil - silicon Dow 200, 100 centistoke	0.98		0.968	0.95	
JA	Oil - silicon Dow 200, 1000 centistoke	0.99		0.972	0.96	
MH	Oil - silicon Dow 704 @ 79°F	1.409		1.02	1.437	
MH	Oil - silicon Dow 705 @ 79°F	1.458		1.15	1.68	

GD	Oil - silicon Dow 710 @ 20°C	1.352		1.11	1.5	8200
JA	Oil - safflower	1.45		0.9	1.3	
JA	Oil - soybean	1.43		0.93	1.32	
M	Oil - sperm	1.44		0.88	1.268	
JA	Oil - sunflower	1.45		0.92	1.34	
M	Oil - transformer	1.391.39		0.92	1.28	
JA	Oil - wintergreen (methyl salicylate)	1.38		1.6	1.6	
DR	Oxygen, O_2, liquid at 90 K	0.9		1.11	1	9.9
M	Paraffin at 15°C	1.3				
	n-Pentane, C_5H_{12}, liquid at 15°C	1.027		0.626	0.642	100
M	Polypropylene oxide (Ambiflo) at 38°C	1.37				
M	Potassium at 100°C, mp = 63.7°C (see 'M' for other temps)	1.82		0.83	1.51	
M	Pyridine	1.41		0.982	1.39	
M	Sodium, liquid at 300°C (see 'M' for other temps)	2.42		8.81	21.32	
M	Solvesso #3	1.37		0.877	1.202	
AS	Sonotrack couplant	1.62		1.04	1.68	
M	Tallow at 16°C	0.39				
M	Thallium, mp = 303.5°C, used in photocells	1.62		11.9	19.3	
M	Trichorethylene	1.05		1.05	1.1	
CRC	Turpentine, at 25°C	1.255		0.88	1.104	
M	Univis 800	1.35		0.87	1.191	
M	Water - heavy, D_2O	1.4		1.104	1.54	
M	Water - liquid at 20°C	1.48		1	1.483	
CRC, DR	Water - liquid at 25°C	1.4967	2.4	0.998	1.494	22
	Water - liquid at 30°C	1.509		1	1.509	19.1
DR	Water - liquid at 60°C (temps up to 500°F listed in 'CRC')	1.55		1	1.55	10.9
M	Water - salt 10%	1.47				
M	Water - salt 15%	1.53				
M	Water - salt 20%	1.6				
CRC	Water - sea, at 25°C	1.531	2.4	1.025	1.569	
DR	Xenon - liquid at 166 K	0.63		2.86	1.8	22
CRC,M	Xylene Hexafloride, $C_8H_4F_6$, at 25°C	0.879		1.37	1.222	
M	m-Xylol, C_8H_{10}	1.32		0.864	1.145	

TABLE B.7

Acoustic Properties of Gases

	Gas	V_L (10^3 m/s)	$\Delta V/\Delta T$ (m/s°C)	ρ (kg/m^3)	Z_L (kRayl)
CRC	Acetone vapor (C_2H_6O) at 97.1°C	0.239	0.32		
CRC	Air - dry at 0°C	0.33145	0.59	1.293	0.4286
M	Air - at 0°C, 25 atm	0.332			
M	Air - at 0°C, 50 atm	0.335			
M	Air - at 0°C, 100 atm	0.351			
M	Air - at 20°C	0.344			
M	Air - at 100°C	0.386			
M	Air - at 500°C	0.553			
CRC	Ammonia (NH_3) at 0°C	0.415		0.771	0.32
CRC	Argon - at 0°C	0.319	0.56	1.783	0.569
CRC	Benzene vapor (C_6H_6) at 97.1°C	0.202	0.3		
CRC	Cardon monoxide (CO) at 0°C	0.338	0.6	1.25	0.423
CRC	Carbon dioxide (CO_2) at 0°C	0.259	0.4	1.977	0.512
M	Carbon disulfate	0.189			
CRC	Carbon tetrachloride vapor (CCl_4) @97.1°C	0.145			
CRC	Chlorine at 0°C	0.206		3.214	0.662
CRC	Chloroform - $CH(Cl)_3$ at 97.1°C	0.171	0.24		
CRC	Deuterium at 0°C	0.89	1.6	0.19	0.1691
CRC	Ethane - C_2H_6 at 0°C	0.308		1.356	0.418
CRC	Ethylene - C_2H_4 at 0°C	0.317		1.26	0.4
CRC	Ethanol vapor - C_2H_5OH at 97.1°C	0.269	0.4		
CRC	Ethyl ether - $C_4H_{10}O$ at 97.1°C	0.206	0.3		
CRC	Helium at 0°C	0.965	0.8	0.178	0.172
CRC	Hydrogen at 0°C	1.284	2.2	0.0899	0.1154
CRC	Hydrogen bromide - HBr at 0°C	0.2		3.5	0.7
CRC	Hydrogen chloride - HCl at 0°C	0.296		1.639	0.485
CRC	Hydrogen iodide - HI at 0°C	0.157		5.66	0.889
CRC	Hydrogen sulfide - H_2S at 0°C	0.289		1.539	0.445
CRC	Methane - CH_4 at 0°C	0.43		0.7168	0.308
CRC	Methanol vapor - CH_3OH at 97.1°C	0.335	0.46		
CRC	Neon - at 0°C	0.435		0.9	0.392
CRC	Nitric oxide - NO at 10°C	0.324		1.34	0.434
CRC	Nitrogen - N_2 at 0°C	0.334	0.6	1.251	0.418
CRC	Nitrous oxide - N_2O at 0°C	0.263	0.5	1.977	0.52
CRC	Oxygen - O_2 at 0°C	0.316	0.56	1.429	0.451
M	Oxygen - O_2 at 20°C	0.328		1.32	0.433
CRC	Sulfur dioxide - SO_2 at 0°C	0.213	0.47	2.927	0.623
M	Water vapor at 0°C	0.401			
M	Water vapor at 100°C	0.405			
CRC	Water vapor at 134°C	0.494			

Abbreviations

AE = *Handbook of Tables for Applied Engineering Sciences*

AH = Andy Hadjicostis, Nutran Company, 206-348-3222.

AJS = A.J. Slobodnik, R.T. Delmonico, and E.D. Conway, *Microwave Acoustics Handbook*, Vol. 3: Bulk Wave Velocities, Internal Report RADC-TR-80-188 (May 1980), Rome Air Development Center, Air Force Systems Command, Griffiths Air Force Base, New York 13441.

AS = Alan Selfridge, Ph.D., Ultrasonic Devices, Inc.

CRC = *Handbook of Chemistry and Physics*, 45th ed., Chemical Rubber Company, Cleveland, OH, p. E-28.

DP = Don Pettibone, Ph.D., Diasonics, Sunnyvale, CA.

FS = Fred Stanke, Ph.D., Schlumberger, Inc., Ridgefield, CT, private communication.

GD = Genevieve Dumas, *IEEE Trans. Sonics Ultrason.*, Mar. 1983.

JF = John Fraser, Ph.D., ATL, Bothell, WA.

KF = Kinsler and Frey, *Fundamentals of Acoustics*, John Wiley and Sons, 1962.

LB = Schaaffs, W., *Numerical Data and Functional Relationships in Science and Technology, New Series Group II*: and *Molecular Physics*, Vol. 5: Molecular Acoustics, K.H. Hellwege and A.M. Hellwege, Eds., Springer-Verlag, Berlin, 1967. (This reference contains velocity and density information for just about any organic liquid. Other volumes in this work contain much information on various anisotropic solids and crystals.)

LP = Laust Pederson

M = MetroTek Inc., Application Note 23.

ME = Materials engineering, Dec. 1982.

RB = Rick Bauer, Ph.D., Hewlett Packard, Page Mill Road, Palo Alto, CA.

RLB = Ram lal Bedi, Ph.D., formerly with Specialty Engineering Associates, Milpitas, CA.

SIM = Simmons, G. and Wang, H., *Single Crystal Elastic Constants and Calculated Aggregate Properties*, 2nd ed., MIT Press, Cambridge, MA, XV, 370, 1971.

© Ultrasonic Devices Inc., 1996.

T_c = Curie temperature

ε_r = Relative dielectric constant, multiply by $8.84 \cdot 10^{-12}$ for MKS units (F/m)

ε_{33} = Unclamped dielectric constant

k_t = Coupling coefficient between E_3 and thickness mode

k_p = Planar (radial) moe coupling coefficient

$\tan\delta$ = loss tangent (dimensionless)

V_3^D = Velocity corresponding to antiresonance (open circuit)

V_3^E = Velocity corresponding to resonance
V_S = Shear velocity
Z_S = Shear impedance times 10^{-6} kg·m^2/s
Z_3^D = Longitudinal wave impedance corresponding to antiresonance times 10^{-6} kg·m^2/s
$\frac{\Delta V}{\Delta T}$ = Change in acoustic velocity per change in temperature in m/s °C.

Loss, or attenuation, is given in several different formats in these tables. The most specific way is with the @ symbol. The number before the @ is the loss in dB/cm, the number after the @ symbol is the frequency at which the attenuation was measured in MHz. For liquids the attenuation is given in Np/cm. To get loss in dB/cm multiply α by 8.686 $* f^2$ where f is the frequency of interest in Hz. This representation obviously assumes that loss increases in proportion to frequency squared, and is most commonly used for low-loss materials such as glass and liquids.

Transducer modeling programs will commonly assume loss increases only in proportion to the first power. If this is the case, then it is appropriate to use the material quality factor, or acoustic Q. To convert between dB/cm and Q, the following equations can be useful:

$$Q = \frac{2 * \pi * (Stored\ energy)}{Energy\ dissipated\ per\ cycle}$$

$$Q = W_0 \frac{Stored\ energy}{Average\ power\ loss}$$

$$Q = \frac{86.9 * \pi * f}{((dB/cm) * velocity)}$$

References

1. Selfridge, A.R., Design and Fabrication of Ultrasonic Transducer Arrays, Ph.D. thesis, Stanford University, Stanford, CA, 1982. Available from University Microfilms, Ann Arbor, MI.
2. Krimholtz, R., Leedom, D.A., and Matthei, G.I., New equivalent circuits for elementary piezoelectric transducers, *Electron. Lett.*, 6, 398, 1970.
3. Mason, W.P., *Electromechanical Transducers and Wave Filters*, Van Nostrand, Princeton, NJ, 1948.
4. Fraser, J.D., The Design of Efficient Broadband Ultrasonic Transducers, Ph.D. thesis, Stanford University, Stanford, CA, 1979.
5. Measured by Alan Selfridge using a vector impedance meter and curve fitting techniques.
6. Vernitron Piezoelectric Division, *Piezoelectric Technol. Data Designers*, 216-232-8600.

7. Private correspondence with Murata.
8. As in [5] though later date.
9. ITT, *Reference Data for Engineers*, 6th ed., H.W. Sams & Co.
10. Kino, G.S., *Acoustic Waves: Devices, Imaging and Analogue Signal Processing*, Prentice-Hall, Englewood Cliffs, NJ, 1987.
11. Same as in [5] except impedance data were measured using a Tektronix 2430 digitizing oscilloscope.
12. Auld, B.A., *Acoustic Fields and Waves in Solids*, Wiley-Interscience, New York, 1973.
13. Ristic, V.M., *Principles of Acoustic Devices*, John Wiley & Sons, New York, 1983.

Appendix C

Complementary Laboratory Experiments

A system of group projects was developed during the evolution of the subject matter of this book when used for teaching purposes. One format involved the use of weekly problem sets for the fundamental part of the material (Chapters 2 through 10), similar in type and level to the questions found at the end of these chapters. During the second part of the course, two alternative schemes were used. One involved the assignment of term papers on a special topic, examples of which are given at the end of this section. The other, and more elaborate approach, consisted of experimental projects. These projects were open-ended as opposed to set-piece laboratory experiments. What was actually done depended on the students' backgrounds, availability of equipment, and qualified instructors. Hence it is stressed that the notes given below should be seen as guidelines or suggestions as to how a suitable laboratory component could be set up and not as formal, ready-to-use laboratory methodology descriptions.

For this second part of the course, students were divided into teams of two or three. A term project was carried out by each team, enabling the students to go more in depth in a given area than they could have done otherwise. Students were asked to divide up tasks in theory/computer calculation on the one hand and experimental testing on the other. Typical subject areas are given below. The approach was very flexible, a particular aspect being worked out in consultation with the teacher, and the actual work carried out under the guidance of a graduate student. The projects were for approximately 1 month, after which the group compiled a single report synthesising the work of all of the participants. The work was then presented in a series of short oral presentations; instruction was given to assist in preparing the report and making the presentation, which was of a length and style similar to that of conference presentations. The advantage of this approach was that students were generally very motivated to learn the theoretical part and to carry out a successful project. Learning to work in a team and acquiring communication skills were other advantages of this approach.

The required material was largely accessible from research laboratories. Computing requirements were modest and in all cases could be met with the departmental PCs. The laboratory equipment available included:

1. HP Model 4195A Network/Spectrum Analyzer
2. One of the following:
   a. MATEC RF tone burst ultrasonic generator and receiver (10 to 90 MHz)
   b. RITEC RAM 10000 tone burst ultrasonic generator and receiver (1 to 100 MHz)
   c. UTEX UT 320/340 Pulser/receiver or equivalent, such as those produced by Panametrics or Metrotek (tone burst systems are ideal for this type of experiment as they allow easy control and variation of the frequency and quantitative verification of frequency-dependent effects)
3. Standard RF attenuators, cables, etc.
4. Laboratory oscilloscope, ideally digital scope with FFT capability, such as the 300 MHz LeCroy digital oscilloscope

A list of typical projects is given below, with notes on particular aspects that can be easily investigated and compared with theory. This list is by no means exhaustive, and it is easy to extend it by the procurement of modest additional resources, such as focusing transducers, additional buffer rods, means of temperature variation and control, magnetic field etc.

1. Transducer characterization

   It is useful to obtain a collection of piezoelectric transducers from various sources. Commercially packaged resonators can easily be obtained in the range 1 to 20 MHz, as can unmounted transducers, longitudinal or transverse, with either fundamental or overtone polish from suppliers such as Valpey Fisher Inc. In the latter case, $LiNbO_3$ transducers with a fundamental in the range of 5 to 15 MHz and with overtone polish are the most convenient choice, typically 5 or 6 mm in diameter.

   Transducer characterization is best made with respect to a well-defined equivalent circuit. This could be a series resonant circuit in parallel with the static capacitance (Butterworth–Van Dyke equivalent circuit for resonators) or the full Mason Model for a loaded transducer. Suggested experiments include:

   a. Characterization of the resonance of an unloaded transducer (resonator) using the network analyzer; determination of transducer parameters by measurement of amplitude and phase response, as well as series and parallel resonant frequencies; identification of harmonic frequencies; effects of liquid loading on the resonance for both longitudinal and transverse polarization.

b. Frequency response of a transducer glued to a buffer rod, with air loading on the opposite face. Points to verify include:
   (i) Frequency response of the odd harmonics.
   (ii) Use of inductors/RF transformers to increase the transducer response.
   (iii) Observation of echoes in the buffer rod.
   (iv) Comparison of shape of the first echo with that of the exciting RF pulse; effect of bond quality on the echo shape.

2. Bulk acoustic wave (BAW) propagation

Experiments in this section are based around the use of a transducer mounted on the end of a buffer rod. Ideally, buffer rods made of materials such as fused quartz, sapphire, etc. can be obtained with end faces optically polished and parallel from suppliers such as Valpey–Fisher. Otherwise, for studies in the low MHz range, it is possible to machine and polish the end faces of materials such as perspex, duraluminium, brass, stainless steel, etc., using standard workshop practices to obtain usable echo trains. Duraluminium is particularly useful due to its low attenuation and its machinability.

The buffer rod should have dimensions of the order of 1 cm in length and 1 cm in diameter; these dimensions are not critical and should be chosen so that the rod diameter is significantly greater than that of the transducer, with the buffer long enough so that clearly separated, nonoverlapping echoes are observed on the oscilloscope. Longitudinal transducers with overtone polish and a fundamental frequency of 5 or 10 MHz are recommended for the experiments of this section. Such experiments include:

a. Mount the transducer on the end of the buffer rod with a suitable ultrasonic couplant; vacuum grease or silicon oil are convenient, as they give a good bond at room temperature which is stable for a few hours and is easily changed. The transducer bond can be improved by wringing it onto the buffer surface using a soft rubber eraser, for example.

b. Tuning the generator to the transducer fundamental frequency; observing echoes. Existence or not of an exponential decay of the echo amplitudes should be registered. Transducer bond can be optimized to give maximum echo amplitude.

c. Estimation of V_L and comparison with the handbook value; estimation of absolute and relative error.

d. Using the same transducer bond as above, steps (b) and (c) should be repeated at odd harmonic frequencies up to the maximum attainable values with the ultrasonic generator used. Variation of the overall modulation of the echo train and the number of echoes is particularly significant. How can these be explained for the particular buffer rod used?

e. For a machined buffer rod, remachine one end face so that there are now nonparallel end faces to within a degree or so. Repeat step (d) and explain any observed variation in the modulation of the echo train.

3. BAW reflection and transmission

These experiments are most conveniently carried out with a buffer rod with the end opposite the transducer partially immersed in a liquid. In this configuration it is possible to measure reflection at normal incidence and transmission and reflection from a plate immersed in the liquid. The appropriate theoretical values can be calculated using the theory of Chapter 7. Recommended experiments are:

a. Use a 5- or 10-MHz longitudinal wave transducer bonded to one end of the buffer rod as in experiment #2; prepare buffer rods of plexiglass, duraluminum, and stainless steel, which form a convenient trio of buffer rods that have low, medium, and high acoustic mismatch to liquids such as water; design and construct sample holders to enable the far end to be immersed in a fluid bath.

b. Pulse echo experiments at low frequency in bare buffer rod; adjustment for obtaining maximum number of echoes.

c. Exposure of the end of the buffer rod to the fluid in question; recording of the echo pattern and comparison with that for the unexposed rod; calculation of the reflection coefficient for each echo; draw conclusions on the accuracy of the method vs. echo number.

d. Systematic study of the three buffer rods against three different liquids with significantly different acoustic impedances; compare with theory.

e. For a given liquid-solid combination at a given frequency, calculate the material and thickness of the layer needed to minimize the reflected signal; attempt to verify this result experimentally.

f. Repeat (c) for the case where there is a reflecting plate immersed in the liquid; trace possible ray paths for various returning echoes in the buffer; compare with experiment to identify all observed echoes; estimate the reflection coefficient at the fluid-plate interface.

4. SAW device fabrication, measurement, and sensor applications

IDTs operating at about 50 MHz can be made very easily in a standard darkroom using photolithography techniques using the following materials; Y-Z $LiNbO_3$ SAW plates, about 15 mm long, 10 mm wide, and 0.5 mm thick; mask for standard transmitter–receiver transducer design, required to have an impedance of 50 Ω when used with the chosen substrate; 10 finger pairs for two transducers about 10 mm apart, aperture approximately 5 mm for Y-Z lithium niobate. The steps for transducer fabrication are as follows:

a. Clean the substrate with acetone and soak in methanol.
b. Deposit approximately 200 nm film of aluminum by flash evaporation.
c. Deposit a photo-resist film by pipette on the substrate in yellow light conditions. Incline the substrate to drain off excess photo-resist.
d. Bake the photo-resist film at 120°C for at least 15 min to harden the film.
e. Clamp the mask on top of the photo-resist film and expose to ultraviolet light for the recommended time.
f. Remove the mask in darkness and dip the substrate for a few moments in NaOH to remove the exposed portions of the photo-resist. The remaining photo-resist protects the aluminum during etching.
g. Etch the plate in a solution of HNO_3, HCl, and H_2O, removing it rapidly at the required moment to avoid overetching.
h. Thoroughly rinse the plate and then remove excess photo-resist with a small amount of NaOH.

If sufficient time and facilities are not available for in-house fabrication, then finished SAW plates with IDTs can be bought from the manufacturer.

A number of instructive experiments can be carried out using the SAW device. These include:

a. Testing the frequency response with the network analyzer: a power splitter can be used to provide a reference signal, enabling tracing of the insertion loss as a function of frequency. The result should be compared with the expected theoretical response.
b. Transducer matching: if the impedance is 50 Ω, then it remains to tune out the static capacitance, here about 0.3 pF. This is most conveniently done with a variable inductance in series with the transducer.
c. Timing flight measurement: the transmitting transducer is excited by a low-amplitude tone burst. To prevent burnout of the IDTs it is advisable to use a fixed attenuator (PAD) of 10 or 20 dB in series with the input if high power sources such as the Matec are used. The source and receiver are tuned to the IDT central frequency. Absolute and relative Rayleigh wave velocity of the substrate can be measured in this way. Compare the measured value with that given in the tables.

d. Liquid loading by leaky waves can be demonstrated very effectively by putting a drop of water on the substrate between the electrodes; the propagated acoustic signal immediately disappears. It is instructive to repeat the experiment with liquids of lower acoustic impedance and increased volatility, such as acetone.

e. Transforming the SAW device into an oscillator is easily accomplished by placing an RF amplifier into a feedback loop connected between the two IDTs, in series with an RF attenuator. The attenuator setting must be low enough so that the loop gain exceeds the losses. Interesting conclusions can be drawn from the behavior of the signal across the device observed on an oscilloscope at high and low values of attenuation. The oscillation frequency should be measured with a frequency counter.

f. Using the SAW device as a temperature sensor is possible due to the temperature dependence of the sound velocity in $LiNbO_3$, which gives rise to a predicted temperature variation of the propagation time as 94 ppm/°C. In light of the discussion in Chapter 13, this can easily be measured as the frequency shift of the oscillator in the preceding section, which is directly proportional to the delay time, hence the velocity variation. The SAW substrate can be placed on a cold plate and then a hot plate to cover a temperature range of about 100°C, around room temperature. A calibrated thermometer should be attached to the SAW substrate, which should then be cycled slowly in temperature. Readings of the frequency shift at various fixed temperatures should be made; the frequency shift vs. temperature should give a linear variation of a value close to that predicted.

5. Advanced experiments

There are a number of more advanced experiments of potential interest, but they rely on the availability of specialized equipment. These possibilities will be mentioned only briefly here; they have been found to be relatively easy to set up and to be instructive, even if carried out at an elementary level.

a. Acoustic radiation measurement by hydrophone and water tank

If an ultrasonic immersion test bath with x-y-z micropositioners is available, then this provides a suitable means for measuring the acoustic radiation patterns of immersion transducers. Immersion transducers can be purchased from vendors such as Panametrics. Detection is carried out by a needle hydrophone which contains a small pointlike piezoelectric detector such that it does not perturb the acoustic field. Measurement of the radiation pattern of a transducer and comparison with theory for both near field and far field is feasible.

b. Acoustic microscopy: if a low-frequency acoustic microscope is available, there are a number of simple experiments that can be performed with few complications. The most direct of these is experimental verification of the resolution of an acoustic lens. The lens is focused on the edge of a plate and scanned in a direction perpendicular to the plate edge at constant height. It is important that the lens axis be vertical and the plate accurately adjusted to be horizontal. Over the plate the reflected amplitude is constant, and it then decreases continuously to zero as the focal point is scanned away from the plate edge into the bulk liquid. The width of the resulting curve gives the resolution. This can then be compared with theory for the lens opening and frequency used.

A second instructive experiment, done in the same configuration as above, is the measurement of a $V(z)$ curve. The lens axis is centered on the middle of the plate, roughly in the focal position. In this case the x,y coordinates of the lens are held fixed, and the plate is scanned along the z axis toward the plate. A series of maxima and minima are observed as described in Chapter 14. The result can be used to deduce the Rayleigh wave velocity in the plate, which can then be compared to the tabulated value.

c. Schlieren imaging: if a Schlieren imaging system is available, then it is the tool of choice to image the propagation paths of ultrasonic waves. Typical operation is at 10 MHz in a water bath. Phenomena such as direct reflection and Schoch displacement are easily observable, as is the imaging of a focused acoustic beam.

6. Topics for term papers

If suitable ultrasonic equipment is not available for experimental projects, then term papers involving literature searches and summaries on specific topics are useful. Possible topics include:

Ultrasonic tomography
Fresnel acoustic lens
SAW biosensors
SAW gas sensors
SAW temperature senors
Acoustic spectrum analyser
Laser generation of ultrasound
Equivalent circuit model of IDTs
Acoustoelectric effect

Index

A

B

C

D

E

F

J

K

L

M

N

Q

R

S

T

U

V

W

X

Y

Z